MW00709821

Bioinformatics

Sequence Alignment and Markov Models

Kal Renganathan Sharma, Ph.D., P.E.
Adjunct Professor
Department of Chemical Engineering
Prairie View A&M University
Prairie View, Texas

New York Chicago San Francisco
Lisbon London Madrid Mexico City
Milan New Delhi San Juan
Seoul Singapore Sydney Toronto

The McGraw-Hill Companies

Library of Congress Cataloging-in-Publication Data

Sharma, Kal Renganathan.
Bioinformatics : sequence alignment and Markov models / Kal Renganathan Sharma.
p. cm.
Includes bibliographical references and index.
ISBN 978-0-07-159306-9 (alk. paper)
1. Bioinformatics. 2. Markov processes. I. Title.
[DNLM: 1. Computational Biology—methods. 2. Markov Chains. 3. Models, Theoretical. 4. Sequence Alignment. QU 26.5 S531b 2009]
QH324.2S53 2009
572.80285—dc22 2008018595

1 2 3 4 5 6 7 8 9 0 DOC/DOC 0 1 4 3 2 1 0 9 8

ISBN 978-0-07-159306-9
MHID 0-07-159306-3

Sponsoring Editor
Taisuke Soda

Production Supervisor
Richard C. Ruzycka

Editorial Supervisor
Stephen M. Smith

Project Manager
Preeti Longia Sinha, International Typesetting and Composition

Copy Editor
James K. Madru

Proofreader
Deepa Pathak

Indexer
WordCo Indexing Services, Inc.

Art Director, Cover
Jeff Weeks

Composition
International Typesetting and Composition

Printed and bound by RR Donnelley.

This work is dedicated to my son
R. Hari Subrahmanyan Sharma
(alias Ramkishan, born August 13, 2001)
with unconditional love.

About the Author

Kal Renganathan Sharma, Ph.D., P.E., has written five books, 12 journal articles, and 448 conference papers. He has earned three degrees in chemical engineering—a B.Tech. from the Indian Institute of Technology, Chennai, and an M.S. and a Ph.D. from West Virginia University, Morgantown. He has held a number of high-level positions at engineering colleges and universities. Dr. Sharma currently teaches at Prairie View A&M University in Prairie View, Texas.

Contents

Preface

I was requested by the former controller of examinations at the University of Madras, India, A. Sivamurthy, to prepare the curriculum and syllabus for a B.Tech. degree in bioinformatics at the Vellore Institute of Technology, Vellore, India, in 2001. I was requested by letter to prepare a project report on a course program for an M.Tech. degree in bioinformatics by Dr. B. Srinivasan, Vice Chancellor at Sri. Chandrasekharendra Saraswati Viswa Mahavidyalaya University, Kancheepuram, India, in 2002. The Vice Chancellor at SASTRA University, Thanjavur, India, R. Sethuraman, chartered me with the task of writing a book entitled *Lecture Notes in Computational Molecular Biology,* to be used for instruction in the newly formed M.Tech. and B.Tech. bioinformatics programs in 2003.

Since I wrote *Lecture Notes in Computational Molecular Biology,* a number of interesting developments in the field of bioinformatics has come about. The Human Genome Project has been completed ahead of time. The biologic databases double in size every 10 months, and the computing speed of microprocessors doubles in speed every 18 months. So a database search that cost $2 today would, two years from now, quadruple in cost to $8 on account of the explosive growth of databases and would be cut back in half to $4 on account of the increase in computing power. There is scope for the development of data search and data storage algorithms and methods. It can be viewed as a marriage between information technology and computational biology. Bioinformatics is emerging as a distinct discipline of its own. Textbooks need to be neither mathematically intimidating nor biologically intensive and laborious. Over 560 end-of-chapter exercises are provided in this book. Appendices with Internet hotlinks to public-domain databases and PERL code commands are given. This book can be used as a textbook for core subjects in a bioinformatics undergraduate program and as an elective for chemical engineering and biotechnology undergraduate and graduate programs.

The dynamic programming methods of Needleman and Wunsch and Smith and Waterman for global, local, and semiglobal alignment of sequences are discussed. The affine gap model and the different scoring schemes to make the alignment more biologically meaningful are treated with worked examples. Further reductions in time and space efficiciency from $O(n^2)$ needed for the dynamic programming algorithms are introduced. These include the greedy algorithms that tap into the existing similarity of biologic sequences that are homologous. The X-drop algorithm for very similar sequences that can be completed in $O(en)$ time, where e is much smaller than n, are discussed. Dynamic array techniques that only need $O(n)$ space for dynamic programming methods are introduced. Sparse dynamic programming table problems are reviewed. Methods discussed in this text feature the software used in industry, such as MUMer, Genie, LAGAN, CHAOS, GLASS, QSAR, AVID, REPuter, CLUSTALW, T-Coffee, DIALIGN, MAFTT, PSI-BLAST, BLAST, FASTA, STAMP, JalView, SAM, HMMER, HMMPRO, Meta-Meme, PFAM, Profile HMMs, GLIMMER, GENEMARK, PROCRUSTES, GRAIL, fGENEH, ROSETTA, GENSCAN, SLAM, HMMSTER, PHDSec, DISULFIND, SAM-T99, JPRED, etc.

Suffix trees can be used to represent sequences. Nineteen string algorithms that search for a pattern in a text that can be completed in $O(n)$ time are discussed. Generalized suffix tree, lazy suffix tree, look-up tables, distributed suffix tree, hash tables, etc., are discussed with examples. The multiple-sequence-alignment problem is shown to be NP complete. A chapter on preliminaries needed to obtain maximum use of the textbook is provided. This contains a bit of molecular biology, computer science, and probability. Approximate multiple-sequence-alignment algorithms are discussed.

Markov models are explained in detail. A genome sequence was obtained from NCBI and modeled using geometric distribution and Markov models. The three questions in hidden Markov models (HMMs), i.e., evaluation, decoding, and learning, are reviewed. The Markov, stationarity, and output independence assumptions are introduced to keep the problems mathematically tractable. The HMM is characterized completely. The number of operations needed to determine the sequence given the HMM, i.e., the evaluation problem that usually takes time $O(N^T)$, where T is the length of the sequence and N is the number of states, can be completed in $O(N^2T)$ time using the forward algorithm. The Viterbi algorithm with optimal path is discussed. HMM applications such as construction of a phylogenetic tree, protein families, wheel HMMs to predict periodicity in DNA, the generalized HMM, database mining, multiple alignments, classification using HMMs, signal peptide and signal anchor prediction by HMMs, and Chargaff parity rule prediction are discussed.

Gene-finding algorithms such as the greedy method of Hertz and Stormo, the Gibbs sampler, the binomial heap method, the interpolated Markov model, the SD site-finding problem, the GPHMM, splice-site VLMMs, Steiner trees, Manhattan distance, the PHMM, and Las Vegas algorithms are reviewed. Protein secondary-structure methods such as the neural networks of Qian and Sejnowski, the PHD architecture of Rost and Sander that provided improved accuracy using evolutionary information, the ensemble method of Riis and Krogh, HMM methods, and DAG-RNNs are reviewed.

Microarray slide preparation methods are discussed. The five steps in the micorray cyle are reviewed. The connection to disease eradication by 2050 is discussed. The confocal scanning microscope used for microarray detection is noted. The fluorescent probe and target optimization to capture gene expression in biochips is outlined. The instrument performance measures are discussed. The four-step process of oligonucleotide synthesis is described. Mechanical microspotting, ink-jet printing, and photlithorgraphy of microarray manufacture are touched on briefly. The normalization of cDNA data using housekeeping genes and the Gosset t distribution is described.

The importance of principles of diffusion in gel acrylamide electrophoresis is shown as a separate chapter. Fick's laws of diffusion and the generalized Fick law of diffusion that can be used to account for finite speed in the propagation of mass are described. Eight reasons are given to seek to generalized Fick's law of diffusion. It is derived using the Stokes-Einstein chemical potential approach. The acceleration term is accounted for as the ballistic term that manifests as damped-wave transport in short-time transient diffusion events. The Taitel paradox is discussed. The final condition in time is used to keep the solution from disobeying the Clausius inequality. The three different regimes of solutions during transient diffusion, conditions where subcritical damped oscillations can occur, are derived. The electrophoretic term is added to the governing equation, and an analytical solution obtained by the method of separation of variables. A new transformation method using a spatiotemporal variable that is symmetric in space and time is used to obtain bounded exact solutions in transient diffusion.

Kal Renganathan Sharma, Ph.D., P.E.

Acknowledgments

I record here with sorrow the demise of my maternal uncle, V. V. Giri, M.D., at the residence of his eldest son, Kartik Giri, M.D., near New York City on June 29, 2007; the death of my great uncle, V. Ramanathan, M.D., in March 2007 in Chennai, India; and the passing of my cousin-aunt, Dr. Renganayaki Mahapatra, Head of the Department of Languages at the University of Calcutta, India, in June 2007. Success profiles of my aunt, Janaki Giri, M.D., a leading oncologist near New York City, and my cousin-uncle, Dr. Viswanathan Bringi, Professor of Electrical and Computer Engineering at Colorado State University, Fort Collins, are inspirational. The courage shown by my first wife, Najma Dalal Sharma, M.D., in battling three cerebral aneurysms is tremendous. I helped her with her medical school fees, and we were married on June 24, 1990. I used to drive between Indian Orchard, Massachusetts, and Morgantown, West Virginia, where we had apartments between 1990 and 1993. She obtained a divorce in 1996. Many thanks to Dr. Vidya, my second wife, and my son, Hari Subrahmanyan Sharma (alias Ramkishan), who I get to see during annual vacations since the marital separation. I am indebted to my parents, S. Kalyanaraman and Shyamala Kalyanaraman, for their unfailing support. We recently celebrated their 70th birthdays on a holy trail in South India.

As author of 448 conference papers, I record with gratitude the financial support received from sources through the years. J. W. Zondlo, Professor of Chemical Engineering at West Virginia University, Morgantown, funded my first paper presentation at Maastricht, Netherlands, in October 1987. Coast-to-coast trips in 1989 for conference paper presentations were funded by Richard Turton, Professor of Chemical Engineering at West Virginia University. Victoria Franchetti Haynes, President, Research Triangle Institute, Research Triangle Park, North Carolina, was forthcoming with aid to attend the American Institute of Chemical Engineers annual meetings in Chicago in 1990, Los Angeles in 1991, and Miami in 1992, and for the annual trips to techonolgy symposiums in St. Louis. Based on my performance reviews, the director told me that when the time comes, I could be nominated for the research fellow program. In October 1995, I was on a transatlantic flight funded by R. Shankar Subramanian, formerly chairman of chemical engineering at Clarkson University, Potsdam, New York, for BDPU test validation in Turin, Italy, and tracer particle technique development at Frieberg, Germany.

Nason Pritchard Funds were granted for my paper presentation at the World Congress of Chemical Engineering in San Diego in the summer of 1996. In 1998 and 1999, I presented 90 conference papers at 19 major conferences. Special thanks to Edward J. Wegmann, Director of the Center for Computational Statistics, George Mason University, Fairfax, Virginia, and N. T. Sivaneri, Professor, Mechanical & Aerospace Engineering, West Virginia University, for their encouragement. Funds were allocated by The Honorable G. Viswanathan, former minister in the state government of Tamil Nadu, for me to travel first class and present papers at national conferences at Kochi, Jodhpur, and Chennai, India. I acknowledge the T. R. Rajagopalan research cell at SATRA University for enabling me to present papers in New Orleans in 2003 and Atlanta in 2006.

Special thanks to Dr. Irvin Osborne-Lee, Head of the Department of Chemical Engineering at Prairie View A&M University, Prairie View, Texas, for financial support to fly from Houston to Salt Lake City in November 2007. Also, thanks to a number of other contributors over the years who cannot be mentioned here on account of space limitations.

CHAPTER 1

Preliminaries

The field of bioinformatics includes algorithms, sequence representation, Markov modeling, neural networks to predict protein secondary structure, and other computational and mathematical modeling methods for analysis and storage of biologic data. It includes the study of structure and function and evolution of genes, protein, and whole genomes. It can be viewed as a marriage between information technology and molecular biology. The Human Genome Project that began in October 1990 was completed years ahead of schedule when the rough draft was presented in June 2000 [1]. The project was planned to last 15 years, but rapid technological advances accelerated the completion to 2003. Project goals were to determine the complete sequence of the 3 billion DNA subunits (bases), identify all human genes, and make them accessible for further biologic study. The sequencing of the mouse genome and rat genome has been completed. The first bacterial genome, *Haemophilus influenzae,* was completely sequenced, annotated, and published in 1995. Since then, more than 200 prokaryote genomes have been sequenced completely, and over 500 prokaryote genomes are at various stages of completion. Seventeen eukaryote genomes and four eukaryote chromosomes have been completed at this writing. The biologic data bank size that is made available in the public domain doubles every 10 months. The number of genes characterized doubles every 2 years. The computing speed of new processors, according to the Moore's law, doubles every 18 months. Thus a data bank search for a gene that would cost $2 today would quadruple in cost to $8 in 20 months owing to the increase in data bank size, and because of the increased speed of the hardware, the cost would be cut in half nearly to $4. Still, a cost increase is seen from $2. There is lot of scope for developing new data structures to store the biologic data and efficient algorithms for conducting data bank searches.

The time taken to align two genome sequences of 3 billion base pairs (bp) in length using an $O(n^2)$ dynamic programming algorithm using a gigahertz personal computer can take about 60 years. On a terra-flop computer, this may come down to half a day. The IBM Blue Gene Project, where peta-flop machines are considered, can further improve the time taken. The storage of data using techniques such as parallel-disk modeling and obtaining approximate alignment in $O(n)$ time solution using suffix-tree representation may be good leads in reducing the time taken and improving the storage efficiency of biologic data. Different data structures, such as the suffix tree, binomial heap, Steiner tree, and Manhattan network, are discussed in the following pages.

The preliminaries needed for getting more use out of the material in this textbook are a bit of molecular biology, computer science, and probability. These are provided in this chapter to make the textbook self-contained.

1.1 Molecular Biology

1.1.1 Amino Acids and Proteins

Discussions in bioinformatics frequently center on two important molecules. These are proteins and nucleic acids. The structures and properties and functions of these molecules in different organisms form the information in the explosive growth of biologic data banks. The name *protein* comes from the Greek word *prota,* meaning "of primary importance." Proteins were first described and named by Berzelius in 1838. However, their central role in living organisms was not fully appreciated until 1926, when Sumner showed that the enzyme urease was a protein. The first protein structures to be solved included insulin and myoglobin; the first was by Sir Frederick Sanger [2–6], who won a Nobel Prize in 1958 for it, and the second by Perutz and Kendrew, also in 1958. Both proteins' three-dimensional structures were among the first determined by x-ray diffraction analysis; the myoglobin structure won the Nobel Prize in chemistry for its discoverers.

Proteins are large bioorganic compounds that are polymeric in nature. They are made of amino acids arranged in a linear chain and joined together between the carboxyl of one amino acid and the amine nitrogen of the other by a bond that is called a *peptide bond.* The sequence of amino acids in a protein is defined by genes and encoded in the genetic code. Although this genetic code specifies the 20 different amino acids, the residues in a protein are often chemically altered in posttranslational modification either before the protein can function in the cell or as part of control mechanisms. Proteins associate to form complexes that are stable. They can work in concert to achieve a particular function, and they participate in every function of the cell. Many proteins are enzymes that catalyze biochemical reactions. They are vital to metabolism. The cell shape is maintained by a system of scaffolding. Proteins in the cytoskeleton form the system of scaffolding. Proteins are also important in cell signaling, immune responses, cell adhesion, and the cell cycle. Protein is also a necessary component in our diet because animals cannot synthesize all the amino acids and must obtain essential amino acids from food. Through the process of digestion, animals break down ingested protein into free amino acids that can be used for protein synthesis.

1.1.2 Structures of Proteins

During formation of the polypeptide polymeric chain, one water molecule is lost per amino acid. This is why the constituents of proteins are called *amino acid residues.* Four different types of protein structures are recognized in the field. These are as follows.

Primary Structure

The primary structure of proteins is the random sequence distribution of the 20 different amino acids concatenated in a polypeptide chain. Each of the 20 different amino acids consists of two parts: (1) the backbone of the protein and (2) the unique side chain, or R group, that determines the physical and chemical properties of the amino acid. Each amino acid consists of an amine (NH_2^+) and a carboxylic acid moiety (COO·). The general formula of the 20 different amino acids can be classified into four categories based on the net charge on the protein molecule. These categories and the amino acids contained in them are as follows:

1. *Positively charged basic amino acids*: lysine (Lys), arginine (Arg), and histidine (His).

   NH_2
   |
   $H_2N\text{-}C_4H_8\text{-}CH\text{-}COOH$ (lysine)

   NH ‖ , NH_2 |
   $NH_2\text{-}C\text{-}NH\text{-}C_3H_6\text{-}CH\text{-}COOH$ (arginine)

   ________ NH_2
   |........ | |
   H_2N $NH\text{-}CH_2\text{-}CH\text{-}COOH$ (histidine)
   \ /
   \ /
   C

2. *Negatively charged acidic amino acids:* aspartic acid (Asp), glutamic acid (Glu).

   NH_2
   |
   $HOOC\text{-}CH_2\text{-}CH\text{-}COOH$ (aspartic acid)

   NH_2
   |
   $HOOC\text{-}C_2H_4\text{-}CH\text{-}COOH$ (glutamic acid)

3. *Polar amino acids:* glycine (Gly), serine (Ser), threonine (Thr), cysteine (Cys), tyrosine (tyr), glutamine (Gln), asparagine (Asn).

CH_2-COOH (glycine)
|
NH_2

OH- CH_2CH-COOH (serine)
|
NH_2

CH_3-CH-CH-COOH (threonine)
| |
OH NH_2

HS-CH_2-CH-COOH (cysteine)
|
NH_2

OH-φ-CH_2-CH-COOH (tyrosine)
|
NH_2

NH_2-C- C_2H_4-CH-COOH (glutamine)
|| |
O NH_2

NH_2-C-CH_2-CH-COOH (asparagine)
|| |
O NH_2

4. *Nonpolar amino acids:* alanine (Ala), valine (Val), leucine (Leu), isoleucine (Ile), proline (Pro), methionine (Met), phenylalanine (Phe), tryptophan (Trp).

CH_3-CHCOOH
| alanine
NH_2

C_2H_6CHCHCOOH
valine |
NH_2

C_2H_6CHCH$_2$CHCOOH
leucine |
NH_2

C_3H_8CHCHCOOH
isoleucine |
NH_2

$CH_3SC_2H_4$CHCOOH
methionine |
NH_2

φ-CH_2CHCOOH
|
NH_2
phenylalanine

CH_2——CH_2
| |
| | (proline)
CH_2 CHCOOH
\ /
\ NH /

φ-e-CH_2CHCOOH (tryptophan)
|
NH_2

Secondary Structure

The polypeptide backbone exists in different sections of the protein either as an α-helix, β-pleated sheet, or random coil. The study of protein secondary structure has attracted a lot attention in the literature. As will be discussed in later chapters, protein secondary structures can be constructed from the primary structure chain sequence distribution. The secondary structure pertains to the *stereoisomerism* exhibited by the polypeptide chain. The problem of secondary structure prediction is one of hydrogen bonding. The polar groups present in the backbone of the polypeptide chain, C=O and N—H, are capable of hydrogen-bond formation. The two structures that solve the problem are the α-helix and β-pleated sheet, in which extended polypeptide backbones are side by side. These structures are stable. They can occur at the exterior of proteins with appropriate hydrophilic side chains or in the hydrophobic interior of proteins with appropriate hydrophobic side chains.

In the α-helix, the polypeptide backbone is twisted into a right-hand helix, called an α-helix. The structure was first recognized in α-keratin by Sanger [2]. For L-amino acids, the right-handed helix is more stable than a left-handed one. The structure has a pitch of three to six amino acids per turn. This results in the C=O of each peptide bond being aligned to form a hydrogen bond with the peptide bond N—H of the fourth distant amino acid residue. The C=O groups point in the direction of the axis of the helix and are aimed at the N—H groups with which they hydrogen-bond, giving maximum bond strength and making the α-helix a stable structure. Thus every C=O and N—H group of the polypeptide backbone is hydrogen-bonded in pairs forming a stable, cylindrical, rodlike structure. Amino acids vary in their tendency to form α-helices.

Proteins are made of mixtures of α-helix and β-pleated-sheet structures. This is also a stable structure in which the polar groups of the polypeptide backbone are hydrogen-bonded to one another. The polypeptide chain lies in an extended or β form with the C=O and N—H groups hydrogen-bonded to those of a neighboring chain. The structure was first recognized in β-keratin. Several chains can form a sheet of polypeptide. It is pleated because successive α carbon atoms of the amino acid residues lie slightly above and below the plane of the β-pleated sheet alternately. The adjacent bonded together polypeptide chains can run in the same direction parallel or in the opposite direction antiparallel. In the latter case, a polypeptide may make tight β turns to fold the chain back on itself.

The *random coil* refers to a section of polypeptide in a protein whose conformation is not recognizable as one of the defined structures of α-helices and β-pleated sheets. It is determined by side-chain interactions and within a given protein is fixed rather than varying in a random way.

One good way to measure protein secondary structure is by x-ray crystallography. In addition, the techniques of neutron diffraction and nuclear magnetic resonance (NMR) can be used to measure protein secondary structure.

Tertiary Structure

The folding of the secondary structure into a macrostructure such as globules is called the *tertiary structure* of a protein. A given protein in a physiologic environment can have a complex three-dimensional structure. The amino acid "backbone" of a protein can rotate freely, allowing amino acids from distal protein domains to come into close contact with each other. As these regions of the protein interact with one another, they will create and stabilize a particular protein conformation. Disulfide bond creation between cysteine residues is one of the primary stabilizing mechanisms.

Quaternary Structure

Two polypeptide chains connected by hydrogen bonding form the *quaternary structure* of a protein.

1.1.3 Sequence Distribution of Insulin

Frederick Sanger was one of the few who won the Nobel Prize a second time. His first Nobel Prize was for his work on protein primary structure in 1958, and the second was for his elucidation of the nucleotide microstructure in 1980. His research work on the structure of insulin took him 12 years. The protein molecule consists of 20 different amino acids. This was known prior to Sanger's work. The microstructure of the protein molecule or the chain sequence distribution of the molecule was not known. Sanger suspected that the differences between the biologic and physical properties of the protein molecule were because of differences in the sequence distribution of the protein molecule(s) [2].

Chain sequence distribution of a copolymer is the relative order of occurrence of the different monomers along a single linear chain. Thus a copolymer with two monomers A and B with random chain sequence distribution may have structures such as

BAABAAABBBABBBBAAAAAB, BBAAABAB, AABABBABABBA, and so on. G. N. Ramachandran had suggested that every third residue in mammalian protein was a certain residue. Some investigators had suggested a periodic microstructure for protein, i.e., ABABABAB for an alternating copolymer with monomers A and B. Some had suggested that protein was a complex mixture.

A range of values from 36,000 to 48,000 for the molecular weight of insulin was reported in the literature. Sanger set out to resolve this discrepancy. He developed a molecular labeling procedure called the *dinitrophenyl* (DNP) *method.* The reagent used was 1,2,4-fluorodinitrobenzene (FDNB). This reacts with the free amino groups of a protein or peptide to form a DNP derivative. The peptide bonds are broken under mild conditions. Hydrolysis of DNP protein results in split of the peptide bonds in the chain bearing the N-terminal residue in the form of a DNP derivative. DNP amino acids are light-yellow substances. They can be extracted from the unsubstituted amino acids using ether as a solvent. Further separation is effected using partition chromatography. The DNP derivatives thus can be fractionated. The chemical structure of the separated compounds is affected by noting the chromatographic rates and comparing them with those of their synthetic analogues. Whereas silica-gel chromatography was used in the initial work on insulin microstructure, paper chromatography has been found to be a satisfactory procedure. On separation and identification, the DNP derivatives could be estimated calorimetrically. When the method was applied to insulin, three yellow DNP derivatives were found in the hydrolysate of the DNP-insulin. One of these was extracted into ether and was identified as ε-DNP-lysine, which was formed by reaction of the FDNB with the free ε-amino group of lysine residues that are bound normally within the polypeptide chain. The others were identified as DNP-phenylalanine and DNP-glycine, and estimation showed that these were two residues of each assuming a molecular weight of 12,000. This lead Sanger to deduce that insulin was composed of four polypeptide chains, two with phenylalanine and two with glycine end groups. The hypothesis that chains of insulin were connected by disulfide bridges was explored by attempting to split the bridges by reduction to SH derivatives. Satisfactory results were obtained by oxidation with performic acid. The cystiene residues are converted to cysteic acid residues, thus breaking the cross-links.

The fractionation of complex mixtures from partial hydrolysis of protein was a technical hurdle. Other investigators have shown that small peptides can be well fractionated by paper chromatography. Tuppy, a postdoctoral associate of Sanger, worked so hard in 1 year that he and Sanger were able to deduce the whole of the sequence of 30 residues. The blueprint was unveiled by Sanger, and the finer details were obtained by his coworkers. The mixture from partial hydrolysis of fraction B was too complex for direct analysis by paper chromatography. They resorted to ionophoresis, ion-exchange chromatography, and adsorption on charcoal. The simplified mixtures then were fractionated by two-dimensional paper chromatography. The peptide spots were cut out, and the material was eluted from the paper, subjected to complete hydrolysis, and analyzed for its constituent amino acids. The analysis from the resulting acidic fraction contained only peptides and cysteic acid. Thus 45 peptides were identified in various fractions of the partial acid hydrolysate, and the following five sequences were found to be present:

1. Phe-Val-Asp-Glu-His-Leu-CysSO$_3$H-Gly (N-terminal sequence)
2. Gly-Glu-Arg-Gly
3. Thr-Pro-Lys-Ala
4. Tyr-Leu-Val-CysSO$_3$H-Gly
5. Ser-His-Leu-Val-Glu-Ala

Proteolytic enzymes are more specific than the acid because only a few of the peptide bonds are susceptible. For example, they considered a peptide Bp3 obtained by the action of pepsin. It had the composition Phe-CysSO$_3$H-Asp-Glu-Ser-Gly-Val-Leu-His, of which the most important components are aspartic acid and serine because they occur only once in the chain. Aspartic acid is present in the N-terminal sequence 1, and serine is in sequence 5. By studying other peptides obtained by the action of pepsin, trypsins, and chymotrypsin, it was possible to find out how the various sequences were arranged and to deduce the complete sequence of the phenylalanyl chain, which is

Phe-Val-Asp-Glu-His-Leu-CysSO$_3$H-Gly-Ser-His-Leu-Val-Glu-Ala-
Leu-Tyr-Leu-Val-CysSO$_3$H-Gly-Glu-Arg-Gly-Phe-Phe-Tyr-Thr-Pro-
Lys-Ala

Paper ionophoresis was required for separation at pH 2.5 for determining the sequence of fraction A by enzymatic hydrolysates. The fraction A sequence was

Gly-Ileu-Val-Glu-Glu-CysSO$_3$H-CysSO$_3$H-Ala-Ser-Val-CysSO$_3$H-
Ser-Leu-Tyr-Glu-Leu-Glu-Asp-Tyr-CysSO$_3$H-Asp

The ammonia generated during hydrolysis with strong acid was used to determine the location of amide groups in the polypeptide chain. Thus the microstructure of insulin was obtained. This was shown by Sanger in his Nobel lecture and is shown below in Fig. 1.1.

Sanger confirmed the random sequence distribution of the 20 different amino acids in the insulin microstructure. Other naturally occurring polypeptides besides insulin are glutathione, carnosine, anserine, oxytocin, vasopressin, bradykinin, and corticotropin. The biologic specificity of a protein is a function of the number of amino acid residues and their sequence.

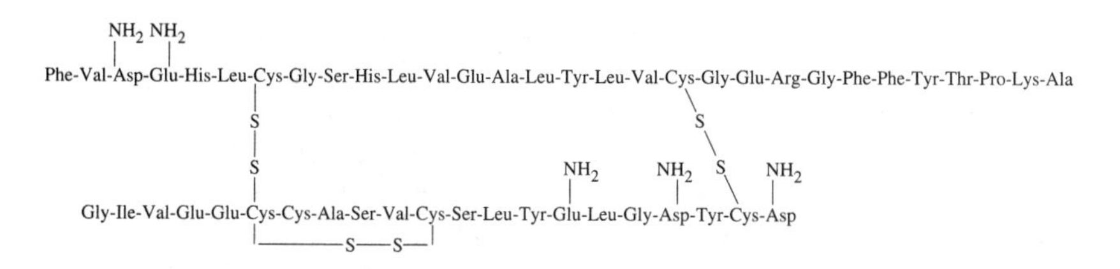

FIGURE 1.1 Microstructure of insulin. Chain sequence distribution of the polypetide chains.

1.1.4 Bioseparation Techniques

The protein chemist is confronted with the problem of isolating, purifying, and characterizing a protein. Identification of a suitable source such as a fresh tissue is the first step for purification of proteins. This tissue is subjected to the action of a blender for grinding to obtain a homogenate that is rich in protein as well as contaminating material. Proteins are temperature-sensitive and fragile. The homogenate is filtered and freed from unwanted material by treatment with suitable solvents. Denaturation of proteins is avoided by control of pH and temperature. As a general rule, purification of proteins is carried out at temperatures close to the freezing point.

Salt precipitation is used to effect separation by addition of ammonium sulfate so that the desired protein remains either in the supernatant or in the precipitate. A mixture of proteins is passed down an ion-exchange column and separated by binding to the column. The bed is regenerated by eluting agents of varying pH. By increasing the pH of the effluent, different fractions of proteins are obtained. The protein solution binds to ion-exchange materials such as cellulosic polymers.

The molecular weight of the proteins can be determined by the use of gels. The discovery of the ultracentrifuge in the early twentieth century was an advancement that allowed precise determination of molecular weights. Svedberg won the Nobel Prize in physics in 1926 for his efforts in the development and use of the ultracentrifuge [7]. Sedimentation is used to measure the molecular weights of proteins and in the study of protein-protein interactions. Sucrose density gradients can be used to separate molecular fragments.

Electrophoresis is a method with superior resolution that is used to separate macromolecules from complex mixtures by the application of an electrical field. The macromolecules, called the *gel,* are placed at one end of the matrix and are subjected to a electrical field. Different macromolecules in the gel will migrate at different speeds depending on the nature of the gel and the characteristics of the macromolecule. Electrophoretic techniques can be used to separate any biomacromolecule such as nucleic acids, polypeptides, and carbohydrates. Tiselius won the Nobel Prize in chemistry in 1948 for his work on the development of electrophoresis as a technique to separate and characterize proteins from complex mixtures [8].

The use of polyacrylamide gel electrophoresis (PAGE) has had a major impact on the ability to isolate and characterize proteins. Polyacrylamide gels are formed by cross-linking an acrylamide monomer with the chemical agent *N,N*-methylene bisacrylamide. The polymerization reaction proceeds as free-radical catalysis with the use of ammonium persulfate and the base TEMED (*N,N,N,N*-tetramethylenediamine) as the initiator. PAGE can be used to resolve the ladders in DNA structure, and the ladders can be used to characterize proteins according to their size or charge. Three methods were developed to measure the primary structure of protein: (1) Sanger's method, (2) the Dansyl chloride method, and (3) the Edman degradation technique.

All three techniques are laborious. It requires lot of material and years of analysis to complete the analysis of even a short protein. In the Edman degradation method, the peptide fragment is treated with phenylisothiocynate at pH 8 to yield phenylthiocarbamyl derivative at the N terminal. The derivative is treated with acid in organic solvent so that the N-terminal amino acid undergoes cyclization to produce phenylthiohydantoin, which is cleaved from the peptide fragment. Thiohydantoin derivative can be identified using paper chromatography, and the peptide is further subjected to the same treatment every time, forming the thiohydantoin derivative from the amino end. Dansyl chloride reagent is used for determination of the N-terminal residue in alkaline conditions. The N-terminal residue forms a yellow fluorescent derivative that can be detected easily. Even small amounts of amino acids can be deduced. Indirect methods can be used to save time. The cDNA responsible for creation of the protein can be cloned and its sequence measured. Then, by deduction, the protein sequence can be obtained.

Mass spectrometry (MS) is a method in which the mass of the molecules that have been ionized can be measured using a mass spectrometer. MS has become a key tool in proteomics research because it can analyze and identify compounds that are present at extremely low concentrations (as little as 1 pg) in very complex mixtures by analyzing their unique signatures. A critical concern in MS is that the methods used for ionization can be so harsh that they may generate very little product to measure at the end. The development of "soft" desorption ionization methods by John Fenn and Koichi Tanaka [9], which allowed the application of MS to biomolecules on a wide scale, earned them a share of the Nobel Prize in chemistry in 2002.

Isoelectric focusing is a variation of electrophoresis that can be used for separating mixtures of protein. A column is used that consists of gel having positive and negative charges. When the protein mixture is injected into the column, the molecules polarize in the electrical field in such a way that the negatively charged ones move toward the anode and the positively charged ones move toward the cathode. At the isoelectric point, i.e., the point in the tube at which each protein attains a neutral pH, the driving force to migration stops. A sharp bend forms at this juncture.

Two-dimensional gel electrophoresis is a way to couple different gel systems with different resolving powers to dramatically improve separation and resolution of complex mixtures of proteins. Two-dimensional gel electrophoresis is an incredibly useful analytic tool that provides a foundation for what is now referred to as *proteomics*.

Chromatography is used extensively for separating different molecular species, including proteins. Different types of chromatography are recognized, such as adsorption chromatography, ion-exchange chromatography, and partition chromatography. Paper chromatography is suitable for separation of small amounts of low-molecular-weight compounds that are soluble in the liquid phase. The liquid phase is water, and the mobile phase consists of a mixture of organic solvents. The paper is spotted with the substance to be separated and immersed in a trough containing the mobile phase. The spots on the paper are developed using a suitable developing reagent. The partition coefficient R_f, is the ratio between the distance run by the compound and the distance traveled by the solvent.

Ion-exchange chromatography also can be employed for separation and purification of proteins. The protein sample is prepared in the right type of buffer and then applied to the ion-exchange column. Molecules possessing no charge will easily pass through, whereas charged molecules will interact with the exchanger and get adsorbed. Proteins then can be eluted from the exchangers. Gel-filtration chromatography is a technique for purification and separation of macromolecules based on their molecular size. Gel permeation, gel exclusion, and molecular sieving are similar methods. Gel-filtration media include polydextrin gels, polyacrylamide gels, agarose gels, and controlled-pore glass beads. The porosity of gel beads in a column is controlled depending on the problem of separation at hand. Larger molecules elute out, and smaller molecules diffuse through the pores in the beads. Later, this can be eluted by using a buffer. Desaltation of a sample can be effected. The use of a biospecific interaction of a protein with a specific ligand is used in affinity chromatography. The chromatographic column uses an inert matrix or support medium that will offer binding sites for the desired protein to be purified. The adsorbate has to be specific, and the adsorbent can be porous, hydrophilic, and capable of covalent binding. Agarose gel, polyacrylamide, and controlled-pore glass beads are examples of adsorbents used by this method.

In thin-layer chromatography (TLC), the chromatogram can be developed a number of times with different solvents with good separation. Quantitative analysis of multiple components can be done in 1 hour using this method. A binding medium such as calcium sulfate is used in TLC. The adsorbent is activated, and then spots are generated and dried. Spots in the form of colored zones also may be observed under ultraviolet light.

High-performance liquid chromatography (HPLC) can be used to isolate, purify, and identify compounds in a mixture. Rapid analysis of nonvolatile, ionic, thermally labile compounds that were previously difficult to separate can be achieved using HPLC. Molecular components of the cell can be determined with high sensitivity, speed, accuracy, and resolution. The mobile phase is a liquid. The solute needs to be soluble in the mobile phase. The mobile phase is forced under high pressure of more than 6000 lb/in^2 into the column. Normal-phase chromatography, bounded-phase chromatography, and reverse-phase chromatography are three kinds of HPLC methods. Separation of ribosomal proteins can be done in this manner. The effect of pore size, pore volume, silica density, and surface area on a given separation is complex.

1.1.5 Nucleic Acids and Genetic Code

Crick, Watson, and Williams were awarded the Nobel Prize in medicine in 1962 for their work in the molecular configuration of nucleic acids, the genetic code, involvement of RNA in the synthesis of proteins, and its significance for information transfer in living material [10]. Nucleic acids are long-chain polymers of nucleotides. Each nucleotide consists of three parts: (1) a sugar, ribose or deoxyribose, (2) phosphoric acid, and (3) a nitrogenous base. The four different nitrogenous bases are adenine, guanine, cytosine, and thymine. Adenine (A) and guanine (G) are purines, and cytosine (C) and thymine (T) are pyrimidines. Uracil (U) is another nitrogenous base found in RNA instead of thymine (T) in DNA. The double-helix three-dimensional structure of DNA was elucidated by Watson, Crick, and Wilkins. They used x-ray crystallography to determine the structure. Rosaland Franklin was one of the pioneers and was deceased by the time the Nobel Prize was awarded. the discovery of DNA is hailed as the most important work in biology in the last 100 years, and the field it opened may be the next scientific frontier for the next 100 years. Outside the helix backbone of the ladder is the sugar-phosphate chain. A complete turn of the ladder is called *pitch* and is about 3.4 Å in length. The space between bases is 2.4 Å, and the diameter of the helix is 20 Å. The sequences of bases in DNA, by a process called *translation,* determine the sequence distribution of protein molecules. All genetic information in living organisms of any kind is carried by the nucleic acids, usually by the DNA. Certain small viruses use RNA as their genetic material. The four bases in DNA can assume $4 \times 4 = 16$ combinatorial forms, 64 triplets, and 256 quartets. The set of bases that code is called a *codon.* The two DNA strands are antiparallel. One strand runs in $5' \rightarrow 3'$ direction and the other strand in a $3' \rightarrow 5'$ direction. The two polynucleotide chains of the double helix interact with each other. The hypothesis that the linear sequence of nucleotides in DNA specifies the linear sequence of amino acids in proteins evolved over a period of time. Kornberg and coworkers [11] discovered and characterized the enzyme polymerase. This was followed a few years later by characterization of RNA polymerase. Kornberg and coworkers clarified the manner by which information in DNA is transcribed into an RNA that is now referred to as *messenger RNA*. Kornberg's son Roger received the Nobel Prize in chemistry in 2006 for his studies of the molecular basis for eukaryotic transcription [12].

Genetic Code

At the time of Crick's Nobel lecture, the genetic code had the following general properties:

- It was fairly certain that codons did not overlap.
- Most, if not all, codons consisted of three adjacent layers.
- Adjacent codons did not overlap.
- The message was read in the correct groups of three by starting at some fixed point.
- Code sequence in the gene was collinear.
- In general, more than one triplet coded each amino acid.
- It was not certain that some triplets may not code more than one amino acid.
- Triplets with code for the same amino acid probably were rather similar.
- It was not known whether there was any general rule that groups code together or whether the grouping was mainly the result of historical accident.
- The number of triplets that do not code an amino acid probably was small.
- Certain codes proposed earlier such as comma-less codes were all unlikely to be correct.
- The code in different organisms probably was similar.

The structural chemistry of the nucleic acids was developed over a period of 70 years in many countries from the chemistry of the constituent purines, pyrimidines, and sugar moieties to work on the nucleosides.

Har Gobind Khorana was awarded the Nobel Prize in medicine in 1968 for his work on nucleic acid synthesis and the genetic code [13]. He looked at the synthesis of short-chain oligonucleotides. The problem he faced was activation of the phosphomonoester group of a mononucleotide, design of suitable protecting groups for the functional groups, and development of methods for the polymerization of specific sequences. Khorana proposed the reaction sequence for the preparation of high-molecular-weight RNA messengers and the subsequent in vitro synthesis of polypeptides of known amino acid sequences (Fig. 1.2). Amplification of the proposed scheme or in vitro studies of the coding problem to produce DNA or RNA products was

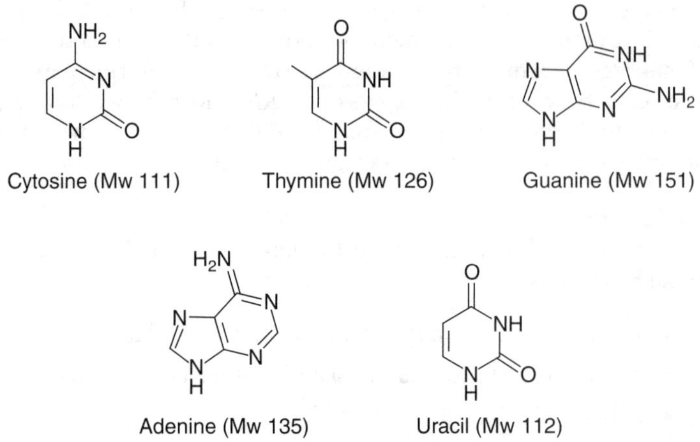

Cytosine (Mw 111) Thymine (Mw 126) Guanine (Mw 151)

Adenine (Mw 135) Uracil (Mw 112)

FIGURE 1.2 Molecular structure of nitrogenous bases in DNA and RNA.

conceived to be a general behavior of the polymerase so long as there was a repeating pattern of nucleotide sequences in the chemically synthesized deoxypolynucleotide templates. Khorana identified the types of reactions catalyzed by DNA polymerase (Fig. 1.3). DNA-like polymers with repeating dinucleotide sequences for polymers with trinucleotide sequences and two polymers with tetranucleotide sequences were prepared. The repeating dinucleotide sequences were TC:GA, trinucleotide TTC:GGA, and tetranucleotide TTAC:GTAA. Khorana identified the transcription of DNA-like polymers by means of RNA polymerase to form single-stranded ribopolynucleotides.

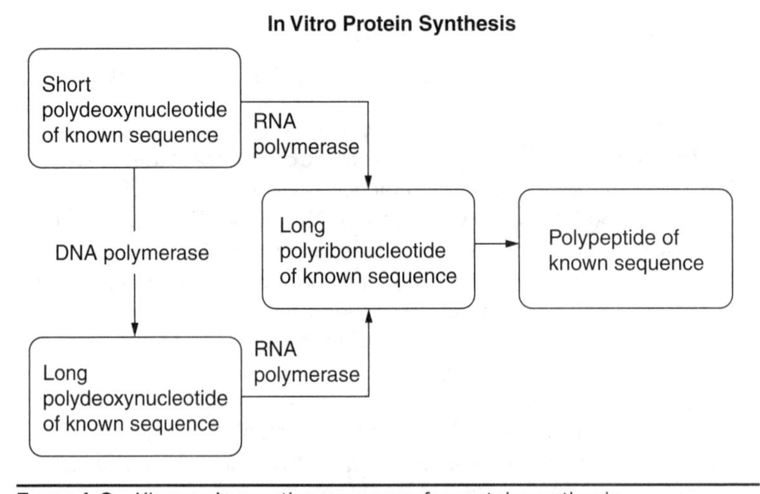

FIGURE 1.3 Khorana's reaction sequence for protein synthesis.

The structure of the genetic code that emerged is shown in Table 1.1. The code is universal. Nirenberg and Leder coshared the Nobel Prize in medicine with Khorana in 1968 for synthesizing trace amounts of protein using artificial RNA molecules. Universality does not mean that all organisms use the same codons for protein synthesis. It means that a trinucleotide codon does not change its meaning from one organism to the next.

The DNA sequence is divided into a series of triplet codes composed of three bases called *codons.* Having more than one codon for one amino acid is called *degeneracy* of the genetic code. Codons that specify the same amino acid are called *synonyms.* Synonyms are usually similar, with variations found only in the third position of the codon. Since the third base of the codon can vary, this base position is called the *wobble position.* Thus the three codons UAA, UAG, and UGA that cause termination of polypeptide chain growth are called *stop codons.* The codons AUG and GUG that stand, respectively, for methionine and valine are also used as signals for initiation of polypeptide chain synthesis and are referred to as *initiation codons.*

1st Letter	2nd Letter				3rd Letter
	U	C	A	G	
	PHE	SER	TYR	CYS	U
	PHE	SER	TYR	CYS	C
U	LEU	SER	C.T.	C.T.	A
	LEU	SER	C.T.	TRY	G
	LEU	PRO	MIS	ARG	U
C	LEU	PRO	HIS	ARG	C
	LEU	PRO	GLN	ARG	A
	LEU	PRO	GLN	ARG	G
	ILEU	THR	ASN	SER	U
A	ILEU	THR	ASN	SER	C
	ILEU	THR	LYS	ARG	A
	MET(C.I)	THR	LYS	ARG	G
	VAL	ALA	ASP	GLY	U
G	VAL	ALA	ASP	GLY	C
	VAL	ALA	GLU	GLY	A
	VAL(C.I)	ALA	GLU	GLY	G

TABLE 1.1 The Genetic Code

The first base of the initiation codon specifies the reading frame of the RNA. In any base sequence, three types of reading frames are possible depending on which base is chosen for the first base. During protein synthesis, only one reading frame is meaningful. The set of codons that runs continuously and is bounded by the initiation codon at one end and the termination codon at the other end is known as an *open reading frame* (ORF) and is used to determine the protein coding regions of DNA.

A number of studies on tRNA structure were conducted on the yeast *Escherichia coli*, rat liver, and wheat germ. Dr. Raj Bhandary and H. Khorana, and colleagues had determined the primary structure of yeast phenylalanine tRNA [14]. All tRNAs whose primary structure is known can adopt the cloverleaf secondary structure (Fig. 1.4). The first general question is, *How are the trinucleotide codons recognized by the protein-synthesizing apparatus?* The answer is the tRNA molecule. The next question is, *What is the evidence that recognition of codons in fact involves nucleotide-nucleotide interaction by virtue of base pairing?* If this is so, then one might expect to find the primary structure of an amino acid–specific tRNA to be three contiguous nucleotide units, complementary to the established codons for the particular amino acid. The concept of anticodons was developed. Only the simplest components of the tRNA structure are shown in Fig. 1.4. Some additional bonds are formed, and the entire structure becomes L-shaped with the 3′ ACCA sequence at one end and the anticodon at the other. The anticodon is complementary to the codon, and 3 bp can form between the codon in the mRNA and the anticodon of the tRNA.

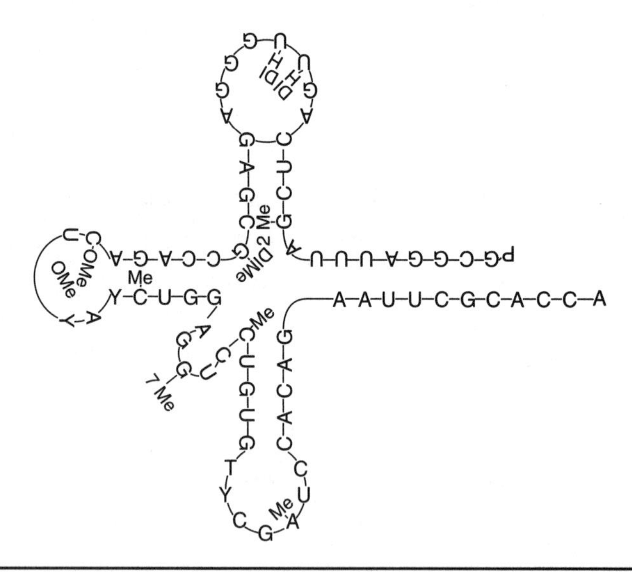

FIGURE 1.4 Cloverleaf model for secondary structure of yeast phenylalanine tRNA.

Determination of Nucleotide Sequences in DNA

A DNA sequence for the genome of bacteriophage φX174 of approximately 5375 nucleotides has been determined using the rapid and simple *plus and minus method* [4]. The sequence identifies many of the features responsible for the production of the proteins of the nine known genes of the organism, including initiation and termination sites for the proteins and RNAs. Two pairs of genes are coded by the same region of DNA using different reading frames.

General methods for determination of DNA sequences have been developed only recently. This is so mainly because of the large size of DNA molecules, the smallest being those of the simple bacteriophages such as φX174, which contains 5000 nucleotides. Initially, the smaller RNA molecules were used for early studies on nucleic acid sequences. Having uncovered the truth about amino acid sequences in protein, Sanger turned his attention to RNA and developed a relatively rapid small-scale method for the fractionation of ^{32}P-labeled oligonucelotides.

The plus and minus method [4] is a relatively rapid and simple technique that has made possible determination of the sequence of the genome of bacteriophage φX174. It depends on the use of DNA polymerase to transcribe specific regions of the DNA under controlled conditions. Another rapid and simple method that depends on specific chemical degradation of the DNA has been described recently by Maxam and Gilbert [15], and this also has been used extensively for DNA sequencing. It has the advantage over the plus and minus method of being applicable to double-stranded DNA, but it requires strand separation or equivalent fractionation of each restriction enzyme fragment studied, which makes it somewhat more laborious.

The general approach used in these studies and in those of proteins depended on the principle of partial degradation. The large molecules were broken down, usually by suitable enzymes, to give smaller products that were then separated from each other and their sequence determined. The separation was done using a gel acrylamide electrophoresis. When sufficient results had been obtained, they were fitted together by a process of deduction to give the complete sequence (Fig. 1.5). Copying procedures were needed for treating large DNA molecules. Pulse labeling with radioactively labeled nucleotides and copying techniques for RNA sequence determination were pioneered by Billeter and colleagues [16]. For DNA sequences, the enzyme DNA polymerase is used. The single-stranded DNA is copied. The enzyme requires a primer, which is a single-stranded oligonucleotide having a sequence that is complementary to and therefore able to hybridize with a region on the DNA sequence. Mononucleotide residues are added sequentially to the 3′ end of the primer from the corresponding deoxynucleoside triphosphates, making a complementary copy of the template DNA. By using triphosphates containing ^{32}P in the α position, the newly synthesized DNA can be labeled. Synthetic oligonucleotides were used as primers initially, but after the discovery of restriction

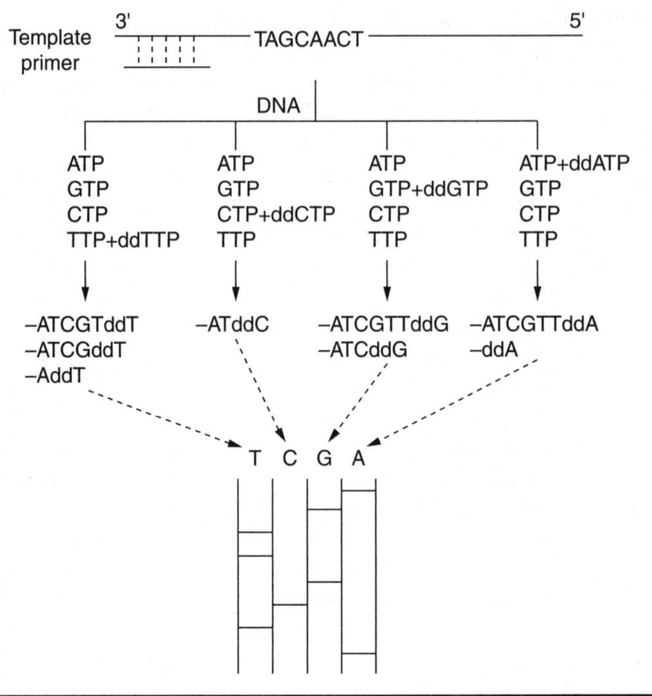

FIGURE 1.5 Principle of the chain terminating for DNA sequencing.

enzymes, it was more convenient to use fragments resulting from their action because they were more readily available. The copying procedure was used to prepare a short specific region of labeled DNA that then could be subjected to partial digestion. One of the difficulties in determining the sequence distribution of DNA was to find specific methods for breaking the strand into smaller fragments. It was found that good fractionations according to size could be obtained by ionophoresis on acrylamide gels. Plus and minus technique, was used to determine the almost complete sequence of the DNA of bacteriophage φX174, which contains 5386 nucleotides.

DNA Transcription, Translation, and Replication

A chromosome contains double-stranded DNA molecules. Its replication is described as semiconservative in that the two original strands called *parental strands* are separated, and each acts as a template for synthesizing a new strand. Each new double helix has one old and one new strand. The basis of the replication is that of complementarity in that a guanine (G) will base pair with a cytosine (C) and an adenine (A) will base pair with a thymine (T) so that a base on the parental strand automatically specifies which base is to be incorporated into the new strand as its partner. This copying process depends on Watson-Crick hydrogen bonding of base pairs. It follows that strand separation is essential to unpair the bases and make them available for base pairing with incoming nucleotides. DNA replication is different in prokaryotes than it is in eukaryotes.

The information in DNA encoded in the sequence of four nitrogenous bases is used to direct the assemblage of 20 different amino acids in the correct sequence so as to produce the protein for which a given gene is responsible. Genes direct protein synthesis by sending out copies of their coded information to the cytoplasm. Messenger RNA is a polynucleotide essentially the same as DNA except for the following differences:

- Sugar is ribose and not the deoxyribose of DNA. An OH is in 2′ position.
- mRNA is single-stranded. Bases are adenine, cytosine, guanine, and uracil.

mRNA is synthesized much like DNA, but the two strands are separated so as to produce a single-stranded template for directing the sequence of nucleotides to be assembled into mRNA. mRNA is made from ATP, CTP, GTP, and UTP by a single enzyme in *E. coli* RNA polymerase. mRNA has a half-life of 20 minutes to several hours in eukaryotes and about 2 minutes in bacteria. Thus, for expression of a gene, a continuous stream of mRNA molecules must be produced from that gene. The flow of information in gene expression from DNA to mRNA is called *transcription* and from mRNA to protein is referred to as *translation*. mRNA production is called *gene transcription*. The RNA molecules produced are called *transcripts*. The synthesis of proteins directed by mRNA is called *translation*.

mRNA in both prokaryotes and eukaryotes is proportionate in length to the size of the protein it codes for. In the DNA from which such RNA molecules are transcribed, the section contents are called *introns*, and the coding stretches are called *exons*. There can be 2–50 introns in human genes, and the lengths of the introns can vary from 50–20,000 bp. Exons usually are less than 1000 bp in length. The primary transcript is processed to eliminate the introns and link together the exons into one mRNA molecule. This is known as mRNA *splicing*.

Example 1.1 The synthetic mRNA has a periodic primary microstructure. Using the two letters U and C, define the synthetic mRNA and its protein products. There are $2^6 = 64$ patterns possible in a member sequence

1. UUU,UUU (Phe, Phe)
2. UCU,CUC (Ser, Leu)
3. UUC,CUU (Phen, Leu)
4. UUU,CCC (Phe, Pro)
5. UUC, UUC (Phe, Phe)
6. CCA, CCA (Pro, Pro)

The protein sequence will depend on the start position.

Case 2

Start position 2, UCU, (Ser, Leu)

Start position 3, CUC, UCU (Leu, Ser)

Case 3

Start position 2, UCC, UCC (Ser, Phe)

Start position 3, CCU, UCC (Pro, Ser)

Start position 4, CUU, UUC (Leu, Phe)

Case 4

Start position 2, UUC, CCU (Phe, Pro)

Start position 3, UCC, CCU (Ser, Leu)

Start position 4, CCU, UUU (Pro, Phe)

Case 5

Start position 2, UCU, UCU (Ser, Ser)

Start position 3, CUU, CUU (Leu, Leu)

Start position 4, UUC, UUC (Phe, Phe)

Case 6

Start position 2, CAC, CAC (His, His)

Start position 3, ACC, ACC (Thr, Thr)

Start position 4, CCA, CCA (Pro, Pro)

1.1.6 Genomes—Diversity, Size, and Structure

Genomes of living organisms are diverse in nature. Some genomes are circular in nature, e.g., in bacteria, whereas other genomes are linear in nature, e.g., in mammals. A 2.91 billion bp consensus sequence of the euchromatic portion of the human genome was generated by the whole-genome shotgun sequencing method [1]. The 14.8.billion bp DNA sequence was generated over 9 months from 27,271,853 high-quality sequence reads (5.11-fold coverage of the genome) from both ends of plasmid clones made from the DNA of five individuals. Two assembly strategies—a whole-genome assembly and a regional chromosome assembly—were used, each combining sequence data from Celera and the publicly funded genome effort. The public data were shredded into 550-bp segments to create a 2.9-fold coverage of the genome regions that had been sequenced, without including biases inherent in the cloning and assembly procedure used by the publicly funded group. This brought the effective coverage in the assemblies to 8-fold, reducing the number and size of gaps in the final assembly over what would be obtained with 5.11-fold coverage. The two assembly strategies yielded very similar results that largely agree with independent mapping data. The assemblies effectively cover the euchromatic regions of the human chromosomes. More than 90 percent of the genome is in scaffold assemblies of 100,000 bp or more, and 25 percent of the genome is in scaffolds of 10 million bp or larger.

Analysis of the genome sequence revealed 26,588 protein-encoding transcripts for which there was strong corroborating evidence and an additional 12,000 computationally derived genes with mouse matches or other weak supporting evidence. Although gene-dense clusters are obvious, almost half the genes are dispersed in low G+C sequences separated by large tracts of apparently noncoding sequence. Only 1.1 percent of the genome is spanned by exons, whereas 24 percent is in introns, with 75 percent of the genome being intergenic DNA. Duplications of segmental blocks ranging in size up to chromosomal length are abundant throughout the genome and reveal a complex evolutionary history. Comparative genomic analysis indicates vertebrate expansions of genes associated with neuronal function, tissue-specific developmental regulation, and the hemostasis and immune systems. DNA sequence comparisons between the consensus sequence and publicly funded genome data provided the locations of 2.1 million single-nucleotide polymorphisms (SNPs). A random pair of human haploid genomes differed at a rate of 1 bp per 1250 on average, but there was marked heterogeneity in the level of polymorphism across the genome. Less than 1 percent of all SNPs resulted in variation in proteins, but the task of determining which SNPs have functional consequences remains an open challenge.

The smallest genomes are found in non-self-replicating suborganisms such as bacteriophages and viruses that piggy-back on the metabolism and replication machinery of free-living prokaryote and eukaryote cells, respectively. The 1.74-Mbp genome of the hypothermophilic *Methanococcus jannaschii* was completely sequenced in 1996. There are 5000 bacterial species per gram of soil. The 3310-Mbp human genome is organized into 22 chromosomes plus the two that determine sex. Chimpanzees, for example, have 23 chromosomes in addition to 4 sex chromosomes. Cats have 38 chromosomes, whereas dogs have 78 chromosomes. The chromosomes in some organisms are not stable. For instance, the *Bacillus cereus* chromosome has been found to consist of a large stable component (2.4 Mbp) and a smaller less stable component (1.2 Mbp) that is more easily mobilized into extra chromosomal elements of different stages. Genomic sequencing is difficult to perform when the chromosomes are not stable. The variation of gene number among different organism is shown in Table 1.2 [19].

The word *gene* was coined in 1909 by the Danish geneticist W. Johannes. Table 1.2 lists the number of genes in organisms with different evolutionary lineages. Gene number identification in organisms is increasing as more accurate methods for their determination become available. The coding and noncoding regions of the genome may be demarcated. The gene is widely recognized as a fundamental hereditary unit of the chromosome that determines the chemical, metabolic, and morphologic characteristics of an individual. The gene is a locus on a chromosome representing a segment of the DNA molecule (cistron) capable of transcription. A eukaryotic gene is a collection of introns and noncoding intervening sequences. Exons are coding regions that give rise to final RNA

Group	Species	No. of Genes
Phages	Bacteriophage MS2	4
Viruses	Cauliflower mosaic	8
Bacteria	*Escherichia coli*	4100
Fungi	*Saccharomyces cerevisiae*	5800
Protoctista	*Oxytricha similis*	12,000
Arthropoda	*Drosophila melanogaster*	15,000
Nematoda	*Caenorhabditis elagans*	19,000
Mollusca plantae	*Nicotine tobacum*	30,000
Chordata	*Homo sapiens*	40,000

TABLE 1.2 Gene Number in Organisms

product, i.e., a protein. In a nutshell, the information transfer from DNA consists of self-duplication of DNA into transcripts called *RNA* by transcription. This encodes the protein sequences by translation.

Genome analysis of the number of genes present in *Homo sapiens* varied considerably from one investigator to another. The number of genes in humans was expected to be around 120,000 [17]. For a complex organism, gene multiplexing makes it possible to produce several different transcripts from many of the genes in the genome, as well as many different protein variants from each transcript. The complex cellular processing of genetic material offers challenges to modeling in bioinformatics.

The human genome data analysis [1] revealed that the gene content in humans may be about 30,000 genes. This is only less than twice the number of genes found in *C. elegans.* The biologic complexity of an organism may be related to the expected number of genes in the organism. Claverie gave an estimate of biologic complexity K in an organism and its relation to the expected number of genes in the genome N [18]. Different functionalities to convert K into N have been suggested [19]:

$$K \approx N \text{ (linear)} \qquad K \approx N^a \text{ (polynomial)}$$
$$K \approx a^N \text{ (exponential)} \qquad K \approx N! \text{ (factorial)}$$

The biologic complexity of an organism may be related to the organism's ability to create diversity in its gene expression, i.e., to the number of theoretical transcription states the organism can achieve.

The human body consists of 10^{12} cells, 23 pairs of chromosomes that consist of 3,310,004,815 bp. The average gene consists of 10,000 bases. The sizes of genes vary. The largest known human gene is the *dystrophin* gene, with 2.4 million bases. The nucleotide sequence is almost exactly the same for the entire human race. For over 50 percent of the genes, function is unknown. In constructing the working draft, 16 genome sequencing centers produced over 22.1 billion bases of raw sequencing data, consisting of overlapping fragments totaling 3.9 billion bases, and provided sevenfold coverage of the human genome. Over 30 percent are high-quality, finished sequences, with 8- to 10-fold coverage, 99.99 percent accuracy, and few gaps. The goals of the projects are to identify approximately 30,000 genes in a human chromosome, determine the sequence of the nucleotide base pairs that make up human DNA, store the information in the databases, improve tools for data analysis, transfer related technologies to the private sector, and address the ethical, legal, and social issues that may arise from the project. Less than 2 percent of the genome codes for proteins. Junk DNA consisting of repeated sequences that do not code for proteins make up at least 50 percent of the human genome. Repetitive sequences are thought to give chromosomes the necessary structure and dynamics. While the genes are randomly distributed in the human genome, the genes of other organism are evenly distributed throughout the genome.

1.2 Probability and Statistics

A gambler's dispute in 1654 led to the creation of the *mathematical theory of probability*. This was accomplished by two pioneers, B. Pascal and P. De Fermat. A French nobleman with an interest in gambling called Pascal's attention to an apparent contradiction concerning a popular dice game. The game consisted of throwing a pair of dice two times. The problem was to decide whether or not to bet even money on the occurrence of at least one double six during 24 throws. A well-established gambling rule led the nobleman to believe that betting on a double six in 24 throws would be profitable, but his own calculations indicated just the opposite. This, among other things, led to a famous exchange of letters between Pascal and Fermat in which the fundamental principles of probability theory were formulated for the first time. The probability of a double six on two throws can be calculated for a fair six-sided pair of dice as

$$P[X = (6, 6)] = 1/6 \times 1/6 = 1/36 \tag{1.1}$$

Thus, in 36 throws, the chance of occurrence of a double six is one. Equation (1.1) takes into account the occurrence of two independent events and that all values will occur with equal likelihood. Bernoulli and de Movrie were big contributors as the subject developed rapidly in the eighteenth century. P. de Laplace published his book, *Theorie Analytique des Probabilities,* in 1912. This widened the scope of probability to many scientific and practical problems. The theory of errors, actuarial mathematics, and statistical mechanics are some examples of applications developed in the nineteenth century. Mathematical statistics in one important branch of applied probability with applications in a wide variety of fields such as genetics, psychology, economics, and engineering. Important contributors to probability since Laplace were Chebychev, Bell, Markov, Von Mises, and Kolmogorov. The search for a widely acceptable definition of probability took nearly three centuries. This was resolved finally by the axiomatic approach developed by Kolmogorov.

1.2.1 Three Definitions of Probability

The *classical definition of probability* states that the probability $P(A)$ of an event A is determined a priori without actual experimentation. It is given by

$$P(A) = \frac{N_A}{N} \tag{1.2}$$

where N is the number of possible outcomes and N_A is the number of outcomes that are favorable to the event A. In the die experiment, A is the double 6 and N is 36.

The *axiomatic definition of probability* uses the set theory. A certain event ρ is the event that occurs in every trial. The union $A + B$ of two events A and B is the event that occurs when A or B both occur. The intersection AB of the events A and B is the event that occurs when both events A and B occur The events A and B are mutually exclusive if the occurrence of one of them excludes the occurrence of the other. Three postulates are given. The probability $P(A)$ of an event A is

$$P(A) \geq 0 \tag{1.3}$$

The probability of the certain event equals 1. If the events A and B are mutually exclusive, then

$$P(A + B) = P(A) + P(B) \tag{1.4}$$

The axiomatic approach is credited to Kolmogorov.

The *relative-frequency approach to the definition of probability* states that probability $P(A)$ of an event A is the limit

$$P(A) = \lim_{n\to\infty} \frac{n_A}{n} \tag{1.5}$$

where n_A is the number of occurrences of A, and n is the number of trials. Probabilities are used to define frequencies and are defined as limits of such frequencies. Both n_A and n must be large. This approach was suggested by von Mises.

1.2.2 Bayes' Theorem and Conditional Probability

The conditional probability of an event A given the event G, denoted by $P(A/G)$ is defined by

$$P(A/G) = P(AG)/P(G) \qquad P(G) \neq 0 \tag{1.6}$$

If G is a subset of A, then $P(A/G) = 1$. If A is a subset of M, then

$$P(A/G) = P(A)/P(G) \geq P(A) \tag{1.7}$$

Bayes proposed a theorem in 1763 that was later named after him. Laplace gave its final form years later, and it can be stated as follows:

$$P(A_i/B) = \frac{P(B/A_i)P(A_i)}{P(B/A_1)P(A_1) + \cdots + P(B/A_n)P(A_n)} \tag{1.8}$$

The conditional probability also can be written in terms of intersection of sets as

$$P(A/G) = \frac{P(A \cap G)}{P(G)} \tag{1.9}$$

$$P(A \cap G) = P(A/G)P(G) = P(G/A)P(A) \tag{1.10}$$

Bayes' theorems then can be stated as

$$P(A/G) = \frac{P(G/A)P(A)}{P(G)} \tag{1.11}$$

1.2.3 Independent Events and Bernoulli's Theorem

Two events A and B are said to be independent if

$$P(A \cap B) = P(A)P(B) \tag{1.12}$$

Suppose that one repeatedly runs independent trials of an experiment in which the probability of success in each trial is p, and the probability of failure in $q = 1 - p$. Then the probability that there are exactly k successes in these n trials is given by ${}^{n}C_{k}p^{k}q^{n.k}$. Let A and B be small positive numbers. Then there is a value of n large enough that the probability that the ratio of the successes in n trials is not within A and p is less than B. In other words, if the experiment is run long enough, the fraction of successes is likely to be close to the correct probability.

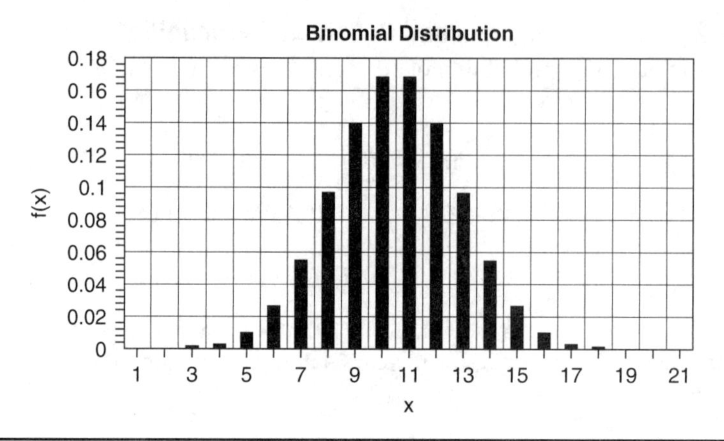

FIGURE 1.6 Binomial distribution with $n = 21$ and $p = q = 0.5$.

1.2.4 Discrete Probability Distributions

Binomial and Multinomial Distributions

The binomial distribution (Fig. 1.6) gives the discrete probability distribution of obtaining n successes out of N Bernoulli trials. The result of each Bernoulli trial is true with probability p and false with probability $q = 1 - p$. Thus

$$\begin{aligned} f(x) = P(X = x) &= {}^{n}C_{x} p^{x} q^{n-x} \\ &= n!/x!(n-x)! p^{x} q^{n-x} \\ &= x = 0, 1, 2, \ldots, n \end{aligned} \tag{1.13}$$

The mean, variance, skewness, and kurtosis of the binomial distribution are given below.

$$\text{Mean } \mu = np$$

$$\text{Variance } \sigma^2 = npq$$

$$\text{Skewness } \alpha_3 = (q - p)/(npq)^{1/2}$$

$$\text{Kurtosis } \alpha_4 = 3 + (1.6pq)/(npq)^{1/2}$$

$A_1, A_2, \ldots, A_k$ events can occur with probabilities $p_1, p_2, \ldots, p_k$, where $p_1 + p_2 + \cdots + p_k = 1$. $X_1, X_2, \ldots, X_k$ are random variables, respectively, giving the number of times that $A_1, A_2, \ldots, A_k$ can occur in a total of n trials so that $X_1 + X_2 + \cdots + X_k = n$. Then the multinomial distribution can be given by

$$P(X_1 = n_1, X_2 = n_2, \ldots, X_k = n_k) = n!/n_1!n_2! \cdots n_k! p_1^{n1} p_2^{n2} \cdots pk^{nk} \tag{1.14}$$

where $n_1 + n_2 + \cdots + n_k = n$. The joint probability function for random variables $X_1 + X_2 + \cdots + X_k = n$. This is the general form of the binomial distribution and the general term in the multinomial expression of $(p_1 + p_2 + \cdots + p_k)^n$.

Poisson Distribution

$$f(x) = P(X = x) = \lambda^x \exp\left(\frac{-\lambda}{x!}\right) \qquad x = 0, 1, 2, \ldots, n \qquad (1.15)$$

$$\text{Mean } \mu = \lambda$$

$$\text{Variance } \sigma^2 = \lambda$$

$$\text{Skewness } \alpha_3 = 1/\lambda^{1/2}$$

$$\text{Kurtosis } \alpha_4 = 3 + 1/\lambda$$

A good example of the Poisson distribution (Fig. 1.7) is the number of typos generated by a good typist. The probability of 1 or 2 typos per page is high, and the probability of generating 10 typos per page is slim. In a similar fashion, the time taken at the teller counter at the bank also can be fit to a Poisson distribution. The probability of the event of the transaction taking 5 or 8 minutes may be high, and the probability of it taking 1 hour will be low. Yet another example is the arrival of students late to class. The probability of students arriving on time or 5 minutes before the hour is high, and the chances that they will arrive ½ hour late will be on the low side.

Hypergeometric Distribution

This is an example in sampling with replacement. Suppose that a box contains b blue marbles and r red marbles. Let us perform n trials of

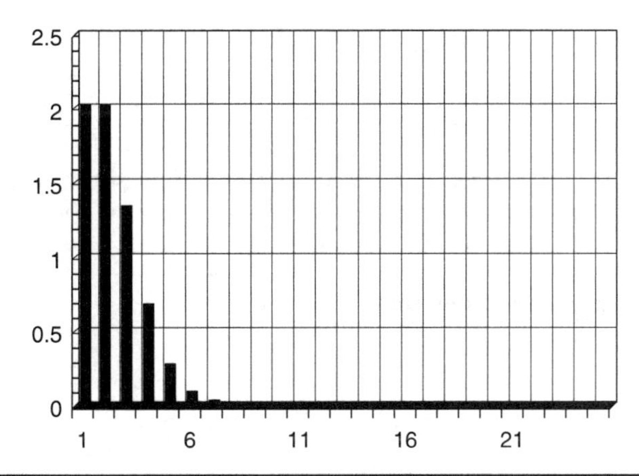

Figure 1.7 Poisson distribution with $\lambda = 2.0$.

an experiment in which a marble is chosen at random and its color is observed and the marble is put back in the box.

$$f(x) = P(X = x) = {}^{b}C_{x}\,{}^{r}C_{n-x} / {}^{b+r}C_{n} \tag{1.16}$$
$$\text{Mean } \mu = nb/(b + r)$$
$$\text{Variance } \sigma^2 = nbr(b + r - n)/(b + r)^2/(b + r + 1)$$

Geometric Distribution

$$f(x) = P(X = x) = pq^{x-1} \qquad x = 1, 2, \ldots \tag{1.17}$$
$$\text{Mean } \mu = 1/p$$
$$\text{Variance } \sigma^2 = q/p^2$$

The chain sequence distribution in copolymers was shown to be modeled using the geometric distribution.

1.2.5 Continuous Probability Distributions

Uniform Distribution and Cauchy Distribution

$$f(x) = \frac{1}{(b-a)} \qquad a \le x \le b = 0 \text{ otherwise} \tag{1.18}$$

$$\text{Mean } \mu = ½(a + b)$$
$$\text{Variance } \sigma^2 = 1/12(b - a)^2$$

A good example of the uniform distribution is the contact time of solid particles at the heat-exchanger surfaces in a circulating fluidized-bed boiler (CFB). The Cauchy distribution is given by

$$f(x) = \frac{a}{\pi(x^2 + a^2)} \qquad a > 0,\ -\infty < x < \infty \tag{1.19}$$

$$\text{Mean } \mu = 0$$

Variance and higher moments do not exist. This distribution is also called the *Lorentz distribution* by physicists. It forms the solution to the differential equation that can be used to describe forced resonance.

Gamma and Chi-Squared Distributions

$$f(x) = x^{\alpha-1} \frac{\exp\left(\frac{-x}{\beta}\right)}{\beta^{\alpha}\Gamma(\alpha)} \qquad x > 0 \tag{1.20}$$

$$= 0 \qquad x \le 0$$

where $\Gamma(\alpha)$ is the gamma function.

$$\text{Mean } \mu = \alpha$$
$$\text{Variance } \sigma^2 = \alpha\beta^2$$

$$\Gamma(n) = \int_0^\infty t^{n-1} \exp(-t)dt \qquad n > 0$$

$$\Gamma(n+1) = n\Gamma(n) \qquad \text{(recurrence formula)}$$
$$\Gamma(1) = 1$$

when n is a positive integer, then

$$\Gamma(n+1) = n!$$

$X_1, X_2, \ldots, X_\gamma$ are γ independently normally distributed random variables with mean 0 and variance 1. Consider the random variable

$$\chi^2 = X_1^2 + X_2^2 + \cdots + X_\gamma^2 \tag{1.21}$$

where γ is the number of degrees of freedom. A special case of the gamma distribution with $\alpha = \gamma/2$, $\beta = 2$.

$$\text{Mean } \mu = \gamma$$
$$\text{Variance } \sigma^2 = 2\gamma$$
$$f(x) = x^{(\gamma/2)-1} \exp(-x/2)/[2^{\gamma/2}\Gamma(\gamma/2)] \qquad x > 0 \tag{1.22}$$
$$= 0 \qquad x \le 0$$

This is the chi-squared distribution.

Student *t* Distribution

$$f(t) = \frac{\Gamma\frac{(\gamma+1)}{2}\left(1+\frac{t^2}{\gamma}\right)^{-(\gamma+1)/2}}{\sqrt{\gamma\pi}\left(\Gamma\frac{\gamma}{2}\right)} \qquad -\infty < t < \infty \tag{1.23}$$

$$\text{Mean } \mu = 0$$
$$\text{Variance } \sigma^2 = \gamma/(\gamma-2) \qquad \gamma > 2$$

where γ is the number of degrees of freedom.

Normal Distribution

$$f(x) = \left(\frac{1}{\sigma\sqrt{2\pi}}\right)\exp - \frac{1}{2}\left(\frac{x-\mu}{\sigma}\right)^2 \tag{1.24}$$

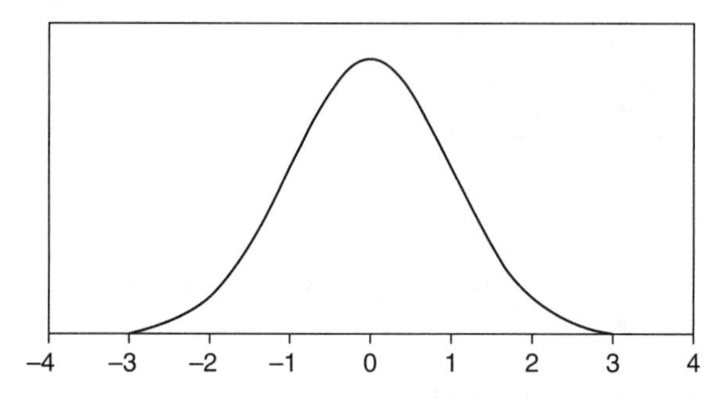

FIGURE 1.8 Normal distribution with zero mean and unit variance.

The normal distribution (Fig. 1.8) was introduced by de Movrie, who approximated binomial distributions for large n. His work was extended by Laplace, who used the normal distribution in error analysis in experiments. Legendre came up with the method of least squares. Gauss by 1809 justified the normal distribution for experimental errors. The name *bell curve* was coined by Galton and Lexis.

Generalized Normal Distribution

The generalized normal distribution was introduced by Sharma to capture the periodicity in pressure fluctuations in addition to the random component [20]. In addition to the mean and standard deviation, the number of saddle points also can be used to characterize the periodicity of pressure fluctuations.

$$f(x) = A \exp(-Bx - Cx^2 - Dx^4) \tag{1.25}$$

where A, B, C, and D are parameters that can be obtained by a least-squares fit of the experimental data. This also can be referred to as a *Sharma distribution.* Periodicity is found in DNA sequences. This periodicity can be represented using the Sharma distribution.

1.2.6 Statistical Inference and Hypothesis Testing

Hypothesis testing is the use of statistics to determine the probability that a given hypothesis is true. The usual process of hypothesis testing consists of four steps:

1. Formulate the null hypothesis H_0 (commonly, that the observations are the result of pure chance) and the alternative hypothesis H_1 (commonly, that the observations show a real effect combined with a component of chance variation).
2. Identify a test statistic that can be used to assess the truth of the null hypothesis.
3. Compute the P value, which is the probability that a test statistic at least as significant as the one observed would be obtained

assuming that the null hypothesis were true. The smaller the P value, the stronger is the evidence against the null hypothesis.

4. Compare the P value with an acceptable significance value α (sometimes called an *alpha value*). If $p \leq \alpha$, then the observed effect is statistically significant, the null hypothesis is ruled out, and the alternative hypothesis is valid.

Type I error is an error in a statistical test that occurs when a true hypothesis is rejected (a false negative in terms of the null hypothesis). *Type II error* is an error in a statistical test that occurs when a false hypothesis is accepted (a false positive in terms of the null of hypothesis).

1.3 Which Is Larger, 2^n or n^2?

During the time and space efficiency analysis of string algorithms, the use of subsequences is encountered. The number of possible distinct subsequences in a sequence of length n can be seen to be

$$^{n}C_0 + {}^{n}C_1 + {}^{2}C_2 + \cdots + {}^{n}C_n \tag{1.26}$$

Expanding $2^n = (1 + 1)^n$ using a binomial expansion:

$$2^n = 1 + n + \frac{n(n-1)}{2!} + \frac{n(n-1)(n-2)}{3!} + \cdots \tag{1.27}$$

Comparing Eq. (1.26) and Eq. (1.27),

$$2^n = {}^{n}C_0 + {}^{n}C_1 + {}^{2}C_2 + \cdots + {}^{n}C_n \tag{1.28}$$

During sequence alignment of two sequences, either global or local by dynamic programming methods, it can be seen that the time taken is usually n^2. It is desirable to evaluate whether 2^n or n^2 is greater for all natural numbers. An attempt is made to prove that $2^n > n^2$ by the principle of induction.

Given: $$2^n \geq n^2 \tag{1.29}$$

To show: $$2^n+1 \geq (n + 1)^2 \tag{1.30}$$

Multiplying both sides of Eq. (1.30) by 2,

$$2^n+1 \geq 2n^2 \tag{1.31}$$

Should $2n^2 > (n + 1)^2$, Eq. (1.31) is shown. Compare $2n^2$ and $(n + 1)^2$, or compare n^2 and $2n + 1$, or compare 1 and $2/n + 1/n^2$. It can be seen that $1 > 1/n^2 + 2/n$.

For $n > 2$, the null case has to be taken because $n = 4$ as at $n = 3$, $2^n < n^2$. Thus 2^n is greater than n^2 for values of n greater than 4. For n less than 4, this is not the case. This can be seen at $n = 3$; in fact, n^2 is greater than 2^n.

1.4 Big *O* Notation and Asymptotic Order of Functions

The big O notation was introduced by E. Landau in 1909 for his discussions of the distribution of prime numbers. The order of growth of the running time of an algorithm gives a simple characterization of the algorithm's efficiency. This allows for comparison of relative performance of different algorithms. When only large input sizes are considered, only the order of magnitude of the running time is important or relevant. This is called the *asymptotic efficiency* of the algorithm. The big O, θ, and Ω notations are introduced to facilitate the analysis of running time and storage space required by computer algorithms.

The worst-case running time function is given by $T(n)$. It is defined on only an integer input size. For a given function $g(n)$, $\theta[g(n)]$ is denoted as the set of functions such that

$$0 \le c_1 g(n) \le f(n) \le c_2 g(n) \tag{1.32}$$

Equation (1.32) is valid for all $n \ge n_0$, and there exist positive constants c_1 and c_2 such that the function is sandwiched between the two function $c_1 g(n)$ and $c_2 g(n)$ for sufficiently large n. It is said of the function $g(n)$ that it is an *asymptotically tight bound* for $f[n - f(n)] = \theta[g(n)]$ denotes the fact a set of functions is involved. The θ notation asymptotically bounds a function from above and below. When there is only an asymptotic upper bound, the O notation is used. For a given function $g(n)$, big O of n is denoted by $O(n)$:

$$O[g(n)] = f(n) \qquad 0 \le f(n) \le cg(n) \tag{1.33}$$

Equation (1.33) is valid for all $n \ge n_0$ and provides an upper bound on a function to within a constant factor. In a similar fashion, the asymptotic lower bound is introduced by $\Omega[g(n)]$:.

$$\Omega[g(n)] = f(n) \qquad 0 \le cg(n) \le f(n) \tag{1.34}$$

Equation (1.34) is valid for all $n \ge n_0$ and provides a lower bound on the function that is considered.

Lemma 1.1 A function $f \in O(g)$ if $f(n)/g(n) = C$ as n goes to infinity and C is less than infinity, including the case where the limit is 0. If the limit of the ratio of f to g exists and is not infinity, then f grows no faster than g. If the limit is infinity, then f does grow faster than g.

For any real number x, the greatest integer less than or equal to x is denoted by $\lfloor x \rfloor$, the floor of x, and the least integer greater than or equal to x by $\lceil x \rceil$, the ceiling of x. For all real x,

$$x - 1 < \lfloor x \rfloor \leq x \leq \lceil x \rceil < x + 1 \tag{1.35}$$

Example 1.1 *Ordering of functions.* Rank the following functions by order of growth:

$$\lg[\lg(n)],\ 1,\ n,\ 2^n,\ e^n,\ n!,\ n^2,\ n^3$$

$$\lg[\lg(n)] < 1 < n < n^2,\ n^3,\ e^n,\ 2^n,\ n! \text{ for large } n \tag{1.36}$$

Summary

The Human Genome Project was completed ahead of time. Bioinformatics involves the algorithms, mathematical models, neural networks, sequence representations, alignment, and other methods of analysis and storage of biologic data. Owing to the doubling of biologic databases every 10 months, there is increased need for inventing new data search and data storage methods. Preliminaries needed for getting more use out of this textbook are a bit of molecular biology, computer science, and probability.

Two important molecules in bioinformatics are proteins and nucleic acids. Insulin was among the first primary protein structure identified. The sequence of amino acids in a protein is defined by genes and encoded by the genetic code. Primary structure, secondary structure, tertiary structure, and quaternary structure are the four different structure types of proteins. The primary structure of a protein consists of a random polymer chain sequence distribution of amino acids. There exist 20 different amino acids. Each amino acid consists of an amine and carboxylic acid group. The 20 different amino acids are lysine, arginine, histidine, which are basic; aspartic acid and glutamic acid, which are acidic; glycine, serine, threonine, cysteine, tyrosine, glutamine, and asparagine, which are polar; and alanine, valine, leucine, isoleucine, proline, methionine, phenylalanine, and tryptophan, which are nonpolar. Owing to hydrogen-bond formation, the secondary structure of protein is formed depending on the primary sequence structure of the protein. Proteins are formed into α-helix, β-pleated sheet, and γ-coil shapes. F. Sanger won two Nobel Prizes—one for discovering the primary chain sequence distribution structure of insulin and another for discovering the chain sequence distribution of nucleic acid. He developed a molecular labeling method. Paper chromatography and gel acrylamide electrophoresis are used to separate the molecular fragments and sequence deduced.

There are a number of bioseparation techniques. These are salt precipitation, ion exchange, ultracentrifuge, sedimentation, polyacrylamide electrophoresis, Sanger's method, the Dansyl chloride method, the Edman degradation technique, paper chromatography, mass spectrometry, isoelectric focusing, two-dimensional gel electrophoresis, adsorption chromatography, partition chromatography, ion-exchange chromatography, gel-filtration chromatography, affinity chromatography, thin-layer chromatography, and HPLC.

Nucleic acids consist of a ribose sugar or deoxyribose, phosphoric acid, and nitrogenous bases. Adenine, guanine, cytosine, thymine, and uracil are nitrogenous bases. DNA has a double-helix structure and two antiparallel strands. The linear sequence of a protein is specified by a linear sequence of nucleic acids by a process of translation and transcription. Har Gobind Khorana received the Nobel Prize for nucleic acid synthesis and the genetic code. The flow of information in gene expression from DNA to mRNA is called *transcription* and from mRNA to protein is called *translation.* Exons are the coding regions of DNA, and intergenic regions are the introns. Splicing occurs when exons are linked together and introns are eliminated.

Genomes vary from one organism to another. About 1.1 percent of the genome is spanned by exons, 24 percent by introns, and 75 percent by intergenic DNA. Gene number of an organism is the number of genes contained in the DNA. The work of Vrenter and colleagues confirmed that the human genome has 30,000 genes. Claverie provided an estimate of biologic complexity K in an organism to the expected number of genes in the genome. Junk DNA consists of repeated sequences that do not code for proteins and make up 50 percent of the human genome.

The famous exchange of letters between Pascal and Fermat on the gambler's dispute gave rise to the concept of probability. Other major contributors to the field of probability are Bernoulli, De Movrie, Laplace, Chebychev, Bell, Markov, von Mises, and Kolmogorov. The three definitions of probability are the classical, axiomatic, and relative-frequency approaches. Conditional probability, Bayes' theorem, and Bernoulli's theorem were discussed. The probability density function, mean, variance, skewness, and kurtosis of discrete distributions such as the binomial, Poisson, hypergeometric, and continuous distributions such as the uniform, Cauchy, gamma, chi-squared, student t, normal, and generalized normal distributions were reviewed. The formulation of null and alternate hypotheses, development of test statistics, and type I and type II errors were reviewed.

$2^n > n^2$ when $n > 4$. The big O notation, floor, ceiling, and asymptotic order of functions O, θ, and Ω were introduced.

References and Sources

[1] J. C. Venter, M. D. Adams, E. W. Myers, et al., "The sequence of the human genome," *Science* 291 (2001), 1304–1351.

[2] F. Sanger, "The chemistry of insulin," Nobel lecture, December 1958. In *Nobel Lectures.* Amsterdam: Elsevier, 1964.

[3] F. Sanger and S. Nicklen Coulson, "DNA sequencing with chain-terminating inhibitors," *Proc. Natl. Acad. Sci. USA* 74 (1977), 5463–5467.
[4] F. Sanger, "Determination of nucleotide sequences in DNA." In *Les Prix Nobel.* Stockholm, Sweden: Nobel Committee, 1980.
[5] F. Sanger and H.Tuppy, "The amino-acid sequence in the phenylalanyl chain of insulin. The investigation of peptides from enzymic hydrolysates," *Biochem J.* 49, (1951), 481–500.
[6] F. Sanger, G. M. Air, B. G. Barell, et al., "Nucleotide Sequence of Bacteriophage X174 DNA," *Nature.* 26, (1977), 687–702.
[7] T. Svedberg, "The Ultracentrifuge." In *Les Prix Nobel.* Stockholm, Sweden: Nobel Committee, 1928.
[8] A. Tiselius, "Electrophoresis and adsorption analysis as aids in investigations of large molecular weight substances and their breakdown products." In *Les Prix Nobel.* Stockholm, Sweden: Nobel Committee, 1948.
[9] K. Tanaka, "The origin of macromolecule ionization by laser irradiation." In *Les Prix Nobel.* Stockholm, Sweden: Nobel Committee, 2002.
[10] F. H. C. Crick,"On the genetic code." In *Les Prix Nobel.* Stockholm, Sweden: Nobel Committee, 1962.
[11] A. Kornberg, J. Bertsch, and H. Khorana, "Enzymatic synthesis of DNA—Oligonucleotides as templates and the mechanism of their replication," *Proc. of National Academy of Sciences.* 61, (1964), 315–323.
[12] R. Kornberg, "The molecular basis of eukaryotic transcription." In *Les Prix Nobel.* Stockholm, Sweden: Nobel Committee, 2006.
[13] H. G. Khorana, "Nucleic acid synthesis and genetic code." In *Les Prix Nobel.* Stockholm, Sweden: Nobel Committee, 1968.
[14] V. L. Raj Bhandary, S. H. Chang, A. Stuart, et al., "t-RNA—Molecular Structure, Sequence and Properties," *Annual Reviews Biochem.* 5 (1976), 805–860.
[15] A. M. Maxam and W. Gilbert, "A New Method for Sequencing DNA," *Proc. of National Academy of Sciences.* 74 (1977), 560–564.
[16] M. A. Billeter, J. E. Dahlberg, H. M. Goodman, et al., "The nucleatide sequence at the 5'-Terminus of the QB RNA minus strand," *Proc. of National Academy of Sciences.* 67 (1970), 921–928.
[17] F. Liang, I. Holt, S. Karamycheve, et al., "Gene index analysis of the human genome estimates approximately 120,000 genes," *Nat. Genet.* 25 (2000), 239–240.
[18] J. M. Claverie, "What if there are only 30,000 human genes," *Science* 291 (2001), 1255–1277.
[19] P. Baldi and S. Brunak, *Bioinfomatics: The Machine Learning Approach.* Boston: MIT Press, 2001.
[20] K. R. Sharma, "Generalized normal distribution to predict periodicity," 93rd AIChE Annual Meeting, Reno, Nev., 2001.

Exercises

1.0 Why does the structure of glycine lead to a polar amino acid and that of alanine a nonpolar amino acid?

2.0 Name some amino acids other than the 20 that constitute the protein primary structure.

3.0 Which amino acids have the largest charge?

4.0 Why is the study of the tertiary structure of proteins important?

5.0 What is the meaning of amino acid residue?

6.0 Compare the principles and process of transcription and translation.

7.0 Explain exons, introns and RNA splicing.

8.0 Show a schematic of operon structure and add a note showing terminators and ribosome-binding sites.

9.0 Display the molecular weights of the 20 different amino acids. What is the molecular weight of insulin? How does this depend on the chain sequence distribution.

10.0 Discuss the differences between

a. nucleoside and nucleotide.

b. codon and anticodon.

c. tRNA and mRNA.

d. amino acid and nucleotide.

e. DNA and RNA.

11.0 Can a codon encode more than one amino acid?

12.0 Do you expect elephants to have more genes than a crocodile. Why?

13.0 Who was credited with the synthesis of a long polyribonucleotide of known sequence?

14.0 Enumerate the process of transcription and translation for CCACGCATGCAGGCGCGCGCGCGGCAT.

15.0 Write a note on denaturization of proteins.

16.0 State true or false.

a. Arginine is a polar amino acid.

b. L-Arnithine is one of the 20 different types of amino acids in protein.

c. Lysine is a basic amino acid.

d. There are sulfide bridges in the insulin molecule.

17.0 Do prokaryotes have genes?

18.0 Where are circular DNA and linear DNA found?

19.0 Write a note on ribosomes.

20.0 What is meant by the DNA shortening problem?

21.0 Which are the start codons and stop codons in the genetic code?

22.0 Why are 61 codons required to code 20 different amino acids?

23.0 Why are there fewer than 61 tRNA molecules?

24.0 What are chaperones and what is their role in protein synthesis?

25.0 Is the mechanism of initiation of translation in eukaryotes not compatible with polycistronic mRNA?

26.0 What disease may be associated with improper protein folding?

27.0 Sketch the lariat formation and splicing of pre-mRNA in eukaryotes.

28.0 How many nucleotides can be handled by the autoradiograph method of directly reading nucleotide sequence distribution?

29.0 A box contains two red and three black marbles. Find the probability that if two marbles are drawn at random (without replacement) both are black. (Ans: 3/10)

30.0 The probability that both are red in problem 29.0 is ________. (Ans: 1/10)

31.0 The probability that one is red and one is blue in problem 30.0 is ________. (Ans: 3/5)

32.0 If at least one child in a family with two children is a boy, the probability that both children are boys is ____________. (Ans: 1/3)

33.0 A shelf contains 6 separate compartments. Show that the number of ways 12 indistinguishable marbles may be placed in the compartment so that no compartment is empty is 462 (*Fermi Dirac Statistics*).

34.0 The probability function of a random variable X is given by

$$
\begin{aligned}
f(x) &= 2p && x = 1 \\
&= p && x = 2 \\
&= 4p && x = 3 \\
&= 0 && \text{otherwise}
\end{aligned}
$$

where p is a constant. What is the probability of $P(0 \leq X \leq 3)$? (Ans: $3p$)

35.0 What is the probability of $X > 1$ in Problem 34.0? (Ans: $5p$)

36.0 Ten percent of the tools produced in a certain manufacturing process turn out to be defective. Assuming that the defective tools can be modeled as rare events and as a Poisson distribution, $f(x) = \lambda^x \exp(-\lambda)/x!)$, with $\lambda = 1$. The probability that in a sample of 10 tools chosen at random 2 are defective is _________. (Ans. 0.1839)

37.0 A typist makes an average of only one error every two pages or 0.5 errors per page. This can be described by the Poisson distribution, $f(x) = \lambda^x \exp(-\lambda)/x!)$ with $\lambda = 0.5$. The probability that the typist will make no error on the next page is ___________. (Ans: 0.607)

38.0 The probability of fewer than two errors in Problem 37.0 is _______. (Ans: 0.986)

39.0 What is the probability of one or more errors in Problem 37.0? (Ans: 0.393)

40.0 Of 80 families with 5 children each, how many would expect to have (a) 3 boys, (b) 5 girls, AND (c) either two or three boys? (Ans: a.10.0; b. 2.5; c. 30)

41.0 During turbulent flow, the age of eddies staying on the wall of the pipe is said to be exponentially distributed. $f(t) = 1/t_{avg} \exp(-t/t_{avg})$, with the mean age being t_{avg} and the variance t_{avg}^2, for $t > 0$. The probability that for a mean time of 100 ms the age of the eddies is greater than 1 second is ____________.
(Ans. 4.5 E.5)

PART 1

Sequence Alignment and Representation

CHAPTER 2

Alignment of a Pair of Sequences

Objectives

The objectives of the chapter are to

- Understand the motivation to study sequence distribution and alignment of sequences.
- Learn to obtain the optimal global alignment of a pair of sequences using dynamic programming (Needleman and Wunsch algorithm).
- Discuss the time taken and space efficiency of global pairwise alignment.
- Learn to obtain optimal local alignment of a pair of sequences using dynamic programming.
- Discuss the time taken and space efficiency of the Smith Waterman algorithm.
- Become familiar with the affine gap model.
- Determine the connection to commercial software packages from techniques discussed in this chapter.

2.1 Introduction to Pairwise Sequence Alignment

Sequence comparison is a field in itself in computer science. It has a lot of interesting applications in bioinformatics. The process of lining up two or more sequences to obtain matches between them is called *sequence alignment.* When two sequences are lined up, it is called a *pairwise alignment* [1], and when more than two are examined, it is referred to as *multiple-sequence alignment.* The sequence distribution can consist of the 20 different amino acids in the protein primary structure of a polypeptide, the 4 nucleotide base pairs in the ribonucleic acid (RNA), or the 4 nucleotide base pairs in the deoxyribonucleic acid (DNA). The similarity among sequences may be based on evolutionary, structural, or functional relationships among them.

Similarities found among nucleotide sequences are also called *identity*. *Conservation* refers to changes at a specific position of an amino acid sequence that preserve the physicochemical properties of the original residue. Similarity attributed to descent from a common ancestor is *homology*. When two or more sequences are aligned and linked to a common ancestor, and when mismatches are found in the alignment, then the mismatches can be detected as *point mutations*.

Gaps in the sequences can be seen as *indels*. Sequence similarity among protein sequences indicates the degree of conservation among them. Conservation in DNA or RNA base pairs can indicate similar functional and structural roles. The objective of sequence alignment is to be able to select two or more sequences and compare them to determine the measure of similarity. The grade of similarity is a measurement used to draw conclusions about whether homology exists between two sequences.

Biomolecules, i.e., DNA and RNA and proteins they encode by gene expression, have been found to be central to the functions of organisms. Their structure and function and their study can be the next frontier in science and is an important fruit of labor from the study of *biochemistry*. These biomolecules can be viewed as polymers—as polynucleotide and polypeptide. The two polymers can be further viewed as random multicomponent copolymers. The polypeptide has a random microstructure with 20^n possible sequence distributions, where n is the degree of polymerization or length of the polypeptide. The 20 comes from the alphabet of amino acids that can be found in the backbone polypeptide chain. The polynucleotide, on the other hand, can have 4^n possible sequence distributions for a length of n of the poynucleotide chain. The 4 comes from the possible base pairs—adenine, guanine, cytosine, and thymine. The RNA would include uracil. The DNA molecule has up to 3 billion base pairs. The sequence distribution microstructures of DNA and insulin were discovered by F. Sanger, who was awarded the Nobel Prize twice, once in 1958 and then again in 1980. The information content in the DNA molecule is high. Most of the information is in the form of random sequences (Fig. 2.1).

As can be seen in the figure, during gene transcription and translation, the random sequences of DNA play an important role in the formation of the copies, and then the codons form the protein molecule by polymerase chain reaction (PCR) in the presence of the polymerase enzyme. The human genome needs to be annotated. Functions of organisms can be linked to the genes at various locations in the genome. The annotation of the genome, i.e., addressing the genes as the originators of such and such functions, is the study of *functional genomics* or *metabolomics*. Sequence information needs to be stored and retrieved from large databases and drives the study of *bioinformatics*. The sequence in itself is not informative. Sequence alignment and analysis are needed to perform these tasks.

```
ORIGIN
   1 tcatagaccg tgccttctag ctgcgacctc acatggtgga aaggggaagg caacctccct
  61 gtagcctctt ttaaaagggc attaatcgca ttcacggggt ctccatcctc ttggcctaac
 121 cacctcccaa aagccctacc ttttagtaat atcacatggg gagttagaat ttcactatat
 181 gaattttggg gggacacaaa catttatgcc acagcagata tctttctacc accttatttg
 241 gtgatttctg ggttttgttt gtttgtttaa gacagagtct cgctctgtcg gccaggctgg
 301 agtgcagtgg caccatctcg gctcaatgca accttcgcct ccccggttca agcgattctc
 361 ctgcctcagc ctcccaagta gctgggatta cggacgtgtg ccaccacgcc tggctaattt
 421 ttgtattttt agtagagact gggtttcacc attttggcca ggctggtccc gaactgctga
 481 gttcaggtga tccacccgcc tcggcctccc aaagttctgg gattacaggc gtgagccacc
 541 atgtccggct ggtgatttct gtttaaaagt tttttcttaa agtgtttttt cccacctagt
 601 ttttcattga atgggtaaaa cattctacat ttgcttttat taaaacaaga aatgaatttt
 661 gctgcatttc aatttataga ttttactatc ctacctcgtg ccaggttctg tgctaagtgc
 721 tgtatatatc tgtgatcaca tttaactttt ataacaagcc aaatgagcag gaactcttat
 781 ctctatctta cagacgaaga atccaaagac cagggacagt aagtaatttg ctcacctggt
 841 ttgccagcct ccatgacaca tcgccgtcca gttctgcctt taattaccaa agcacaacac
 901 gctgctttga ttcccctctc ctcggcgcca gaattcaaga gtgaagttaa accgcaaggg
 961 ctgagttaga agattggcct cagttccctg ttcccaccag caggtggcac cgtctcctag
1021 cggaattctt acttgaacgt tttgcttcca tttctgcaga ggcatggtga acacagttac
1081 accaccaaag tgttcctcct ggctgagttt gcctatcttg ttcagtgaag acaacccatg
1141 aggacaaatg gtgttaatga gaagcttttg cggagttaca gagatcctcg tatttcttta
1201 aaatacacct aataacgtta actctgcaat aatttgtaga tcatgttaaa tcttagctat
1261 cttcctcttg ccacccagtg tgcttcaagc cacatggttc agagcaccat ttaatgtgaa
1321 actccaattt taaaacaaag tgaaccttcc ttttacaaaa ccatgagaca agttacagag
1381 taatgaccac ccacatgacc ttgaagtgat tttgagtgag tgagtgtaac ttccgtggct
1441 gccatttaaa ttggattcaa atccaaatgg ctccacctcc atgtcatcag acctcttgtg
1501 ccctgattcc cttggctaag ttcacagtac cttccacatc aggttgtggc aatgattacc
1561 tgaggttaat acgataaaag cacatggtaa gcactcctaa atgatagcca atataaagac
1621 tcagttctcc caattccaag ggtccccacc atgatagaaa aggatctttt ggtaaataga
1681 gtatgtttag ctcttgctag gtctttaaat actttgctgg gggccaggca ccatggctca
1741 cacctgtaat cccaccgcct taggagactg aggctggagg atcctttgcg gccaagagtt
1801 tgagaccagc ctgggcaaca cagcaagacc ctatttctac aaaaataaaa ataaaaatta
1861 accaggcttt gtacacactt gtagtcccat tacttgggag gctgaggcag gaggatccct
1921 caagcccaag agttcaaagc tgtagtgagc tatgattgcg ccactgcact ccagcctggg
1981 tgacagagta agactctgtt tcaaaacaac aacaacaaac aaaaacctca aaacctcttt
2041 gttggactta acttccagct cctccatgta gtaccttagt acccttgcag cccgtttctc
2101 ttttacaaga caacaatgtt gttataaact catttggatg tggtcccgtg gaggagtatt
```

FIGURE 2.1 Nucleotide sequence of *Homo sapiens* base pairs 1–3001 (From NHLBI Resequencing and Genotyping Service, N01.NV.48196, J. Craig Venter Institute, Rockville, MD, http://rsng.nhlbi.nih.gov [2].)

2.2 Why Study Sequence Alignment

Sequence alignment can be the key to finding a cure for autoimmune disorders. Autoimmune disorders are those in which harm is inflicted on a patient's cells by signals from within the patient himself or herself by mistake. Rather than targeted annihilation of the culprit virus by the immune system, the signal from within the patient triggers the attack of the patient's cells. This is a case of mistaken identity. Thus a double deleterious effect is in place—i.e., failure of the immune system and damage done to the cells.

Researchers have shown that the protein signal in the patient is specific to the sequence distribution in the cells. When the sequence distribution of the culprit virus and the sequence distribution in the cells at the site of the disease in the patient are identical, the harm is done by signal action directed at the common sequence distribution, and then the cells of the patient are harmed. Thus, when the signal from the protein to attack the infecting virus is confused and the cells in the body with the identical sequence distribution are attacked instead, autoimmune disorder sets in. Once this match is determined, then drugs can be designed whose therapeutic action can alter the gene expression, thus effecting a cure. In multiple sclerosis, the immune system's T cells attack the patient's nerve cells. In a similar fashion, bone cells are attacked during the onset of rheumatoid arthritis. The infection is still unchecked, and this results in a double deleterious effect on the patient. In multiple sclerosis, it was conjectured that the myelin sheath proteins that were sequenced were matched in a protein database with similar bacterial and viral sequences, and tests were conducted to determine whether the T cells attacked the myelin sheath proteins with the same sequence as the bacterial and virus proteins. The result was identification of certain bacterial and viral proteins that were confused with myelin sheath proteins. Thus autoimmune diseases arise from an overactive immune response of the human anatomy against substances and tissues usually present in the human anatomy—i.e., the human anatomy attacks its own cells. There are more than 80 autoimmune disorders reported today. They afflict 5.7 percent of the population. Some of the known autoimmune disorders are listed in Table 2.1.

Recently, a key set of genes that can be used to manipulate immune system activity was discovered [3]. This discovery may lead to new therapies for autoimmune disease. The immune system is often described as a kind of military unit, a defense network that

Addison's disease	Aplastic anemia	Autoimmune hepatitis
Celiac disease	Crohn's disease	Diabetes mellitus
Gestational pemphigoid	Goodpasture's syndrome	Graves' disease
Kawasaki's disease	Multiple scelorisis	Myasthenia gravis
Opsoclonus myoclonus syndrome	Optic neuritis	Ord's thyroiditis
Pemphigus	Pernicious anemia	Primary biliary cirrhosis
Rheumatoid arthritis	Reiter's syndrome	Sjögren's syndrome
Takayasu's arteritis	Temporal arteritis	Hemolytic anemia
Wegener's granulomatosis	Primary thyroiditis	Ulcerative colitus
Hashimoto's thyroiditis	Systematic lupus	Dermatomyositis

TABLE 2.1 List of Autoimmune Disorders

guards the body from invaders. White blood cells, or T cells, serve as frontline soldiers of immune defense, engaging invading pathogens head on. T cells are commanded by regulatory T cells. Regulatory T cells are themselves controlled by a master gene regulator called Foxp3. Master gene regulators bind to specific genes and control their level of activity, which, in turn, affects the behavior of cells. In fact, when Foxp3 stops functioning, the body can no longer produce working regulatory T cells. When this happens, the frontline T cells damage multiple organs and cause symptoms of type 1 diabetes and Crohn's disease. Researchers have scanned the entire genome of T cells and have located the genes controlled by Foxp3. Roughly 30 genes were found to be controlled directly by Foxp3, and one, called *Ptpn22,* showed a particularly strong affinity. The list of the genes that Foxp3 targets provides an initial map of the circuitry of these cells, which is important for understanding how they control a healthy immune response. Autoimmune diseases on a molecular level can be considered in a "black box." The molecular mechanisms of these diseases can be understood using sequence alignment.

Sequence alignment is usually attempted in terms of sequence database searching. The sequence is analyzed by comparative methods against existing databases to *develop hypotheses* concerning relatives and function. For example, an abundant message in a cancer cell line may bear similarity to protein phosphates genes. This relationship would prompt experimental scientists to investigate the role of phosphorylation and dephosphorylation in the regulation of cellular transformation.

The *common inheritance* can be found and the *evolutionary tree* constructed from the knowledge gained by sequence alignment. Evolution is considered at the molecular level in such projects. Chimpanzee and *Homo sapiens* were found to have a common ancestor recently. The wings of bats and those of butterflies have evolved independently. Evolution can be linked to changes in DNA. Molecular evolution is the study of the history of changes in an organism during evolution and its relation to changes in DNA. For example, cytochrome C and hemoglobin were sequenced. Family trees were constructed based on the assumption that closely related organisms have similar sequences. Thus chimpanzee was found to be closer to *Homo sapiens* than to rattlesnake. Sequence comparisons thus are motivated by the study of evolution at a molecular level.

Prior to analyses using DNA sequences, it is first necessary to determine the actual sequence itself. The length of DNA can be as much as 3 billion base pairs. Practical considerations limit the sequencing of DNA all at once because of this length. Via Sanger's plus-minus method and other methods, about 450 to 500 base pairs can be sequenced at a time. Many overlapping small pieces are sequenced. Then these fragments are assembled into one long contiguous sequence. One problem is that the location of the fragments within the genome and with respect to each other is not generally known. Enough fragments are sequenced so that there will be many overlaps between them, and the fragments can be matched up and assembled. This process is called *shotgun sequencing*. Sequence similarity is used to obtain the overlaps needed in shotgun sequencing. DNA sequence can be used to obtain the translated *polypeptide microstructure* using the genetic code.

Gene finding and its role in disease mechanisms have been receiving increased attention in recent years. These can be achieved by sequence alignment. For example, genes responsible for longevity have been discovered recently by the scientists at the National Institute of Aging. These genes can be searched for in sequence databases.

The genomes of various organisms have been sequenced in their entirety and the information stored using computer resources world over. *Sequence database searches* can be conducted depending on the problem at hand. For this, reliable sequence alignment methods are needed. In order to reduce database search costs, more research is being undertaken in this area. The databases have doubled in size because of the advent of high-throughput automated fluorescent DNA sequencing technology. Analyses of DNA sequences are used in the construction of phylogenetic trees, in genetic engineering using restriction site mapping, in determining gene structure through intron/exon prediction, in making inferences about protein coding sequences through open-reading-frame (ORF) analysis, etc.

Drugs can be designed based on the sequence distribution of the nucleotides or protein in culprit viruses. Examples of viruses for which this has been done include influenza virus, Japanese yellow fever virus, measles virus, rabies virus, TA coliphase virus, cauliflower mosaic virus, human immune deficiency virus (HIV) type 2, vaccinia virus, polio virus, serum hepatitis virus, etc. The drugs interact with the protein in the virus and changes the protein signaling that originally caused the disease, leading to a cure. On the other hand, the gene expression can be altered by therapeutic action, leading to a change in the protein signal, effecting a cure.

The *protein secondary structure* can be deduced from the sequence distribution of the polynucleotide. There are three different types of protein secondary structures—α-helix, β-pleated-sheet, and γ-coil/loop conformations. The 1997 Nobel Prize for medicine went to S. B. Prusiner for his work on prions [4]. Prion proteins have been associated with so-called mad cow disease and its human variant, Creutzfeldt-Jakob syndrome. In these proteins, the same sequence may adopt different stable conformations—a bad conformation with a mixture of helices and sheets and a normal conformation with a bundle of helices. The bad conformation prions were shown to have an autocatalytic effect and may be responsible for the transformation of normal conformation prions into bad ones. Based on local sequence information, such conformational conflicts those in prion proteins will be difficult to solve by any prediction method. However, a local method may be able to report that a piece of a sequence may have a higher potential for both helix and sheet as opposed to coil.

The *protein folding problem* can be viewed as given the primary protein microstructure, what is the final three-dimensional (3D) folding of the protein? It was shown by Anfinsen and colleagues [5] that ribonuclease could be denatured and refolded without loss of enzymatic activity. This showed that all the information that a protein needs resides in its primary structure. Hence it is possible to derive the rules for protein folding from analyses of sequences with known structures. These rules can be applied to prediction of the 3D structure of protein given only a linear sequence of amino acids.

The DNA sequence of a *clone* can be obtained from the study of biologic sequences. In an experiment to clone a specific gene whose sequence is known, it is necessary to check and validate that the cloned sequence is identical to the published one. Should the results of sequence similarity reveal misaligned or mismatched sequences, the experiments must be designed to correct those sequences. For example, cloning errors can result from using inappropriate primers at the cloning step. The use of a low-fidelity enzyme in a PCR experiment can produce errors during cloning.

Proteins are *classified* according to their sequence distribution. Multiple sequence alignment (MSA) is used in the study of genetic diseases.

2.3 Alignment Grading Function

A *string* is an ordered sequence of characters or symbols more generally. These characters or symbols in particular are usually drawn from a set called the *alphabet.* The alphabet is a set of characters or symbols from which the strings are constructed. Here is an example: Consider a section of the chain sequence distribution of DNA of *Homo sapiens* as shown in Eq. (2.1):

$$[S]: \quad \text{cttgatctta} \tag{2.1}$$

$[S]$ is a string. It can be seen that the string contains the characters c, t, g, and a only. Thus the alphabet Σ for string $[S]$ is (a, c, g, t). The set of all strings over Σ of any length is the *Kleene closure* of Σ and is denoted Σ^*. The length of the string is the number of characters contained in the string. String length can be fixed or variable. Now consider another string $[T]$ drawn from the chimpanzee:

$$[T]: \quad \text{cttaatcaaa} \tag{2.2}$$

In order to measure how similar the strings $[S]$ and $[T]$ are or to quantitate the similarity of the two strings $[S]$ and $[T]$, an alignment grading function is introduced.

The simplest events that occur during the course of molecular evolution are substitution of one base for another and insertion or deletion of a base pair. Radiation can cause these changes. When a is deleted in string [S],

$$[S_a]: \quad \text{c \_ t g a t c t t a} \tag{2.3}$$

$$[S]: \quad \text{c t t g a t c t t a} \tag{2.4}$$

An insertion of R between t and a is shown by

$$[S_a]: \quad \text{c t t g a t c t t \_ a} \tag{2.5}$$

Two letters arranged one over another are called *matched.* If two matched letters are equal, then the match is called an *identity.* Otherwise, the match is called a *mismatch.* An insertion or deletion (*indel*) is one or more letters aligned against a_. A mismatch is generally a substitution. When only the matches and not the details of the indels are specified, the resulting arrangement is called a *trace.* Consider one possible alignment between [S] and [T]:

$$[S']: \quad \text{c t \ t g a t \ c \ t t a} \tag{2.6}$$

$$[T']: \quad \text{c t \ t a a t \ c \ a a a} \tag{2.7}$$

This alignment has 7 identities, 3 mismatches, and 0 indels. This alignment represents a certain hypothesis about the evolution of the sequences. Seven of the nucleotides have not changed, and three nucleotides have been inserted, deleted, or substituted. In order to evaluate the goodness of an alignment, an identity is given a value +2, and a substitution and an indel are given values of –1 and a mismatch a value of –1. The result for alignment [S′] and [T′] is $7 \times 2 - 3 - 1 = 11.0$.

Grading Function $\sigma(x, y)$ denotes the grade of alignment of aligning x and y, where x and y are each a single character or space. Thus, in the examples shown in Eqs. (2.6) and (2.7), the grading function can be written as $\sigma(\text{a, a}) = 2$, $\sigma(\_, \text{g}) = -1 = \sigma(\text{c}, \_)$; $\sigma(\text{a, t}) = -1$.

Length of String If S is a string, then $|S|$ denotes the length of the string, and $S[i]$ denotes the ith character of S. For example, from Eq. (2.7), $|S| = 10$, $S[2] = \text{t}$, and $S[8] = \text{t}$.

Alignment of Strings An alignment A maps S and T into strings S' and T' that may contain indels or space characters, where $|S'| = |T'|$, and removal of indels from S' and T' leaves S and T, respectively. The grade of the alignment A is given by

$$A = \Sigma_L^1 \; \sigma(S'[i], T'[i]) \tag{2.8}$$

where $1 = |S'| = |T'|$.

Optimal Alignment An optimal alignment of S and T is one that has the maximum possible alignment grade for these two strings.

Given two strings $[S]$ and $[T]$ of lengths n and the other sequence m, the *number of possible alignments* of the two strings can be estimated as follows: The length of the new sequences after introducing indels is

$$\text{Max}[n, m] \leq L \leq n + m \tag{2.9}$$

Let $q(i, j)$ = number of alignments of i letters of S with j letters of T. It can be shown using advanced combinatorics that

$$q(n, m) = q(n-1, m) + q(n-1, m-1) + q(n, m-1) \tag{2.10}$$

$$q(n,n) = \frac{(1+\sqrt{2})^{2n+1}}{\sqrt{n}} \tag{2.11}$$

This comes from the fact that the beginning of the alignment can be only one of three things—an identity, a substitution, or an indel.

In the *brute-force method* for finding alignments, all the possible alignments are searched, and the output is the maximum grade. A *subsequence* of a string S means a sequence of characters of S that need not be consecutive in S but do retain their order as given in S. For instance, aatt is a substring of S = aattcctc. Assuming that $[S] = [T] = n$ and that the grading function restricted to $\sigma(\_,\_) \leq 0$, the algorithm for finding all the possible alignments of S and T is given in [6]. The running time and storage space needed for this algorithm are calculated as follows: String of length n has ${}^nC_i$ subsequences of length i, that is,

$${}^nC_i = \frac{n!}{(n-i)!i!} \tag{2.12}$$

The number of pairs of subsequences of length $i = ({}^nC_i)^2$.

$$\text{Alignment length} = n + n - i = 2n - I \tag{2.13}$$

With i matched, $(n - i)$ mismatched = number of blanks. Grade of alignment of each pair in alignment is calculated as follows:

$$\text{Total grade of basic operations} = \sum_{0}^{\infty} ({}^nC_i)^2(2n-i) \tag{2.14}$$

It can be shown by induction that

$$n^{2n}C_n > 2^{2n} \qquad \text{for } n > 4 \tag{2.15}$$

$$\text{Null case, } n = 4; \qquad \frac{4(8!)}{(4!)^2} > 2^8 \tag{2.16}$$

$$280 > 256 \tag{2.17}$$

$$\frac{(2n)!n}{(n!)^2} > 2^{2n} \tag{2.18}$$

Show that

$$\frac{(n+1)(2n+2)!}{(n+1)!(n+1)!} > 2^{2(n+1)} \tag{2.19}$$

$$\frac{2(n+1)(2n)!n}{n(n!)^2} > 2.2^{2n} \tag{2.20}$$

Given $\frac{2n!n}{(n!)^2} > 2^{2n}$, show that

$$\frac{2n+1}{n} > n \tag{2.21}$$

or show that

$$2 + 1/n > 2 \tag{2.22}$$

This is true for any natural n.

$$\sum_{i=0}^{\infty} {}^nC_i^2(2n-i) > n\sum_{i=0}^{\infty} {}^nC_i^2 \tag{2.23}$$

$$n\sum_{i=0}^{n} {}^nC_i^2 = {}^nC_0^2 + {}^nC_1^2 + {}^nC_2^2 + \cdots \tag{2.24}$$

Now $a^2 + b^2 + c^2 + d^2 + \cdots = (a + b + c + d + \cdots)^2 - 2(abc\leftarrow)$ and (2.25)

$${}^nC_0{}^2 + {}^nC_1{}^2 + {}^nC_2{}^2 + \cdots = ({}^nC_0 + {}^nC_1 + {}^nC_2 + \cdots)^2 - 2(\leftarrow) \tag{2.26}$$

$$= 1^{2n} + {}^{2n}C_1 + {}^{2n}C_2 + \cdots \tag{2.27}$$

or

$$(1 + 1)^n = {}^nC_0 + {}^nC_1 + {}^nC_2 + \cdots \tag{2.28}$$

or

$$O(n\Sigma\, {}^nC_i^2) = O(n2^{2n}) \tag{2.29}$$

Running time is at least $n2^{2n}$ for string of length n. The rate of growth of the higher-order term is the most important determinant of how long an algorithm runs on large inputs independent of constant of proportionality and any lower-order terms.

2.4 Optimal Global Alignment of a Pair of Sequences

2.4.1 Needleman and Wunsch Algorithm

A computer-adaptable method using dynamic programming algorithm was suggested for *optimal global alignment of two sequences* by Needleman and Wunsch [7]. In global alignment, the two sequences are aligned end to end. Needleman and Wunsch developed a computer-based statistical and general method applicable to the search for similarities in the amino acid sequences of two proteins. A number of authors have studied the question of how to construct a good grading function for sequence comparison, including Altschul and colleagues [8], Altschul [9,10], and Altschul and Gish [11]. From these findings it is possible to determine whether significant homology exists between the proteins. Another goal for seeking alignment is to establish full genetic relationships between proteins. This information is used to trace their possible evolutionary development. The *maximum match* can be defined as the largest number of amino acids of one protein that can be matched with those of another protein while allowing for all possible deletions.

Prior to automation of sequence alignment, it was done by hand—by eyeballing them. Plots were generated by creating a grid with one sequence on top and another on the side. Red dots were placed wherever two sequences matched. Diagonal lines could be found in the graph. When connected, they formed the optimal alignment of the pair of sequences. One method to arrive at the optimal alignment is to calculate all possible alignments and assign each alignment a grade of alignment with penalties for the gaps—both for quantity and size of gaps—and then to choose the alignment with the highest grade. This is computationally prohibitive.

Needleman and Wunsch [7] were the first to devise a computationally feasible method for automated sequence alignment. Their application is built on an analogy with the old visual comparison method of alignment. They used a Pascal array that maps directly to the sequence alignment plot, and their algorithm then operated within this context. They describe the alignment process as "pathways through the array" that are evaluated to find the "maximum match." The algorithm works by progressively building a path through the array, gaining rewards for obtaining matches and incurring penalties for gaps. This kind of approach constitutes dynamic programming, a common method for optimizing computer algorithms. Needleman and Wunsch obtained alignment of whole myoglobin and human β-hemoglobin and alignment of bovine pancreatic ribonuclease and hen's egg lysozyme.

Given strings S and T with $|S| = n$ and $|T| = m$, an optimal global alignment of S and T can be obtained using dynamic programming. A grade of alignment $G(i, j)$ of string $S(i)$ and $T(j)$ is defined. The grade of optimal alignment of S and T is $G(m, n)$. The dynamic programming method is used to solve for the general problem of computing all grades $G(i, j)$ with $0 \le i \le n$ and $0 \le j \le m$ in order of increasing i and j.

Algorithm 2.1 *Global Alignment*
Basis:

$$G(0, 0) = 0 \tag{2.30}$$

$$G(i, 0) = G(i-1, 0) + \sigma[S(i), \_] \qquad \text{for } i > 0 \tag{2.31}$$

$$G(0, j) = G(0, j-1) + \sigma[\_, T(j)] \qquad \text{for } j > 0 \tag{2.32}$$

Recurrence formula:

$$\begin{aligned} G(i, j) = \max\{&G(i-1, j-1) + \sigma[S(i), T(j)]\}, \\ &G(i-1, j) + \sigma[S(i), \_], \\ &G(i, j-1) + \sigma[\_, T(j)] \end{aligned} \tag{2.33}$$

The interpretation of the alignment is as follows: Consider the optimal alignment of the first i characters from S and the first j characters from T. In particular, consider the last aligned pair of characters in such an alignment. This last pair must be one of the following:

1. $[S(i), T(j)]$, in which case the remaining alignment excluding this pair must be an optimal alignment of $S(1), \ldots, S(i-1)$ and $T(i), \ldots, T(j-1)$ (i.e., It must have grade $G(i-1, j-1)$ or
2. $[S(i), \_]$, in which case the remaining alignment excluding this pair must have grade $G(i-1, j)$ or
3. $[\_, T(j)]$, in which case the remaining alignment excluding this pair must have grade $G(i, j-1)$.

A traceback procedure is used to obtain all the alignments.

Example 2.1 *Global Alignment of Two Sequences by Dynamic Programming*
Demonstrate the dynamic programming method to obtain the global alignment of the two strings with the following sequence distribution:

S: a c g t t t g c a

T: c c a t g c g a

Solution
The grading function used was +2 for a match, –1 for a mismatch, and –1 for an indel.

Recovering the alignments:

```
I. a  c  g  t  t  g  c  _  a
   c  c  a  t  _  g  c  g  a
```

$$\text{Grade of alignment: } 5 \times 2 - 4 = +6 \tag{2.34}$$

S[i] T[j]		a	c	g	t	t	g	c	a
	0	−1	−2	−3	−4	−5	−6	−7	−8
c	−1	−1	+1	0	−1	−2	−3	−4	−5
c	−2	−2	+1	0	−1	−2	−3	−1	−2
a	−3	0	0	0	−1	−2	−3	−2	+1
t	−4	−1	−1	−1	+2	+1	0	−1	0
g	−5	−2	−2	+1	+1	+1	+3	+2	+1
c	−6	−3	0	0	0	0	+2	+5	+4
g	−7	−4	−1	+2	+1	0	+2	+4	+4
a	−8	−5	−2	+1	+1	0	+1	+3	+6

TABLE 2.2 Global Alignment of Two Sequences—Dynamic Programming

```
II. a  c  g  t  t  g  c  _  a
    c  c  _  a  t  g  c  g  a
```

Grade of alignment: $5 \times 2 - 4 = +6$ (2.35)

Example 2.2 *When the Grade of Alignment is 0 and When the Grade of Alignment Is Less Than 0*

In Example 2.1, what is the meaning when the grade of alignment is (a) 0 and (b) negative (or less than zero)? From Table 2.2, an alignment of two sequences in which the grade of alignment is 0 can be selected as follows: A 0 grade of alignment can be expected when these two sequences are aligned:

```
S:  a  c  g  t  t  g
T:  c  c  a  t
```

The tracebacks for alignment of *S* and *T* would be

```
I. a  c  g  t  t  g
   _  c  c  a  t  _
```

Grade of alignment: $2 \times 2 - 2 \times 1 - 2 \times 1 = 0$

```
II. a  c  g  t  t  g
    c  c  _  a  t  _
```

Grade of alignment: $2 \times 2 - 2 \times 1 - 2 \times 1 = 0$

```
III. a  c  g  t  t  g
     c  c  a  _  t  _
```

Grade of alignment: $2 \times 2 - 2 \times 1 - 2 \times 1 = 0$

Compared with the alignments in Example 2.1, where the number of characters that were aligned was 5 out of a length of 9 of the mapped string, in Example 2.2, the number of characters that were aligned was 2 out of a length 6 of the mapped string. Most of the mapped string in Example 2.1 was matched, and a minority of characters was aligned in Example 2.2. Maybe a positive grade of alignment can signify that more characters are aligned in the string, and a 0 grade of alignment can denote that only few characters are aligned.

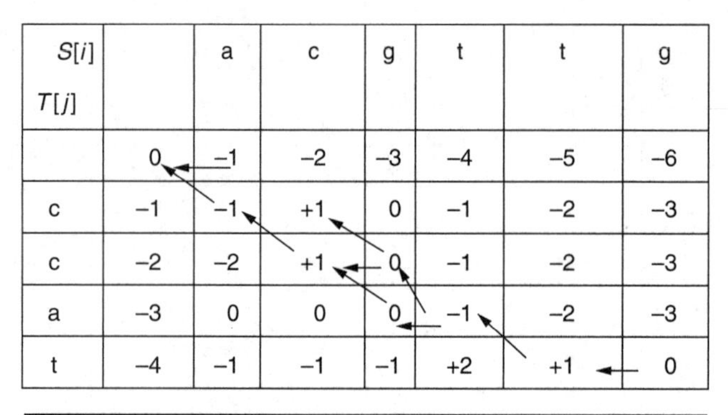

S[i] / T[j]		a	c	g	t	t	g
	0	–1	–2	–3	–4	–5	–6
c	–1	–1	+1	0	–1	–2	–3
c	–2	–2	+1	0	–1	–2	–3
a	–3	0	0	0	–1	–2	–3
t	–4	–1	–1	–1	+2	+1	0

Table 2.3 Global Alignment of Two Sequences with Grade of Alignment = 0

From Table 2.3, two strings can be selected in which the grade of alignment can be expected to be negative. These strings are as follows:

S: a c g t
T: c c a

The traceback procedure can be used, and the alignments recovered for a grade of alignment –1 would be

IV. a c g t
_ c c a

Grade of alignment: $1 \times 2 - 2 \times 1 - 1 \times 1 = -1$

V. a c g t
c c _ a

Grade of alignment: $1 \times 2 - 2 \times 1 - 1 \times 1 = -1$

VI. a c g t
c c a _

Grade of alignment: $1 \times 2 - 2 \times 1 - 2 \times 1 = -1$

In the tracebacks of the alignments shown in IV, V, and VI recovered from the alignments shown in Table 2.3, only one character is aligned compared with a length of 4 of the mapped string. The number of misaligned characters is larger than the aligned characters.

Thus the interpretation of the grade of alignment can be made as follows based on the preceding calculations:

1. When the grade of alignment is greater than 0, the number of characters aligned is greater than the number of mismatched characters in the sequence.
2. When the grade of alignment is 0, the number of matched characters and the number of mismatched characters are equal to each other.
3. When the grade of alignment is less than 0, the number of mismatched characters is greater than the number of matched characters.

Example 2.3 *Semiglobal Alignment*
Consider the two strings gatcatcgcagcgttagtagc and gctgcg. An optimal global alignment returns the following:

g a t c a t c g c a g c g t t a g t a g c

g _ _ c _ t _ g c _ g _ _ _ _ _ _ _ _ _

A biologically more meaningful alignment would be

g a t c a t c g c a g c g t t a g t a g c

_ _ _ _ _ _ _ g c t g c g _ _ _ _ _ _ _ _

How would you change the grading scheme to return the latter alignment compared with the former?

Solution
This can be done by awarding no penalty to end gaps or allowing free end gaps. This would be *semiglobal alignment.* All gaps inserted before or after the alignment will not be penalized. The traceback procedure is similar to that of the global alignment procedure.

2.5 Dynamic Programming

Bellman [12] began the systematic study of dynamic programming in 1955. He used a tabular solution method that was called *dynamic programming*. Prior to Bellman's work, dynamic programming was used in optimization techniques such as in finding the optimal reactor heat-exchanger network of the sulfur trioxide–forming step in the oleum process to manufacture sulfuric acid. Bellman was the first to provide the approach with a solid mathematical basis. The time taken for the longest common subsequence problem, as suggested by Smith and Waterman [13] using dynamic programming methods, is $O(mn)$. Knuth [14] posed the question of whether subquadratic algorithms for the longest common subsequence (LCS) problem exist. Masek and Paterson [15] answered this question in the affirmative by giving an algorithm that runs in $O(mn)/\lg(n)$ time, where $n \leq m$, and the sequences are drawn from a bounded size. For the special case in which no element appears more than once in an input sequence, Szymanski [16] showed that the problem can be solved in $O(n + m)\lg(n + m)$ time. In 1970, Knuth conjectured that a linear time algorithm for the problem of finding the LCS would be impossible. It will be shown in a subsequent chapter that the LCS of two strings can be found in linear time using a generalized suffix tree.

The solutions to subproblems are combined to solve a given problem in the dynamic programming method. This is similar to the divide-and-conquer principle used a lot in computer algorithms. In the divide-and-conquer strategy, the problem is split into subproblems. The subproblems are solved recursively, and the solutions are combined to form the solution of the original problem. When the subproblems are not independent of each other, the dynamic programming method may be applicable. Every subproblem is solved only once, and the results are saved in the tabular form in the dynamic programming method compared with the divide-and-conquer, where more work is done than necessary. Optimization problems can use the dynamic programming method. Many solutions are possible for such problems. Each solution has a grade, and the extremum is of interest as the optimal solution. The development of a dynamic programming algorithm can be broken into four steps [17]:

1. Characterize the structure of an optimal solution.
2. Recursively define the grade of an optimal solution.
3. Compute the grade of an optimal solution in a bottom-up fashion.
4. Construct an optimal solution from the computed information.

Two key ingredients to the application of dynamic programming problems are identification of the optimal substructure and overlapping subproblems.

2.6 Time Analysis and Space Efficiency

The optimal global alignments can be obtained using dynamic programming in $O(mn)$ time. When $m = n$, the time taken becomes $O(n^2)$.

Proof: An $(m + 1)(n + 1)$ table needs to be filled. Each and every entry is computed with a maximum of six table look-ups, three additions, and a three-way maximum in time c.

Complexity of the algorithm $= c(n + 1)(m + 1) = O(mn)$. Reconstructing a single alignment $= O(n + m)$ time. The space required for retaining the grades of alignment in the table is mn also.

2.7 Dynamic Arrays and $O(n)$ Space

The space required in dynamic programming during global alignment of two sequences can be reduced from $O(n^2)$ as follows: It can be realized that only the global optimal grade of alignment is needed. Thus two dynamic rows at any given time will be sufficient. In order to construct the next row of alignments, the previous row is sufficient. Thus space required will be $O(2m)$ or $O(m)$, (Hirschberg, [18]). Reconstructing an alignment is somewhat more complicated but can be achieved in $O(n + m)$ space and $O(nm)$ time with a divide-and-conquer approach [18,19].

2.8 Subquadratic Algorithms for Longest Common Subsequence Problems

Hunt and Szymanski [20] introduced a fast algorithm for computing LCSs. They provided a running time of $O(r + n)\lg(n)$, where r is the total number of ordered pairs of positions at which the two sequences match. In the worst case, the algorithm has an $O[n^2 \lg(n)]$ running time. However, for applications where most positions of one sequence have few matches in the other sequence, a running time of $O[n \lg(n)]$ can be expected.

Let S be a finite sequence of elements chosen from some alphabet Σ. The length of the sequence S is $|S|$. $S[i]$ is the ith element of S, and $S[i{:}j]$ is the sequence $S[i]$, $S[i + 1]$, $S[i + 2]$, . . . , $S[j]$. If U and V are finite sequences, then U is said to be a subsequence of V if there exists a monotonically increasing sequence of integers $r_1, r_2, \ldots, r_{|U|}$ such that $U[i] = V[r_i]$ for $1 \leq i \leq |U|$. U is a common subsequence of S and T if U is a subsequence of both S and T. A longest common subsequence is a common subsequence of greatest possible length. Both sequences are assumed to have the same length n. The number of elements in set $\{(i, j)$ such that $S[i] = T[j]\}$ is denoted by r.

The data structure used in the algorithm of Hunt and Szymanski is $G_{i,k}$, an array of threshold grades defined by the smallest j such that $S[1{:}i]$ and $T[1{:}j]$ contain a common subsequence of length k. Each $G_{i,k}$ may be considered as a pointer that signifies how much of the T sequence is needed to produce a common subsequence of length k with the first i elements of S. Each row of the G array can be seen to be increasing; i.e.,

Lemma 2.1 $G_{i,1} < G_{i,2} < \cdots < G_{i,p}$, as defined earlier.

Lemma 2.2 $G_{i,k.1} < G_{i+1,k} \leq G_{i,k}$

Lemma 2.3 $G_{i+1,k}$ = smallest j such that $S[i + 1] = T[i]$ and $G_{i,k}$ if no such j exists $G_{i,k-1} < j < G_{i,k}$

Lemmas 2.1, 2.2, and 2.3 are stated, and the proofs are available in [21]. This algorithm can be completed with an $O[n^2 \lg(n)]$ time efficiency to determine the length of the common subsequence. This can be refined to improve the running time to $O(r + n)\lg(n)$, and the longest common subsequence can be recovered.

Algorithm 2.2 *Length of Longest Increasing Subsequence*

```
G[0] = 0
Recurrence formula:
        For i = 1 to n,
                G[i] = n + 1
                    For i = 1 to n
```

```
                    For j = n to 1, step 1
                    If S[i] = T[j], then
        Begin
        Find k such that G[k - 1] < j ≤ G[k]
                G[k] = j
                End
                Print largest k such that G[k] ≠ n + 1
```

A small amount of preprocessing will improve the performance of Algorithm 2.2 in great measure. The main source of inefficiency in Algorithm 2.2 is the inner loop *j*, in which the elements are searched for repeatedly in *T* sequences that match *S*[*i*]. A linked list can be used to eliminate this search step. For each *I*, a list of corresponding *j* positions is needed such that *S*[*i*] = *T*[*j*]. These lists must be retained in decreasing order in *j*. All positions of the *S* sequence that contain the same element may be set up to use the same physical list of matching *j*'s.

Algorithm 2.3 *Find and Print Longest Common Subsequence of S and T*

```
   Initialize Arrays S[1; n], T[1; n], G[0, n], MATCHLIST
[1, n], LINK[1, n], PTR.
Build Linked Lists
                For i = 1 to n
                        MATCHLIST[i] = <j_1, j_2, . . . , j_p>
such that j_1 > j_2 > · · · > j_p and S[i] = T[j_q] for 1 ≤ q ≤ p.
Initialize Threshold Array:    G[0] = 0
        For i = 1 to n
                G[i] = n + 1
                LINK[0] = null
Compute Successive Threshold Grades
     For i = i to n
          For j on MATCHLIST[i]
          Find k such that G[k - 1] < j ≤ G[k];
            If j < G[k], then
                 G[k] = j
                 LINK[k] = new node (I, j, LINK[k - 1])
                    End:        End
Recover Longest Common Subsequence in Reverse Order
        k = largest k such that G[k] ≠ n + 1
        PTR = LINK[k]
        While PTR ≠ null do
                Print (I, j) pair pointed to by PTR
        Advance PTR;    End
```

An LCS of the sequences *S* and *T* is found and printed by Algorithm 2.3. The time efficiency of the algorithm is $O(r + n)\lg(n)$, and the space required is $O(r + n)$. The key operations in the implementation of Algorithm 2.3 are the operations of inserting, deleting, and testing membership of elements in a set where all elements are restricted to the first *n* integers. van Emde Boas [21] has shown that each such operation can be performed in $O\{\lg[\lg(n)]\}$ time. Time taken for initializing using this data structure is $O\{n \lg[\lg(n)]\}$.

2.9 Optimal Local Alignment of a Pair of Sequences

2.9.1 Smith and Waterman Algorithm

Smith and Waterman [13] introduced the local alignment problem and proposed an $O(mn)$ time to solve it. The recurrence formula used for local alignment is similar to that used for global alignment except for an additional term in the max function while obtaining the grade of alignment and is given below:

Algorithm 2.4 *Local Alignment of Two Sequences*

```
Basis:    G(i, 0) = 0                          (2.36)
          G(0, j) = 0                          (2.37)
          σ(a, -) = σ(_, a) ≤ 0                (2.38)

Recurrence Formula:

          G(i, j) = Max(0),
   G(i - 1, j - 1) + σ(S|i|, T|j|),
      G(i - 1, j) + σ(S|i|, _),
       G(i, j -1) + σ(_, T|j|)                 (2.39)
```

Consider an optimal alignment of a suffix α of $S(1), S(2), \ldots, S(i)$ and a suffix β of $T(1), T(2), \ldots, T(j)$. There are four possible cases:

1. $\alpha = \lambda$ and $\beta = \lambda$, in which case the alignment has grade 0.
2. $\alpha \neq \lambda$, $\beta \neq \lambda$, and the last matched pair in A is $(S|i|, T|j|)$, in which case the remainder of A has grade $G(i-1, j-1)$.
3. $\beta \neq \lambda$, and the last matched pair in A is $(\_, T(j))$, in which case the remainder of A has grade $G(i, j-1)$.
4. $\alpha \neq \lambda$, and the last matched pair in A is $(S(i), \_)$, in which case the remainder of A has grade $V(i-1, j)$.

Example 2.4 *Optimal Local Alignment of Two Sequences*

Find the optimal local alignment of *S:* a c g t t g c a and *T:* c c a t t g c. The grading function has no gap penalty or different grade of alignment for mismatch.

Solution

$$\sigma(-1, a) = \sigma(a, \_) = \sigma(a, b) = -1 \tag{2.40}$$

$$\sigma(a, a) = +2$$

The grade of alignment is shown in Table 2.4. Retracing the path from any maximum entry to zero entry:

```
c  g  t  t  g  c
c  a  t  t  g  c
```

Grade of alignment: $5 \times 2 - 1 \times 1 = 9$

S[i] / T[j]		a	c	g	t	t	g	c	a
	0	0	0	0	0	0	0	0	0
c	0	0	2	1	0	0	0	2	1
c	0	0	2	1	0	0	0	2	1
a	0	2	1	1	0	0	0	1	4
t	0	1	1	0	3	2	1	0	3
t	0	0	0	0	2	5	4	3	2
g	0	0	0	2	1	4	7	6	5
c	0	0	2	1	1	3	6	9	3

TABLE 2.4 Local Alignment of Two Sequences

2.10 Affine Gap Model

A gap in an alignment of S and T is a maximal substring of either S' or T' consisting only of spaces. The motivation is that for certain applications, the penalty proportional to the length of a gap is not needed. For instance, a mutation causing insertion or deletion of a large substring may be considered a single evolutionary event and may be nearly as likely as insertion or deletion of a single residue. In cDNA matching, biologists are interested in learning which genes are expressed in which types of specialized cells and where those genes are located in the chromosomal DNA. To study gene expression within specialized cells, one way is to first capture the mRNA as it leaves the nucleus. Then the complementary DNA is made from the mRNA using an enzyme called *reverse transcriptase.* The cDNA is thus a concatenation of the gene's exons. Then the cDNA is sequenced. The sequenced cDNA then is matched against chromosomal DNA to find the region of chromosomal DNA from which the cDNA derives. In this process, the introns are not heavily penalized, which will match the gaps in the cDNA.

In general, the gap penalty may be some arbitrary function $\phi(q)$ of the gap length q. The best choice of their function depends on the application. In the cDNA matching application, what is known about the common length of introns is reflected in the penalty grades/scores. There are programs in the literature in which gap penalties are precise linear functions. There exist $O[nm \lg(m)]$ time algorithms for the case when $\phi(q)$ is concave downward [19,22]. An arbitrary function can be selected as a gap penalty function, but this requires cubic time [7]. The affine gap model can be developed in which the penalty for a gap has two parts—one for inserting a gap and another that depends linearly on the length of the gap. That is, the gap penalty is $W_g + qW_s$, where W_g and W_s are both constants and $W_g \geq 0$, $W_s \geq 0$, and $q \geq 1$ is the length of the gap. When a model has a constant penalty regardless of gap length, this is the special case of $W_s = 0$. The global alignment algorithm and local alignment algorithm presented earlier can be suitably modified to include the affine gap penalty.

Algorithm 2.5 *Local Alignment with Affine Gap Penalty*

$$\sigma(a, -1) = \sigma(-1, a) = 0 \tag{2.41}$$

$$\text{Maximize } \sum\sigma(S'[i], T'(j) - W_g(\#\text{gaps}) - W_s(\#\text{spaces}) \tag{2.42}$$

Where S' and T' are S and T strings with spaces inserted in them and $|S'| = |T'| = 1$.

1. $G(i, j)$ is the grade of an optimal alignment of S and T.
2. $G'(1, j)$ is the grade of an optimal alignment of S and T whose last pair matches S and T.
3. $F(i, j)$ is the grade of an optimal alignment of S and T whose last pair matches $S(i)$ with space.
4. $E(i, j)$ is the grade of an optimal alignment of S and T whose last pair matches a space with T.

Basis:

$$G(0, 0) = 0 \tag{2.43}$$

$$G(1, 0) = -W_g - iW_s, \text{ for } i > 0 \tag{2.44}$$

$$G(0, j) = -W_g - jW_s, \text{ for } j > 0 \tag{2.45}$$

$$E(i, 0) = -\infty, \text{ for } i > 0 \tag{2.46}$$

$$F(0, j) = -\infty, \text{ for } j > 0 \tag{2.47}$$

Recurrence Relation:

$$\text{For } i > 0 \text{ and } j > 0 \tag{2.48}$$

$$G(i, j) = \max[G'(i, j), F(i, j), E(i, j)] \tag{2.49}$$

$$G'(i, j) = G'(i-1, j-1) + \sigma[S(i), T(j)] \tag{2.50}$$

$$F(i, j) = \max[F(i-1, j) - W_s, V(i-1, j) - W_g - W_s] \tag{2.51}$$

$$E(i, j) = \max[E(i, j-1) - W_s, V(i, j-1) - W_g - W_s \tag{2.52}$$

Time taken by affine gap model can be seen to be $O(nm)$.

Example 2.5 Illustrate use of the affine gap model in finding the optimal local alignment between *S:* acgucguagg and *T:* uaggaugcgcau.

Solution

$$\sigma(-1, a) = \sigma(a, \_) = -1; \ \sigma(a, a) = +2; \sigma(a, b) = -1 \tag{2.53}$$

The grading function in the first solution procedure has no separate grades for mismatches from indels. No gap penalty is levied. The recurrence formula as described by Eq. (2.44) is used. This is applicable for optimal local alignment. The zero is an added term in the max function compared with global alignment. Retracing the local alignments (Table 2.5) from maximum grade of alignment to zero, two possibilities can be identified:

$$\begin{aligned} &S: \quad \text{u a g g} \\ &T: \quad \text{u a g g} \end{aligned} \tag{2.54}$$

Grade of alignment: 4 × 2 = 8.0

$$\begin{aligned} &S: \quad \text{a c g \_ u \_ c g u a} \\ &T: \quad \text{a g g a u g c g c a} \end{aligned} \tag{2.55}$$

With six identities, two mismatches, and two indels, the optimal grade of alignment would be 6 × 2 − 2 × 1 − 2 × 1 = 8.0. In some applications, the alignment shown in Eq. (2.54) is preferred to the alignment shown in Eq. (2.55), especially in gene finding and cDNA matching. This can be factored into the program as follows: Let $W_s = -1$, $W_g = -1$. The grade of alignment will be unchanged. Only one local alignment is obtained under the new scheme (Table 2.6). Another grading scheme that achieves the same result is to change the grading function to $\sigma(a, a) = 2$, $\sigma(a, b) = -2$.

		A	C	G	U	C	G	U	A	G	G
		0	0	0	0	0	0	0	0	0	0
U	0	0	0	0	2	1	0	2	1	0	0
A	0	2	1	0	1	0	0	0	4	1	0
G	0	1	1	2	1	1	2	1	1	6	3
G	0	0	0	3	2	1	2	1	0	3	8
A	0	2	1	2	2	1	1	1	3	2	5
U	0	1	1	1	4	3	2	3	2	2	4
G	0	0	0	3	3	3	5	4	2	4	4
C	0	0	2	1	2	5	4	4	3	3	3
G	0	0	1	4	3	4	7	6	3	5	5
C	0	0	2	3	3	5	6	6	5	4	4
A	0	2	1	2	2	4	5	5	8	7	6
U	0	1	1	1	4	3	4	7	7	7	6

TABLE 2.5 Local Alignment of a Pair of Sequences with an Affine Gap Penalty

		A	C	G	T	C	G	T	A	G	G
		0	0	0	0	0	0	0	0	0	0
T	0	0	0	0	2	0	0	2	0	0	0
A	0	2	0	0	0	0	0	0	4	2	0
G	0	0	0	2	0	0	2	0	2	6	4
G	0	0	0	2	0	0	2	0	0	4	8
A	0	2	0	0	0	0	0	0	2	2	6
T	0	0	0	0	2	0	0	2	0	0	4
G	0	0	0	2	0	0	2	0	0	2	2
C	0	0	2	0	0	2	0	0	0	0	0
G	0	0	0	4	2	0	4	2	0	2	2
C	0	0	2	2	2	4	2	2	0	0	0
A	0	2	0	0	0	2	2	0	4	2	0
T	0	0	0	2	2	0	0	4	2	2	0

TABLE 2.6 Local Alignment of a Pair of Sequences—Dynamic Programming: Grading Function of Highest Penalty for Indel

2.11 Greedy Algorithms for Pairwise Alignment

When the two sequences that are aligned differ only by sequencing errors, a greedy algorithm [23] can be used that is much faster than traditional dynamic programming approaches and guarantees an optimal alignment. Chao and colleagues [24] presented greedy algorithms for solving a simple formulation of the alignment problem called the *longest common subsequence problem.* This problem is equivalent to finding the fewest one-character insertion and deletion operations that will convert one sequence into another. Let S and T be two sequences with sequence lengths m and n, and let e denote the minimum number of operations. e is the *edit distance* between the two sequences. When two DNA sequences are considered, for example, in which the shorter sequence is very similar to some concatenated region of the longer sequence, a similar region of the longer sequence is determined, and then an optimal set of single-nucleotide changes such as insertions, deletions, or substitutions is computed that will convert the shorter sequence to that region. The grade-of-alignment scheme is developed to model sequencing errors rather than evolutionary processes.

Greedy alignment algorithms presented by Ukkonen [25] and Miller and colleagues [23] are best used when e is a lot smaller than m or n. The time efficiency of the algorithms is a worst case of $O[\min(m, n)e]$ and space $O(m + n)$. The space needed is an order of magnitude smaller than that required by the dynamic programming approaches presented by Smith and Waterman [13] and Needleman and Wunsch [7]. The expected-case time efficiency of greedy algorithms can be $O(ne^2)$.

Greedy algorithms for sequence alignment are implemented in the assembly of the Unigene database maintained by the National Center for Biotechnology Information (NCBI). The algorithm suggested by Chao and colleagues [24] consists of two phases. The interval in the longer sequence that should be aligned with the shorter sequence is located during phase I. A divide-and-conquer approach is employed to obtain the alignment in phase II. The end gaps are then added to the alignment.

Algorithm 2.7 *Tool for Aligning very Similar DNA Sequences [26]*

Input: $S: a_0\, a_1\, a_2 \cdots a_{m-1}$; $|S| = m$

$T: b_0\, b_1\, b_2 \cdots b_{m-1}$; $|T| = n \qquad n \geq m$

The edit graph for sequences S and T is a directed graph with a vertex at each integer grid point (x, y), $0 \leq x \leq m$ and $0 \leq y \leq m$. Let $I(k, c)$ denote the x grade of the farthest point in diagonal k that can be reached from the source [i.e., grid point (0, 0)] with cost c and that is free to open an insertion gap. The grid point can be (1) reached by a path of cost c that ends with an insertion or (2) reached by path of cost c –1 and the gap-open penalty of 1 can be "paid in advance." Let $D(k, c)$ denote the grade of the farthest point in diagonal k that can be reached from the source with cost c and is free to open a deletion gap. Let $S(k, c)$ denote the x grade of the farthest point in diagonal k that can be reached from the source with cost c. With proper initializations, these vectors can be calculated by the following recurrence relation:

$$I(k, c) = \max[I(k-1, c-1), S(k, c-1)]$$

$$D(k, c) = \max[D(k+1, c-1) + 1, S(k, c-1)]$$

$$S(k, c) = \text{snake}\{k, \max[S(k, c-1) + 1, I(k, c), D(k, c)]\}$$

where $\text{snake}(k, x) = \max[x, \max(z: a_{x}, \ldots, a_{z-1} = b_{x+k, \ldots,}\, b_{z-1+k})]$.

Since the vectors at cost c depend only on those at costs c and $c - 1$, a linear-space version of the preceding relationship can be derived.

Exact phase I: Phase I can be accomplished by applying the recurrences for I, D, and S where all costs in row 0 are initialized to 0. Once row m is reached, the desired interval has been located. Although the worst-case running time for this approach is $O(mn)$, the average running time is $O(n \times \text{dist})$, where dist is the distance of S and T. The average length of a snaked fragment is a small constant.

Phase II: Backward vectors $I^*(k, c)$ denotes the x grade of the farthest I node in diagonal k that can reach the sink [i.e., grid point (m, n)] with cost c. $D^*(k, c)$ is the x grade of the farthest D node in diagonal k

that can reach the sink with cost c. Let $S^*(k, c)$ denote the x grade of the farthest S node in diagonal k that can reach the sink with cost c. After initializations, these vectors can be computed by the following recurrence relation:

$$S^*(k, c) = \text{snake}^*\{k, \min[S^*(k, c-1) - 1, D^*(k, c-1), I^*(k, c-1)]\}$$

$$D^*(k, c) = \min[D^*(k-1, c-1) - 1, S^*(k, c)]$$

$$I^*(k, c) = \min[I^*(k+1, c-1), S^*(k, c)]$$

where $\text{snake}^*(k, x) = \min[x, \min(z{:}\ a_z, \ldots, a_{x-1} = b_{z+k}, \ldots, b_{x-1+k})]'$

A linear-space version of the recurrence relation can be derived. The pseudocode for the linear space algorithm for alignment is as follows:

```
Procedure path(I1, J1, Type1, I2, J2, Type2, Dist)
{     if boundary cases then
      {Output the edit script; return
      Else
      {
        Mid ← Dist/2
        Mid ← Dist -mid
   A Linear Space Forward Pass Computes S(k, mid), D(k,
mid), and I(k, mid) for J1 - I1 - mid ≤ k ≤ J1 - I1 + mid.
   A Linear Space Backward Pass Computes S*(k, mid), D*(k,
mid), and I*(k, mid) for J2 - I2 - mid* ≤ k ≤ J2 - I2 + mid*.
   Let K be the diagonal such that that X(K, mid) ≥ X*(K,
mid*) where X is S, D, or I.
Path[I1, J1, Type1, X(K, mid), X(K, mid) + K, X, mid]
Path[X(K, mid), X(K, mid) + K, X, I2, J2, Type2, mid*]}}
```

2.12 Other Alignment Methods

Altschul and colleagues [8] developed an algorithm in which the sequences are searched for on diagonals of length k (k-tuples) ahead of time. The k-tuples then are evaluated, and groups of continuous tuples are labeled *significant diagonals*. A *window space* can be identified in the grid as the region around the most significant diagonals that represent partial matches. In this method, diagonals and regions are used instead of the traceback used in Needleman and Wunsch's and Smith and Waterman's algorithms. The window size is controllable and hence can speed up the computations. The time taken did not better the $O(n^2)$ needed for dynamic programming methods. Sharma [27,29,30] suggested a heuristic algorithm for approximate global alignment of a pair of sequences with less time efficiency than the

$O(n^2)$ time needed for dynamic programming. The two sequences are parsed by generating a random index i. Depending on match or mismatch, the indel and gap are introduced. The grade of alignment is calculated. The next random index is called for. This procedure is repeated until the maximum grade of alignment is reached. The maximum grade of the alignment is reached in $O(en)$ time efficiency, where e is the number of indels and gaps called for. For some sequences, this may be $O(n)$, but the alignment is approximate. The worst-case time efficiency will revert to $O(n^2)$.

2.13 Pam and Blosum Matrices

Protein sequence alignments have become an important tool for molecular biologists. Local alignments are frequently constructed with the aid of a *substitution grade of alignment matrix* that specifies a grade for aligning each pair of amino acid residues. Over the years, many different substitution matrices have been proposed, based on a wide variety of rationales. Statistical results, however, demonstrate that any such matrix is implicitly a log-odds matrix with a specific target distribution for aligned pairs of amino acid residues. In the light of information theory, it is possible to express the grades of alignments of a substitution matrix in bits and to see that different matrices are better adapted to different purposes. The most widely used matrix for protein sequence comparison has been the PAM-250 matrix [9]. It is argued that for database searches, the PAM-120 matrix generally is more appropriate, whereas for comparing two specific proteins with suspected homology, the PAM-200 matrix is indicated. Altshcul [9] discussed the lipocalins, human α_1, β-glycoprotein, the cystic fibrosis transmembrane conductance regulator, and the globins.

In protein sequence comparison, the conservative substitutions are given a different weighting for good reason. Matches in amino acids or identities should be given greater weight than substitutions. Among substitutions, it is desirable that more conservative substitutions should be given higher grade than less conservative and nonconservative substitutions. The PAM-250 matrix was constructed with certain substitution grades. The larger the number, the more common is a particular substitution. For example, glycine is commonly regulated and replaced by alanine, and vice versa. This is sensible because they are the amino acids with the smallest side chains. Similarly, aspartic acid and glutamic acid frequently substitute for each other. Serine and proline and glutamic acid and alanine substitute for each other. PAM represents 1 unit of evolutionary divergence. After 250 cycles of change have taken place in 100 amino acids, 80 still may have considerable similarity to the original sequence. The PAM-250 matrix (Fig. 2.2)

	Ala	Arg	Asn	Asp	Cys	Gln	Glu	Gly	His	Ile	Leu	Lys	Met	Phe	Pro	Ser	Thr	Trp	Tyr	Val
Ala	13	6	9	9	5	8	9	12	6	8	6	7	7	4	11	11	11	2	4	9
Arg	3	17	4	3	2	5	3	2	6	3	2	9	4	1	4	4	3	7	2	2
Asn	4	4	6	7	2	5	6	4	6	3	2	5	3	2	4	5	4	2	3	3
Asp	5	4	8	11	1	7	10	5	2	3	2	5	3	1	4	5	5	1	2	3
Cys	2	1	1	1	52	1	1	2	7	2	1	1	1	1	2	3	2	1	4	2
Gln	3	5	5	6	1	10	7	3	6	2	3	5	3	1	4	3	3	1	2	3
Glu	5	4	7	11	1	9	12	5	5	3	2	5	3	1	4	5	5	1	2	3
Gly	12	5	10	10	4	7	9	27	5	5	4	6	5	3	8	11	9	2	3	7
His	2	5	5	4	2	7	4	2	15	2	2	3	2	2	3	3	2	2	3	2
Ile	3	2	2	2	2	2	2	2	2	10	6	2	6	5	2	3	4	1	3	9
Leu	6	4	4	3	2	6	4	3	5	15	34	4	20	13	5	4	6	6	7	13
Lys	6	18	10	8	2	10	8	5	8	5	4	24	9	2	6	8	8	4	3	5
Met	1	1	1	1	0	1	1	1	1	2	3	2	6	2	1	1	1	1	1	2
Phe	2	1	2	1	1	1	1	1	3	5	6	1	4	32	1	2	2	4	20	3
Pro	7	5	5	4	3	5	4	5	5	3	3	4	3	2	20	6	5	1	2	4
Ser	9	6	8	7	7	6	7	9	6	5	4	7	5	3	9	10	9	4	4	6
Thr	9	5	6	6	4	5	5	6	4	6	4	6	5	3	6	8	11	2	3	6
Trp	0	2	0	0	0	0	0	0	1	0	1	0	0	1	0	1	0	55	1	0
Tyr	1	1	2	1	3	1	1	1	3	2	2	1	2	15	1	2	2	3	31	2
Val	7	4	4	4	4	4	5	4	15	10	4	10	4	10	5	5	5	v	4	17

FIGURE 2.2 PAM-250 matrix.

is derived from 71 sets of sequences aligned and extrapolated to 250 cycles of mutations per 100 residues. In PAM, point-accepted mutations and well-tolerated mutations are given certain meaningful grades based on the observed mutation frequencies in several thousand proteins. The PAM-250 matrix is the log probability of one amino acid changing into another amino acid. A grade of alignment above 0 indicates that such an amino acid change is more than expected by chance. Grades less than 0 denote pairs of amino acids that seldom undergo interchange.

The BLOSUM substitution matrix is constructed in a similar fashion to PAM (Fig. 2.3). Target frequencies of mutations out of background mutations are used. A blocks database is used for deriving the mutation frequencies. Blocks contain local multiple alignment of distantly related sequences. BLOSUM has an evolutionary model in its matrix formulation. Since it is derived from direct date rather than from extrapolated grades as in PAM, BLOSUM 62 means that sequences having 62 percent similarity are merged into a single sequence for detecting the matrix grade. No gap penalty is considered in either BLOSUM or PAM.

	C	S	T	P	A	G	N	D	E	Q	H	R	K	M	I	L	V	F	Y	W
C	**9**	−1	−1	−3	0	−3	−3	−3	−4	−3	−3	−3	−3	−1	−1	−1	−1	−2	−2	−2
S		**4**	1	−1	1	0	1	0	0	0	−1	−1	0	−1	−2	−2	−2	−2	−2	−3
T			**4**	1	−1	1	0	1	0	0	0	−1	0	−1	−2	−2	−2	−2	−2	−3
P				**7**	−1	−2	−1	−1	−1	−1	−2	−2	−1	−2	−3	−3	−2	−4	−3	−4
A					**4**	0	−1	−2	−1	−1	−2	−1	−1	−1	−1	−1	−2	−2	−2	−3
G						**6**	−2	−1	−2	−2	−2	−2	−2	−3	−4	−4	0	−3	−3	−2
N							**6**	1	0	0	−1	0	0	−2	−3	−3	−3	−3	−2	−4
D								**6**	2	0	−1	−2	−1	−3	−3	−4	−3	−3	−3	−4
E									**5**	2	0	0	1	−2	−3	−3	−3	−3	−2	−3
Q										**5**	0	1	1	0	−3	−2	−2	−3	−1	−2
H											**8**	0	−1	−2	−3	−3	−2	−1	2	−2
R												**5**	2	−1	−3	−2	−3	−3	−2	−3
K													**5**	−1	−3	−2	−3	3	−2	−3
M														**5**	1	2	−2	0	−1	−1
I															**4**	2	1	0	−1	−3
L																**4**	3	0	−1	−2
V																	**4**	−1	−1	−3
F																		**6**	3	1
Y																			**7**	2
W																				**11**

FIGURE 2.3 Log-odds matrix for BLOSUM 62.

Summary

Sequence alignment is a process of lining up two or more sequences to obtain matches among them. Sequence alignment can be used to develop cures for autoimmune disorders, to accomplish phylogenetic tree construction, to identify polypeptide microstructure, during shotgun sequencing, in gene finding, in restriction site mapping, in ORF analysis, in genetic engineering, during drug design, in protein secondary structure determination, and in protein folding, clone analysis, protein classification, etc. An alignment grading function is introduced to keep track of the degree of alignments and pick the optimal alignment.

Optimal global alignment of a pair of sequences can be achieved in $O(n^2)$ time using Needleman and Wunsch's dynamic programming algorithm. A dynamic programming table is filled, and the optimal alignment falls out of the procedure. The different alignments can be identified using trace-back procedures. Penalty and rewards are selected such that when the grade of alignment is greater than 0, the number of characters aligned is greater than the number of mismatched characters in the sequence; when the grade of alignment is 0, the number of mismatched characters is equal to each other; and when the grade of alignment is less than 0, the number of mismatched characters is greater than the number of matched characters. Semiglobal alignment is obtained by awarding no penalty to end gaps or allowing free end gaps. Development of the dynamic programming algorithm consists of characterizing the structure of an optimal solution, recursively defining the grade of an optimal solution, computing the grade of an optimal solution in a bottom-up fashion, and constructing an optimal solution from the computed information. The space requirement of $O(n^2)$ can be reduced to $O(n)$ using Hirschberg's dynamic array method. Algorithms for finding longest common subsequence in less than quadratic time are discussed. The Smith and Waterman algorithm can be used to obtain the optimal local alignment between a pair of sequences using the dynamic programming method in $O(n^2)$ and $O(n^2)$ space efficiency. The affine gap model can be used to define penalties for gaps and gap lengths in order to obtain biologically meaningful alignments.

Greedy algorithms can be used for aligning sequences that differ only by a few errors. Miller and colleagues have developed a method that can guarantee optimality in $O(en)$ time, where e is much less than n, and in $O(m + n)$ space. These are implanted in the Unigene database by NCBI. Other methods for obtaining sequence alignment include method of significant diagonals, the heuristic method, approximate alignments, hamming, etc. The PAM and BLOSUM matrices are provided, and the benefits of using them for alignments are outlined. The methods described were applied to sequences with varying microstructures, such as alternating, random, and block distribution.

The concept of supersequence was introduced. The X-drop algorithm for global alignment was touched upon. The effect of repeats in a sequence on dynamic programming procedures is explored. The antidiagonal was defined, and banded diagonal methods were explored. The implications of what would happen when the dynamic programming table is sparse were explored. The stability of global and local alignment was touched on. The staircase table, inverse dynamic programming, consensus sequencing, and sequencing errors and their ramifications were introduced.

References

[1] J. E. Hopcroft and J. D. Ullman, *Introduction to Automata Theory, Languages and Computation*. Reading, MA: Addison-Wesley, 1979.

[2] T. B. Stockwell, D. A. Busam, S. M. Ferriera, et al., "Nucleotide sequence in *Homo sapiens*," *J. Craig Venter Institute* (submitted 2006).

[3] A. Marson, K. Kretschmer, G. M. Framton, et al., "Foxp3 occupancy and regulation of key target genes during T-cell stimulation," *Nature* 445 (2007), 931–935.

[4] S. B. Prusiner, "Prions," Nobel Lecture, http://nobel.se, 1997.

[5] C. B. Anfinsen, E. Haber, M. Sela, and F. H. White, Jr. "The kinetics of formation of native ribonuculease during oxidation of the reduced polypeptide chain," *Proc. Natl. Acad. Sci. USA* 47, (1961), 1309–1314.

[6] M. Tompa's Course Notes, Computational Biology, CSE 527, University of Washington, Seattle, winter 2000.

[7] S. B. Needleman and C. D. Wunsch, "A general method applicable to the search for similarites in the amino acid sequence of two proteins," *J. Mol. Biol.* 48 (1970), 443–453.

[8] S. F. Altschul, W. Gish, W. Miller, et al., "Basic local alignment search tool," *J. Mol. Biol.* 215 (1990), 403–410.

[9] S. F. Altschul, "Amino acid substitution matrices from an information theoretic perspective," *J. Mol. Biol.* 219 (1991), 555–565.

[10] S. F. Altschul, "A protein scoring system sensitive at all holutiny distances," *J. Mol. Evol.* 36 (1993), 290–300.

[11] S. F. Altschul and W. Gish, "Local alignment statistics," *Methods Enzymol.* 266 (1996), 460–480.

[12] R. Bellman, *Dynamic Programming*. Princeton, NJ: Princeton University Press, 1957.

[13] T. F. Smith and M. S. Waterman, "Indentification of common molecular subsequences," *J. Mol. Biol.* 147 (1981), 195–197.

[14] R. E. Knuth, *The Art of Computer Programming*. Reading, MA: Addison-Wesley, 1997.

[15] W. J. Masek and M. S. Paterson, "A faster algorithm computing string edit distances," *J. Comput. Syst. Sci.* 20 (1980), 18–31.

[16] T. G. Szymanski, "A special case of the maximal common subsequence problem," Technical Report TR-170, Princeton University Computer Science Laboratory, Princeton, NJ, 1985.

[17] T. H. Cormen, C. E. Leiserson, R. L. Rivest, and C. Stein, *Introduction to Algorithms*. Boston, MA: MIT Press, 2001.

[18] D. S. Hirschberg, "A linear-space algorithm for computing maximal common subsequences," *Commun. ACM* 18 (1975), 341–343.

[19] E. W. Myers and W. Miller, "Optimal alignments in linear space," *Comput. Appl. Biosci.* 4 (1988), 11–17.

[20] J. W. Hunt and T. G. Szymanski, "A fast algorithm for computing longest common subsequences," *Commun. ACM* 20 (1977), 350–353.

[21] P. van Emde Boas, "Preserving order in a forest in less than logarithmic time." In *Proceedings of the 16th Annual Symposium on the Foundations Computer Science*, October 1975, pp. 75–84.
[22] Z. Galil and R. Giancarlo, "Speeding up dynamic programming with applications to molecular biology," *Theoretical Comput. Sci.* 64 (1989), 107–118.
[23] Z. Zhang, S. Schwartz, L. Wagner, and W. Miller, "A greedy algorithm for aligning DNA sequences," *J. Comput. Biol.* 7 (2000), 203–214.
[24] K. M. Chao, J. Zhang, J. Ostell, and W. Miller, "A tool for aligning very similar DNA sequences," *CABIOS* 13 (1997), 75–80.
[25] E. Ukkonen, "Algorithms for approximate string matching," *Information Control* 64 (1985), 100–118.
[26] N. L. Howard, F. J. Joubert, and D. J. Strydom, "The amino acid sequence of ostrich (*Struthio camelus*) cytochrome C," *Comp. Biochem. Physiol. [B]* 48 (1974), 75–85.
[27] K. R. Sharma, "Seeking opitmal hamming distance between nucleoide sequence using @RAND key," 58th Northwest Regional Meeting of the Americal Chemical Society, Bozeman, MT, June 2003.
[28] W. L. Gray, B. Starnes, M. W. White, and R. Mahalingam, "The DNA sequence of the simian varicella virus genome," *Virology* 284 (2001), 123–130.
[29] K. R. Sharma, "New data structures for increased efficiency of database search in genomics," 227th ACS National Meeting, Anaheim, CA, March–April 2004.
[30] K. R. Sharma, "New data structures in bioinformatics to improve search cost," AIChE Spring Meeting, New Orleans, April 2004.

Further Reading

S. Henikoff and J. G. Henikoff, "Amino acid substitution matrices from protein blocks," *Proc. Natl. Acad. Sci. USA* 89 (1992), 10915–10919.
W. R. Pearson and D. J .Lipman, "Improved tools for biological sequence comparison," *Proc. Natl. Acad. Sci. USA* 85 (1988), 2444–2448.
D. Sankoff and J. B. Kruskal, *Time Warps, String Edits and Macromolecules: The Theory and Practice of Sequence Comparison.* Reading, MA: Addison-Wesley, 1983.
K. R. Sharma, *Lecture Notes in Computational Molecular Biology.* Kumbakonam, Tamil Nadu, India: SASTRA University Press, 2005.

Exercises

1.0 *Amino acid sequence of ostrich* [28]. Phylogenetic studies were carried out to compare the positioning of chicken, turkey, duck, penguin, pigeon, and ostrich according to their cytochrome structures. The amino acid sequence of *Struthio camelus* cytochrome C was derived by sequencing tryptic peptides. The sequence differs from that of the typical bird cytochrome C in a single position. What would be a good grading scheme to obtain the local alignment between the sequences using affine gap model?

2.0 What is the alphabet of (a) protein sequences, (b) DNA sequences, and (c) RNA sequences?

3.0 What is an indel, and what is a gap?

4.0 Why are mutations, conservation, and homology important in sequence alignment?

5.0 What is the increase in database search cost expected with time and why?

6.0 What is the difference between genomics, proteomics, and metabolomics?

7.0 How many different sequences can exist with a length of 10 and an alphabet of 4 letters?

8.0 Sketch the disease mechanism of an autoimmune disorder?

9.0 Give two examples of autoimmune disorders?

10.0 What is the connection to sequence alignment during shotgun sequencing?

11.0 What is Kleen closure?

12.0 When seeking an alignment of two DNA sequences, can you use the genetic code to obtain the translated protein sequences, obtain their alignment, and then map the alignment to the original DNA sequences. How close to the optimal alignment is the alignment of DNA sequences. Discuss the time and space efficiency of this approach and compare it with obtaining a direct alignment between the DNA sequences.

13.0 What is the expected role of sequence alignment in *personalized medicine?*

14.0 How are better drugs designed using sequence alignment methods?

15.0 What does sequence distribution have to do with prion proteins and the 1997 Nobel Prize given to Prusiner?

16.0 What is the significance of a positive grade, negative grade, and 0 in the grading function to obtain the grade of alignment when sequences are aligned?

17.0 What is a trace of alignment?

18.0 What does it mean when the optimal grade of alignment is 0?

19.0 Szymanski showed that for the special case when no element appears more than once in the input sequence, the alignment problem can be solved for in $O(m + n)\lg(n + m)$. This is lower than the $O(n^2)$ time efficiency needed for any two general sequences. Why does the time increase when the characters repeat in the sequences?

20.0 What is the LCS between a pair of sequences S and T?

21.0 *Sequence distribution microstructure.* What would be different about the grade of alignment during pairwise global alignment of two sequences when one of the sequences is (a) randomly distributed, (b) alternating distributed, or (c) block architecture?

22.0 *Optimal global pairwise alignment.* Find the optimal global alignment between

S: cccaaggtacg
T: acacacacaca

23.0 Other than the optimal grade of alignment, how many alignments come to within 1 of the grade of optimal alignment in Exercise 3.0.

24.0 Can the optimal global alignment in Exercise 3.0 be improved on by using an affine gap penalty model?

25.0 *Grading functions during optimal global alignment.* Choose the appropriate grading functions to align the following strings globally:

S: uucgauugu
T: cccggguga

26.0 Prove that ${}^{n+m}C_n = \sum {}^{n}C_k\, {}^{m}C_k$ for $k \geq 0$.

27.0 *Second-best grade of alignment.* Declump and find the second-best local alignment in Example 2.3.

28.0 *Reverse of sequence.* Consider the string *S* with the sequence distribution shown below:

S: gcuauaauauu

Construct a string *T* with the sequence architecture that is the reverse of the sequence distribution of *S*. Using a global optimal alignment using dynamic programming, show that the optimal grade of the alignment is 11. Retrieve the optimal alignments. How many alignments come within ±1 of the global maximum grade of alignment. Is there another grade during the computation that is greater than the optimal grade of alignment. What is the significance of this?

29.0 *Cell grades during local alignment vs cell grades during global alignment.* An additional term 0 is used in the maximum term when calculating the grade of alignment when seeking an optimal local alignment compared with the maximum term when calculating the grade of alignment used when seeking a global alignment of two sequences. Why is this?

30.0 *Affine gap penalty.* Repeat Example 2.5 with $W_g = W_s = 2.0$.

31.0 *Interpretation of grade of alignment.* Can the optimal grade of alignment take on negative numbers.

32.0 *Gap penalty.* Obtain the global optimal alignment between the pair of sequences *S* and *T*:

S: attagacttaag
T: agctagg

The suggested grading scheme is 2 for an identity (match), –1 for mismatch, and –2 for the gap penalty and shows the initialization, matrix fill, and traceback steps. Recover all the possible alignments.

33.0 *Longest common subsequence.* Determine the LCS of strings *S* and *T* with the sequence distributions shown below:

S: dcdbddb
T: dcbdcdba

34.0 *Optimal local alignment.* Find the best local alignment between the pair of sequences *S* and *T*.

S: ggatgaaccgd
T: agtatgcgagcad

35.0 *Optimal Local Alignment of Protein Sequences.* Show the local alignment between protein sequences and between the pair of sequences *S* and *T:*

S: kcitgtnvtqdigrad
T: qmlhatndvacd.

36.0 *Pair of sequences with no repetitions of characters.* What can be achieved during global alignment of two sequences when both sequences have no repetitions of characters in them.

37.0 *BLOSUM and PAM matrices.* What are the strength and weaknesses of BLOSUM and PAM matrices.

38.0 *Affine gap penalty affixation.* What are the drawbacks of affine gap penalty affixation?

39.0 The best substitution matrix for Smith-Waterman comparisons of distant homologues is often BLOSUM-45. The best matrices for BLAST are different. Why?

40.0 When will the optimal alignment not be sought by the FASTA and BLAST software?

41.0 When is *t*blast*x* preferred to blast*n*?

42.0 What are the advantages of using multiple sequence alignments of genomic DNA sequences and a multiple sequence alignment of a group of homologous proteins?

43.0 What are the advantages of using multiple sequence alignment instead of pairwise sequence alignments.

44.0 Can you have palindromes in DNA sequences?

45.0 *PAM-250 grading matrix.* Align the sequences

S: eehgwagaeh
T: eaehwap

using the PAM-250 grading matrix and a gap penalty of –8. Seek the following: (a) global alignment; (b) local alignment; (c) global alignment with the end-gap penalties. For all alignments, provide the complete dynamic programming matrix. Use SSEARCH, which can be run on the Internet from http://workbench.sdsc.edu, to align these sequences and compare their alignment with your results.

46.0 *Affine gap penalty with translation, gaps, and transfers.* Given the following strings:

S: cgccautacgcgaatttta
T: catataaacgct

seek a global alignment using the following parameters: identity = +4; translation = –2; gap = –8 fixed; and transfers = –4. Align the two sequences, and report their grade of alignment. Revise the algorithm to produce a local alignment.

47.0 Give examples in computational molecular biology where each of the following alignment strategies would be appropriate.

(a) Global alignment with no end-gap penalties

(b) Global alignment

(c) Local alignment

(d) Spaces penalty

48.0 Use dotplot analysis and view the alignment of sequences given in Exercise 46.0.

49.0 Match 1-tuples and 2-tuples and diagonal sum for the sequences in Exercise 46.0. Develop a sequence comparison by the method of hashing.

50.0 *Award for matches, penalty for mismatch and gaps.* Use a match grade of alignment of +5, a mismatch penalty of –4, and a gap penalty of –3 and develop a dynamic programming algorithm for aligning two DNA sequences.

S: acugacgagcaucaucgaugcac
T: gaagacaucgucgau

51.0 *Escherichia coli promoter sequences.* Align the –10 signal in *E. coli* promoter sequences TATAAT with the sequences GTTACGTAA. Use the grading function 2 for a match, –1 for a mismatch, and –3 for a gap. Does the complementary sequence of *S* match better. What is the time-taken efficiency?

52.0 *Sequence distribution with high degree of alternation.* The local alignment of sequences *S:* u c u u c a a and *T:* c c a u u c are shown in Table 2.7. Recover the alignments using the traceback procedure from a local maximum grade of alignment to a minimum grade of alignment. What is the meaning of two local maximum grades of alignments of 5?

		u	c	u	c	a	a
		0	0	0	0	0	0
c	0	0	2	1	2	1	0
c	0	2	2	1	3	2	1
a	0	2	1	1	2	5	4
u	0	2	1	3	2	4	4
u	0	2	1	3	2	3	3
c	0	1	4	3	5	4	3

TABLE 2.7 Local Alignment of Sequences with Alternating Sequence Distribution

53.0 *Both strings with alternating sequence distribution.* Consider two strings *S* and *T* with alternating sequence distribution shown below:

S: ugugugugugugugug
T: gugugugugugugugu

Use the Needleman and Wunsch global alignment method with a grading function of +2 for matches and –1 for mismatches, indels, and gaps to show that the optimal alignment of the two sequences would be in a fashion that the traceback path would be along the diagonal from the right bottom cell to the left top cell of the dynamic programming array.

54.0 *One string with block sequence distribution architecture.* Consider the two strings *S* and *T* with one of them, *S*, having a chain sequence distribution with a block architecture.

S: uuuucccc
T: ucauuccc

Obtain the optimal local alignment between the two strings. Show that the optimal grade of alignment for the strings is +10. Find the second-best alignment.

55.0 *Oligonucleotides.* PBMCs are peripheral blood mononuclear cells. Single-stranded DNA (ssDNA) can be synthesized that ligands to human PBMCs. PBMCs are isolated from whole blood and contain a complex mixture of cell types of B-lymphocytes, T-lymphocytes, and monocytes. Ligands to PBMCs have many uses, including imaging lymph nodes for cancer screening and flow cytometry for AIDS monitoring. A library of synthetic DNA oligonucleotides containing 40 random nucleotides was created. The sequences of two clones are given below:

S: aguuuggau
T: gugagaaau

Using a grading scheme of +1 for a match, –1/3 for a mismatch, –1 for a gap opening, and –1/3 for a gap extension, obtain a global sequence alignment of strings *S* and *T*.

56.0 What are the differences in alignment of the strings *S* and *T* in Exercise 55.0 when the grading scheme shown in Exercise 37.0 is used and when the following grading scheme is used:

$$\sigma(a, a) = +2; \sigma(a, b) = -1; \sigma(a, \_) = -1; \sigma(\_, a) = -1$$

57.0 *Hamming.* Obtain an approximate global alignment of two sequences *S* and *T*. Using an @RAND key, go to the *k*th position. Compare $S[k]$ and $T[k]$. If there is a match, skip; if not, introduce an indel. Call for the next RAND(k). Compute grade of alignment. Stop when grade of alignment increases to a maximum. What is the time-taken efficiency for this procedure? Is there a reduction in time-taken efficiency from $O(n^2)$ to $O(en)$, where *e* is the number of indels called for? How close does this alignment come to the optimal alignment?

58.0 *Distance metric.* Consider two strings *S* and *T* with the following sequence distribution obtained from Fig. 2.1:

S: gttcaggtga
T: gggacacaaa

Define a distance between the two strings in such a fashion that when they are parsed, if there is a match, then the distance $D(a, a) = 0$, and if there is a mismatch, $D(a, b) = +1$. Thus, when the distance between two strings is large, they are misaligned. What is the distance between strings *S* and *T* shown above. Using the Needleman and Wunsch algorithm, obtain a global alignment between strings *S* and *T*. What happens to the distance between the mapped strings when the optimal alignment is obtained?

59.0 *Nonaligned sequences.* One way to speed up the $O(n^2)$ time efficiency needed for optimal global alignment between two strings *S* and *T* is to identify regions of sequences that are misaligned. Suppose that there are two sequences *S* and *T* with no characters in common or only one or two characters in common. Would it be better to define a distance metric and maximize the distance between the mapped strings. In this way, the optimal nonalignment can be obtained. What is the biologic significance of nonalignment. When would this be preferred to the alignment schemes discussed in this chapter.

60.0 *DNA sequence of simian varicella virus.* In nonhuman primates, simian varicella virus (SVV) causes a natural disease that is clinically similar to human varicella-zoster virus (VZV) infections. The SVV and VZV genomes are similar in size and structure and share extensive DNA homology. SVV DNA is 124,138 bp in size, 746 bp shorter than VZV DNA, and 40.4 percent G + C. The viral genome includes a 104,104-bp unique long component bracketed by 8-bp inverted repeat sequences and a short component composed of a 4904-bp unique short region bracketed by 7557-bp inverted repeat sequences. A total of 69 distinct SVV open reading frames (ORFs) were identified, including three that are duplicated within the inverted repeats of the short component. Each of the SVV ORFs shares extensive homology with a corresponding VZV gene. The only major difference between SVV and VZV DNA occurs at the leftward terminus. SVV lacks a VZV ORF 2 homologue. In addition, SVV encodes an 882-bp ORF A that is absent in VZV but has homology to the SVV and VZV ORF 4. The results of this study confirm the relatedness of SVV and VZV. This provides further support for simian varicella as a model to investigate VZV pathogenesis and latency. What grading scheme would you suggest for seeking an alignment between SVV and VZV viral genomes?

61.0 *Greedy algorithm to align DNA sequences.* Obtain the optimal global alignment of two sequences *S*: ccatacgtggttggtt and *T*: acgg using the greedy method. How is this an improvement over the dynamic programming method of Needleman and Wunch?

62.0 *Supersequence for global alignment.* Consider two sequences *S* and *T*

S: tgttgtcccc
T: cttgccttcc

Define a supersequence *U* where *S* and *T* are subsequences of *U*. Let the supersequence *U* be given by

U: tgtttgtccccttggcttc

How can U be used in obtaining the optimal global alignment of sequences S and T?

63.0 *Supersequence for local alignment.* How can the supersequence construction shown in Exercise 62.0 be used to obtain the optimal local alignment between sequences S and T.

64.0 *X-drop algorithm for global alignment.* The procedure to obtain the global alignment between two sequences S and T by the method of dynamic programming calls for the creation of table of the size of $m \times n$ at a space efficiency of $O(n^2)$. In order to save space and time, a method is developed where most of the cells in the array are not filled. Since only the traceback from the right bottom cell to the left top cell is important, an X-drop procedure can save time and space. The grade of alignment is calculated across the diagonal of the table. When there is a match, the next cell diagonal down can be called for. When there is a mismatch, x cells vertically can be traversed from the diagonal until a match is found. All along indels/gaps can be called for. On finding the match, the procedure can continue either from the diagonal cell branched off from or from the matched cell to the diagonal cell down. In this way, the space required would be $O[k \max(m, n)]$, where k is the departure from the diagonal in steps, and the time taken would be $O[k, \max(m, n)]$. For nearly aligned sequences, k can be small, and the best-case space and time can be linear, $O(n)$. Show an example of nearly aligned sequences S and T and the advantages of the X-drop method compared with the recurrence relation discussed by Eqs. (2.31) through (2.34).

65.0 *X-drop algorithm for local alignment.* How suitable is the X-drop method outlined in Exercise 46.0 to obtain local alignment between sequences S and T? What happened to the guaranteed optimality? Should the entire table be filled to confirm local maxima?

66.0 *X-drop algorithm with gap penalty.* For biologic applications where the gap penalty needs to be levied, as shown in Example 2.4, how do you expect the X-drop method outlined in Exercise 46.0 to fare?

67.0 *Repeats in a sequence.* Consider two sequences S and T as follows:

S: acgtacgtacgt
T: ccgatca

It can be seen that acgt repeats three times in the sequence S. When asked to obtain the global alignment between the two sequences S and T by the method of dynamic programming, as shown in Eqs. (2.31) through (2.34), how can you use the knowledge of the repeats to cut down the time taken and space efficiency from $O(n^2)$. Filling which cells and what cell calculations can be cut down to increase the time and space efficiency?

68.0 *Hirschberg array for local alignment.* Can the dynamic array method suggested by Hirchberg and discussed in the chapter be used to obtain the optimal local alignment using the dynamic programming method of Smith and Waterman. Why not?

69.0 *Antidiagonal* [24]. Antidiagonal k is the set of all points (i, j) such that $i+j=k$. Thus antidiagonal $k=2$ would mean the cells (2, 0), (0, 2) and (1, 1). Antidiagonal computation and half-nodes can speed up the sequence alignment process. Given an example of alignment of two sequences, where does antidiagonal computation speed up the time taken? What is the payoff ?

70.0 *Edit distance* [25]. Edit distance between two sequences S and T is the minimum cost c of a sequence of editing steps such as insertions, deletions, and changes that convert one sequence into another. A tabulating method was developed to compute c as well as the corresponding editing sequence in time efficiency of $O[c \min(m, n)]$ and space efficiency of $O[c \min(c, m, n)]$, where all editing steps have the same cost independent of the characters involved. If the editing sequence that gives cost c is not required, the algorithm can be implemented in space efficiency of $O[\min(c, m, n)]$. Consider two sequences S and T

S: aacaaagtta
T: attgaaacaa

Convert sequence S to T, and confirm the time and space efficiency of the editing method shown in [26].

71.0 *Band across diagonal.* In the method of dynamic programming, to align two sequences and to obtain the optimal global alignment grade, filling a table of grades with m rows and n columns is called for. It can be seen that the alignments can be recovered using a traceback procedure. The alignments are close to the diagonal of the table. A lot of cells in the table that are far from the diagonal are needed to recover the optimal global alignment. Thus a procedure can be developed that calls for computations only across the diagonal in the table from (0, 0) to (m, n) and a few cells from the top and bottom of the main diagonal. Thus confinement to within a band of the main diagonal can reduce the number of computations needed to obtain the optimal grade of alignment and recover the alignments. What is the speedup expected as a function of the width of the band? What is the space reduction achieved as a function of the width of the band?

72.0 *Tradeoff between time efficiency and optimality.* Suppose that a tradeoff is allowed between time efficiency and optimality. In order to obtain an optimal global alignment between two sequences within twice the optimal grade of alignment, what is the speedup and space reduction that can be expected?

73.0 *Global alignment of three sequences.* Show that a dynamic programming method can be used to obtain optimal global alignment of three sequences S, T, and U. The time-taken efficiency would be $O(n^3)$, and the space required would be $O(n^3)$. A cube of cells with m rows, n columns, and o floors has to be filled to complete the procedure.

74.0 *Local alignment of three sequences.* Show that a dynamic programming method can be used to obtain optimal local alignment of three sequences S, T, and U. The time-taken efficiency would be $O(n^3)$, and the space required would be $O(n^3)$. A cube of cells with m rows, n columns, and o floors has to be filled to complete the procedure.

75.0 *Affine gap penalty.* Consider the local alignment of three sequences S, T, and U. How would you modify the procedure developed in Exercise 74.0 to incorporate the affine gap penalty.

76.0 *Hirschberg space array for global alignment of three sequences.* Show that the dynamic array concept developed by Hirschberg and discussed in Sec. 2.7 can be extended to obtaining the optimal grade of alignment during global alignment of three sequences S, T, and U in space efficiency of $O(n^2)$.

77.0 *Hirschberg space array for local alignment of three sequences.* Can the dynamic array concept developed by Hirschberg and discussed in Sec. 2.7 be extended to obtaining the optimal grade of alignment during local alignment of three sequences S, T, and U? Why?

78.0 *Hamming for three sequences* [28]. Can an approximate global alignment be obtained of the sequences S, T, and U using the @RAND key? What is the tradeoff between speed and optimality? What is the space required?

79.0 *Dynamic programming table for global alignment.* In the method of dynamic programming, to obtain optimal global alignment of two sequences S and T, a table of grades has to be generated with mn cells. Are there two sequences S and T for any set of cell grades in the table? Discuss.

80.0 *Dynamic programming table for local alignment.* In the method of dynamic programming, to obtain optimal local alignment of two sequences S and T, a table of grades has to be generated with mn cells. Are there two sequences S and T for each and every set of cell grades in the table? Discuss.

81.0 *Sparse table.* Obtain the optimal local alignment of two sequences S and T given below:

S: acgtt
T: acaaa

Show that the dynamic programming table will be filled as in Table 2.8. What is the unique feature of Table 2.8? Is it a sparse matrix or sparse table? Once the sparse matrix is recognized, can the time efficiency be increased and space required cut down?

82.0 Recover the local alignments from Table 2.8 using the traceback procedure.

83.0 Note that the grading function had a +2 for matches in Table 2.8. If the grade of alignment for a match is –1, can a table with sparse matrix property such as the one in Table 2.8 be generated for any two sequences S and T? Why?

84.0 *Sparse table.* During the local alignment procedure of Smith and Waterman, using dynamic programming depending on the nature of the sequences S and T, some tables can be seen to be sparse; i.e., a lot of the cells have zero grade. For example, only half Table 2.8 is filled, and the rest in not filled. How can this knowledge be used to decrease the space required and increase the time efficiency of the method?

S(i) T(j)		a	c	g	t	t
	0	0	0	0	0	0
a	0	2	0	0	0	0
c	0	1	4	0	0	0
a	0	1	3	3	0	0
a	0	1	2	2	2	0
a	0	1	1	1	1	1

TABLE 2.8 Local Alignment of Two Sequences *S* and *T* in Exercise 81.0

85.0 *Stability of global alignment.* In the dynamic programming method, to obtain the global and optimal alignment between two sequences *S* and *T*, the traceback procedures originate from the right bottom cell and end up at the left top cell. What is it in the procedure that keeps it from taking a detour to the right top cell or a wavy path?

86.0 *Stability of local alignment.* In the dynamic programming method, to obtain the local and optimal alignment between two sequences *S* and *T*, the traceback procedures originate from the cell with the local maximal and end up at the cell with the local minima. What is it in the procedure that keeps it from taking a wavy path?

87.0 *Smith and Waterman's seminal article* [13]. Smith and Waterman presented their dynamic programming work in 1981. The two sequences that they considered in their article in the *Journal of Molecular Biology* were as follows:

S: acagccuccgcuuag
T: aaugccauugacgg

Obtain the optimal local alignment of sequences *S* and *T*.

88.0 Is there a need for use of affine gap penalty for the sequences *S* and *T* given in Exercise 86.0.

89.0 *Needleman and Wunsch's seminal article* [8]. Needleman and Wunch presented their dynamic programming work in 1970. The two sequences they considered in their article in the *Journal of Molecular Biology* were as follows:

S: abcnjrqclcrpm
T: ajcjnrckcrbp

Obtain the optimal global alignment of two sequences *S* and *T*.

90.0 What is the biologic significance of the weighting factors used in the grading scheme in Exercise 89.0.

91.0 *Staircase table* [29,30]. Consider the dynamic programming table shown in Table 2.9. When the two sequences are aligned using the dynamic

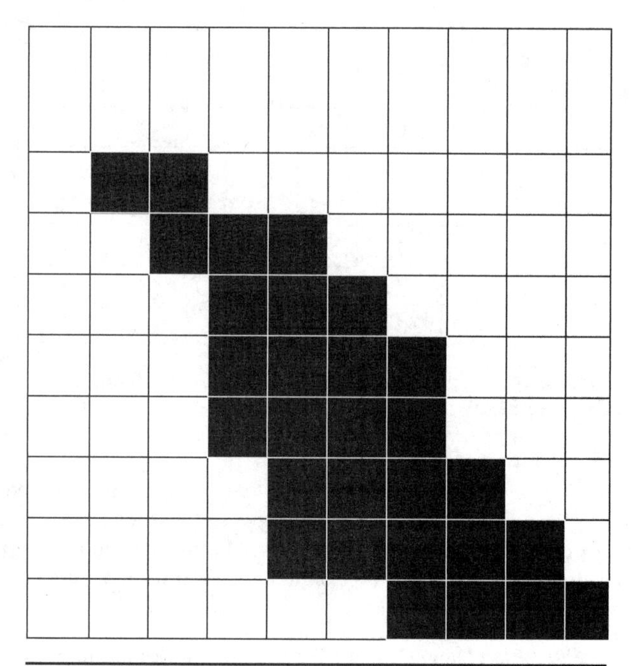

TABLE 2.9 Staircase Table

programming method for global optimal alignment, consider the special case where the grades of alignment for the sequences S and T form a staircase such as the one shown in the table. The unshaded regions are insignificant grades, and the significant grade of alignment only falls in the shaded region. For such cases, can the time efficiency of alignment be speeded up and the space required cut down?

92.0 If the local optimal alignment of two sequences results in a dynamic programming table such as Table 2.9, how would you modify the procedure? What is the speed-up that can be expected and space savings that can result from realizing the nature of the particular sequences that form a staircase region in the dynamic programming table.

93.0 It is generally agreed that in problems such as matrix multiplication, dynamic programming, and greedy algorithms where there is an optimal structure to the subproblems, the time taken cannot be better than $O(n^2)$. Reports in the literature for a linear time solution for the longest common substring problem can be seen and implemented in commercial software successfully. How should you interpret these two observations?

94.0 *Inverse dynamic programming for global alignment.* Define a formal inverse problem of conversion of a filled dynamic programming table (with m rows and n columns) with grades of alignment in each cell to obtain the optimal global alignment into two sequences S and T.

95.0 *Inverse dynamic programming for local alignment.* Define a formal inverse problem of conversion of a filled dynamic programming table (with m rows and n columns) with grades of alignment in each cell to obtain the optimal local alignment into two sequences S and T.

96.0 *Inverse dynamic programming for local alignment with affine gap penalty.* Define a formal inverse problem of conversion of a filled dynamic programming table (with m rows and n columns) with grades of alignment in each cell to obtain the optimal local alignment with affine gap penalty into two sequences S and T.

97.0 *Global alignment to grading scheme.* Consider a global alignment of a pair S' and T' that originated from the sequences S and T. For a desired alignment, how would you devise a grading scheme?

98.0 *Local alignment to grading scheme.* Consider a local alignment of a pair S' and T' that originated from the sequences S and T. For a desired alignment, how would you devise a grading scheme?

99.0 *Local alignment with affine gap penalty to grading scheme.* Consider a local alignment of a pair S' and T' with affine gap penalty that originated from the sequences S and T. For a desired alignment, how would you devise a grading scheme?

100.0 Consider a desired alignment such as the one shown below:

S': a c _ c g t _ _ c a a
T': _ c g c t t a a c a t

For the global alignment shown for sequences S: accgtcaa and T: cgcttaacat, what grading scheme is used?

101.0 Consider a desired local alignment such as the one shown below:

S': t a t a t a t a t e e
T': _ a _ a t a _ _ t e e

What would be the grading scheme used be for aligning the sequences S: tatatatatee and T: aatatee?

102.0 How would you change the grading scheme obtained in Exercise 101.0 to obtain the following local alignment:

S': t a t a t a t a t e e
T': _ a _ a t a _ a t e e

103.0 When you align two sequences S and T using dynamic programming, suppose that the indels/gaps are not allowed for one sequence and are allowed for the other sequence. Would this change the time-taken efficiency and space required to fill the table?

104.0 *Sequencing errors.* The experimental procedures used to obtain the sequence distribution of DNA usually have a 5 percent error from the wet laboratory and 7 percent during shotgun sequencing owing to extrapolation. Approximate global alignment of DNA sequences may even be preferred as opposed to optimal global alignment that takes $O(n^2)$ time and $O(n^2)$ space. What is the tradeoff in speed versus accuracy when a sublinear time algorithm is developed for an approximate global alignment.

105.0 *Consensus sequence.* A consensus sequence *C* of *S* and *T* sequences can be defined as follows:

S: a c g t t t g c g g c
T: a a t g t a g c a g a
C: a a g g t a g c g g a

How can the consensus sequence *C* figure in obtaining alignment between *S* and *T*?

106.0 *Geometric distribution.* The dynamic programming methods to align any two sequences *S* and *T* require $O(n^2)$ time and $O(n^2)$ space for guaranteed optimality. When DNA sequences are aligned, a geometric distribution model can be developed to characterize the two sequences *S* and *T*. When the sequences are parsed for a match, matched regions are given a positive weight, and when a mismatch is encountered, the model can be called for to check whether the mismatch resulted from some experimental error or from some biologic phenomenon. Search across a diagonal is performed, and excursions as in the X-drop method shown in Exercise 64.0 are allowed. Show that this can result in less time and space taken and that a closer to optimal grade of alignment can be achieved.

107.0 *$O(mn/K)$ time taken.* In the BLAST software, alignment strategies employed are segments. Show what a segment is by an example.

108.0 *Polymorphism.* Polymorphism is the differences in DNA among individuals of the same species. Given long DNA sequences, the optimal solution of dynamic programming will assign a penalty for mismatches that result from polymorphism. Given the biologic interpretation, how would you devise a grading scheme to reduce the penalty only in the case of polymorphism?

CHAPTER 3

Sequence Representation and String Algorithms

Objectives

The objectives of this chapter are to

- Be able to store a sequence in a suffix tree data structure.
- Be able to construct a suffix tree, suffix array, and suffix links in $O(n)$ time and space.
- Be able to use a suffix tree to align sequences.
- Be able to search for a pattern p in a text t.
- Be able to learn string matching algorithms such as Knuth-Morris-Pratt (KMP), Boyer-Moore (BM), and finite automaton (FA).
- Be able to solve problems with variations in the KMP, BM, and FA algorithms.
- Be able to achieve sublinear time taken efficiency in pattern matching algorithms.

3.1 Suffix Trees

3.1.1 Overview of Suffix Trees in Sequence Analysis

The paper that introduced the suffix tree was awarded the "Algorithm of the Year" award by Knuth in 1973. This feat was achieved by Weiner. Prior to that, the data structure was called different names, such as *bi.tree, prefix tree, PAT tree, position tree,* the *repetition finder,* and the *sul tree.* The construction of the suffix tree was improved by Ukkonen [1] to give a linear time algorithm for the construction of a suffix tree.

A suffix tree of length n over a constant alphabet can be constructed in $O(n)$ time. It can be used to find patterns in DNA and protein sequences [2].

Any string of length m can be degenerated into m suffixes, and these suffixes can be stored in a suffix tree. Creating this structure requires time $O(m)$, and searching for a pattern in it requires time $O(n)$, where n is the length of the pattern. The internal structure of a DNA sequence can be exposed by storing it in a suffix tree. Patterns that exist are identified and saved accordingly. These two properties make the suffix tree an appealing structure for a diverse range of bioinformatics applications, including multiple-genome alignment [3], selection of signature oligonucleotides for DNA arrays [4], and identification of sequence repeats [5]. The search engine REPfind of REPuter software uses a compact implementation of the suffix tree to locate exact repeats in linear time and space. Up to the size of human genome, the sequence can be stored in a suffix tree in linear time and space. Exact repeats are used as seeds from which significant degenerate repeats are constructed, allowing for mismatches, insertions, and deletions. Degenerate palindromic repeats also can be detected. A suffix array is an array of all suffixes that are sorted. Suffix arrays and suffix trees can be derived from each other in $O(n)$ time.

The *generalized suffix tree (GST),* can be used to represent a set of strings and stores all the suffixes of all the strings. The i from the leaf label (i, j) denotes that the suffix is from string S_i, and j represents the starting position of the suffix in S_i. An edge label in GST is represented by three integers (i, j, l) and is a substring of one of the sequences. i is the string number, j and l are the starting and ending positions of the substring in S_i. N is the total number of characters of all strings in set S. The GST can be constructed in $O(N)$ space.

At least two programs based on suffix trees are available for whole-genome alignment: MUMer [6] and Multiple Genome Aligner (MGA) [3]. MUMmer and MGA both use common subsequences as anchors for the alignment. While they both use the same data structure, "a suffix-tree data structure", which permits very fast and memory-efficient comparison (of the genomes) [6], the two applications take different approaches to genome alignment. MUMmer extracts Maximal Unique Matches (MUMs)—sequences that occur exactly once in each genome—sorts these sequences to find the longest set of MUMs occurring in the same order in both sequences, and uses this set of sequences to anchor the multiple alignment. The gaps between anchors are filled using the Needleman and Wunsch [7] dynamic programming alignment algorithm. Since the Needleman and Wunsch algorithm does not scale well for multiple sequences (its time and space requirements increase exponentially with the number of strings), MUMmer is restricted to comparing two genomes. MGA computes the longest nonoverlapping sequence of Maximal Multiple Exact Matches (multiMEMs) and uses these to guide the multiple alignment. A MEM is defined as a (k + 1)-tuple ($l, p\_0, p\_1, \ldots, p\_k -1$) such that l indicates the length of the MEM and p_i indicates the start coordinate of the exact match in genome i. A maximal MEM cannot be extended to the left or the right and is referred to as a *multiMEM*. Gaps are shortened by recursively extracting multiMEM sequences and finally are filled using ClustalW—a progressive/iterative multiple-alignment method.

A long MUM can figure in the optimal alignment of a pair of sequences. Consider two sequences S and T. First, all the MUMs between the two sequences S and T are found. The longest sequence of MUMs that occurs in the same order in either sequence is found. The regions between the MUMs are aligned.

Suffix tree can be used to represent the DNA sequence and is an interesting data structure. Consider the string S:

$$\begin{array}{lcccccccc} S: & T & T & A & T & T & A & C & G \\ i & 1 & 2 & 3 & 4 & 5 & 6 & 7 & 8 \end{array} \quad (3.1)$$

For every i, I = 1 to k; let the substring a be the shortest substring beginning at i that does not occur elsewhere in S.

Position i	**Substring**
1	TTAT
2	TAT
3	AT
4	TTAC
5	TAC
6	AC
7	C
8	C

The suffix tree representation of the sequence in string S given in Eq. (3.22) is shown in Fig. 3.1.

The n terminal nodes of the suffix tree $S = S_1, S_2, \ldots, S_n$ consist of $1, 2, 3, \ldots, n$. The sequence of labels on the edges from the root to the terminal node i is the identifying substring for position i. The suffix tree for the S, the sequence of length 8, is shown in Fig. 3.1. The largest matching regions between two sequences S and T can be found using the concept of a suffix tree. The string concatenation of letters from root to leaf will give the associated suffix. The longest repeat in sequence S is TTA. This is obtained simply by reading the labels at the tips of the longest branches of the tree. TTA begins at positions 1 and 4. Contained in the suffix tree is all the repeat information of sequence S.

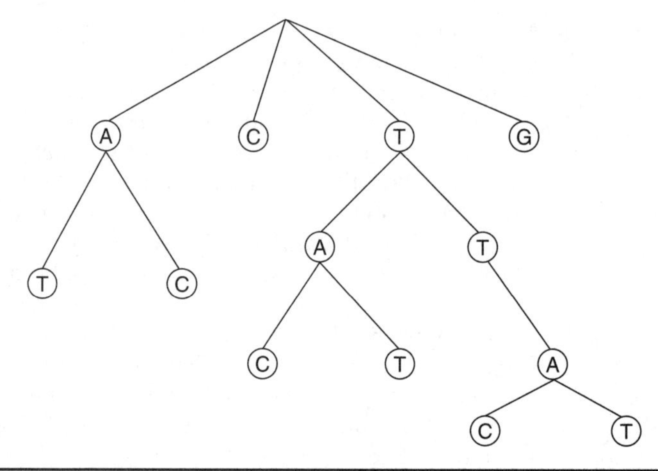

FIGURE 3.1 Suffix tree representation of sequence TTATTACG.

3.2 Algorithm for Suffix Tree Representation of a Sequence

Each internal node has at least two children. An n leaf suffix tree has at most $n - 1$ internal nodes. The maximum number of children is bounded by $|\Sigma| + 1$. However, for the edge labels, the size of the tree is $O(n)$. Each edge label is denoted by two numbers for the starting and ending positions.

Algorithm 3.1 *Suffix Tree Representation of a Sequence*

```
Input:    (B, depth)                                         (3.2)
Output:   (list, depth) for α ∝ Σ                            (3.3)
          for all α∝Σ                                        (3.4)
          list(α) = ϕ                                        (3.5)
          for all i ∝ B                                      (3.6)
          List(a_i+depth) ← list(a_i+depth) ∪ (i)            (3.7)
```

Algorithm 3.2 *Find Repeats in a Sequence*

```
Input:    S_1, S_2, . . . , S_n                              (3.8)
          for node = top
List node ← (1, 2, 3, . . . , n)
          for all nodes with (list node) > 1                 (3.9)
          Suffix [list(node, depth)]
```

Tandem repeats are short segments of DNA that occur more than once in the DNA sequence consecutively and participate in the polymerase chain reaction (PCR) gene expression reactions. A tandem repeat can be defined as $\tau = s's'$. This is a repeated occurrence of string s' twice. τ is also called a *primitive tandem repeat*. A *square* is a two-repeat tandem sequence. A *tandem array* is when substring s' repeats more than two times consecutively. Thus $\tau = s'^k$. s' is denoted by (i, s', k). It also can be represented as a tuple. Crochemore [8] developed an algorithm that computes all occurrences of primitive tandem repeats in $O[n \lg(n)]$ time. All occurrences of tandem repeats can be found in $O[n \lg(n) + occ]$, where the number of occurrences of tandem repeats in the string is given by *occ*.

The complete system code for MUMmer 2 is freely available from the TIGR Web site at www/tigr.org/software/mummer. MUMmer uses suffix trees to create an internal representation, and based on this representation, two genomes can be *approximately aligned* in linear time and space [7]. The error from optimal alignment can be calculated. For example, MUMmer 1.0 aligns the 4.7-Mbp genome of *Escherichia coli* and the 3.0-Mbp large chromosome of *Vibrio cholerae* in 74 seconds on a 1-GHz desktop computer, requiring 293 MB of memory. The memory requirement of 38 bytes/bp, although it grows only linearly with the size of the input sequences, is still a limitation of original system. This has been reduced dramatically in MUMmer 2.0. For the same two genomes, the new system computes the alignment in only 27 seconds and requires only 100 MB of memory. Both speed and memory usage have been improved by a factor of nearly 3.

There are three significant technical improvements in the core algorithms of MUMmer 2.0. The first is the reduction in the amount of memory used to store suffix trees by employing techniques described by Kurtz and Schleiermacher [5]. At most, 20 bytes/bp for amino acids is used. The maximum memory usage occurs in the case where each internal node in the suffix tree has only two children. In practice, however, many nodes have more than two children, particularly in the case of polypeptide sequences, which reduce the actual memory requirement. The second significant core improvement is an alternative algorithm to find exact matches. The original algorithm built a suffix tree containing two input sequences and then found all MUMs between them.

3.3 Streaming a Sequence Against a Suffix Tree

Given a pair of sequences, one sequence is stored in the suffix tree. The second sequence or the query is *streamed* against the suffix tree, exactly as if it were being added but without actually adding it. This procedure was introduced by Chang and Lawler [9] and fully described by Gusfield [10]. Using this process, where the query sequence would branch off from the tree can be identified. In this way, all matches to the reference sequence are determined. For example, the query ATGTCC is streamed against the string S in Fig. 3.2. Wherever a branch occurs at a tree position with just a single leaf beneath it, the match is unique in the reference sequence. By checking the character immediately preceding the start of this

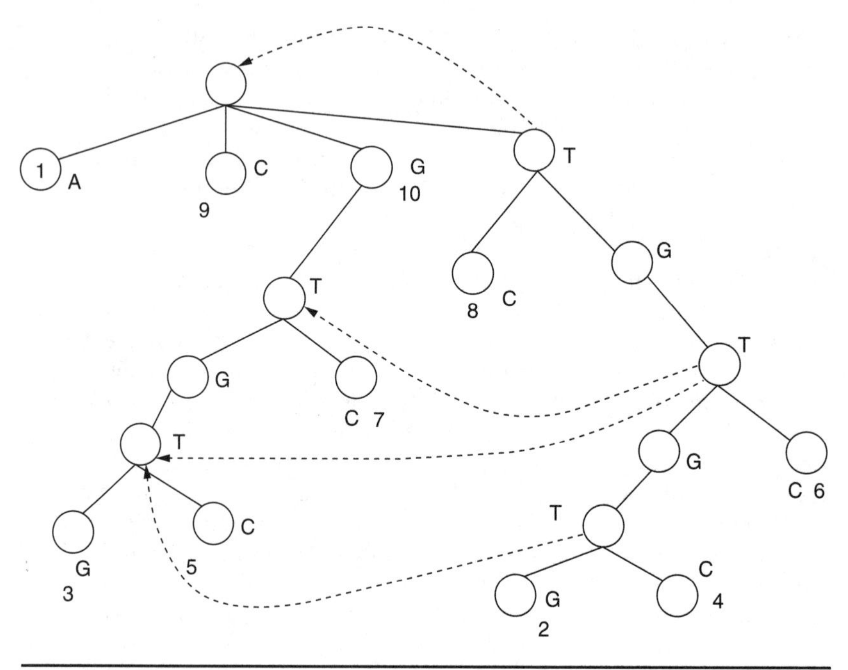

FIGURE 3.2 Streaming a sequence against a suffix tree.

match, it can be determined whether it is a maximal match. Thus, in time proportional to the length of the query sequence, all maximal matches between it and a unique query string in the reference sequence can be identified. Because the query is streamed through outputting matches as they are found, it is not known which sequence will occur later in the query. The advantage of this method is that only one of the two sequences can be streamed against the reference sequence that is stored as a suffix tree. Delcher and colleagues [6] have used this technique to compare two assemblies of the entire genome, each approximately 2.7 billion characters, using each chromosome as a reference and then streaming the entire genome past it.

S:	A	T	G	T	G	T	G	T	C	(3.10)
	1	2	3	4	5	6	7	8	9	

Position	Subsequence
1	A
2	TGTGTG
3	GTGTG
4	TGTGTC

5	GTGTC
6	TGTC
7	GTC
8	TC
9	C

How a string is streamed against a suffix tree is shown in Fig. 3.2. Leaf 7 represents GTC, and leaf 5 represents GTGTC. At the point shown in the figure, the input stream starting at position i is matched as indicated by the arrow. The match extends to the corresponding arrow position in the tree. In this case, the match can be seen to be unique because there is a single leaf below this position in the tree. The number label of the leaf gives the starting position of the match in the suffix tree string. To find the next match, the suffix links in the tree are used. These are indicated by arrows at the ends of the curvilinear dashed lines. These links are constructed for each internal node in the tree. Because the match is made as far as possible in the tree, the matches are maximal on the right hand side (RHS) of the strings being compared. The one-sided uniqueness property of MUMmer 2.0 can be an advantage when comparing queries that represent only a partial genome assembly. The third technical improvement is the addition of a new module to cluster matches. The original version of MUMmer computed a single longest alignment between the sequences.

3.4 String Algorithms

String algorithms can be used to find patterns in DNA sequences. The strategy used is similar to that used to find a pattern in a text, such as the Find tool used in word-processing software. The string-matching problem can be formalized as text in an array is $T(1, 2, \ldots, n)$ of length n and the pattern in an array is $P(1, 2, \ldots, m)$ of length m, $m \leq n$. The elements of P and T are drawn from a finite alphabet Σ. The character arrays P and T are called *strings of characters.* The preprocessing time and matching times for different string-matching algorithms are shown in Table 3.1. The problem is to find pattern P in text T.

Algorithm	Preprocessing Time	Matching Time
Rabin-Karp	$\theta(m)$	$[(n - m + 1)m]$
Knuth Morris Pratt	$\theta(m)$	$O(n)$
Boyer Moore	$O(m + \sigma)$	$O(n)$
Finite automaton	$O(m\|\Sigma\|)$	$O(n)$

Table 3.1 Preprocessing and Matching Times for Different String-Matching Algorithms

3.4.1 Rabin-Karp Algorithm

The Rabin-Karp algorithm [11] performs well in practice and can be generalized to two-dimensional pattern matching. A hashing function is used in the algorithm. A quadratic number of comparisons is avoided. Instead of checking at each letter of the text, it is sufficient to look through a window to see whether it looks like the pattern that is searched for. The hashing function has to be efficiently computable and highly discriminating for strings.

Algorithm 3.3 *Rabin-Karp Matcher*

```
Length(T) = n; length of (P) = m                                  (3.11)
        d^(m-1) mod q = h                                         (3.12)
        p = 0; t_0 = 0
          For i = 1 to m,
             p = [dp + P(i)] mod q                                (3.13)
             t_0 = [dt_0 + T(i)] mod q
          For s = 0 to n - m,
            do if p = t_s
              then if P(1, . . . , m) = T(s + 1, . . . , s + m)
                then print "Pattern occurs with shift" s
                                                                  (3.14)
                   if s < n - m
        then t_(s+1) = {d[t_s - T(s + 1)h] + T(s + m + 1)} mod q
                                                                  (3.15)
```

The Rabin-Karp matcher takes $O(m)$ preprocessing time, and its matching time is $O[m(n - m + 1)]$ in the worst case.

3.4.2 Knuth-Morris-Pratt (KMP) Algorithm

This algorithm can be used for pattern-matching problems. The straight solution in simple string matching can be archived in worst-case time of $O(mn)$ and space required $O(nm)$. For example, let the P = AGAGU and T = AGAGAGUUA. A set Σ is defined to be the alphabet or set of characters from which the characters in P and T may be chosen from, and let $\alpha = |\Sigma|$. The flowchart or finite automaton (Fig. 3.3) has two types of nodes j.

Some are read nodes, which mean, "Read the next character. If there is no further character in the text string, halt; there is no match." One read node is designated the start node.

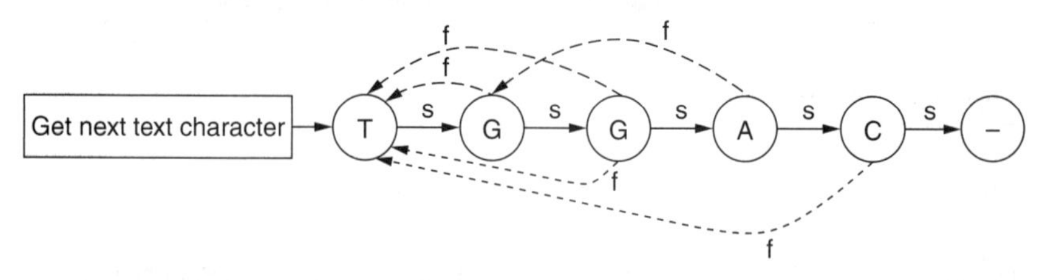

Figure 3.3 KMP flowchart for TGGAC.

A stop node, which means, "Stop; a match was found." It is marked with *a*. The flowchart has α arrows leading out from each read node. Each arrow is labeled with a character from Σ. The arrow that matched the text character just read is the arrow to be followed; that is, it indicates which node to go to next. The read nodes serve as sort of memory. For instance, if execution reached the third read node, the last two characters read from the text were *A*'s. What preceded them is irrelevant. The time taken for KMP algorithm is $O(n + m)$, an improvement over the $O(nm)$ for straightforward matching [12].

Example 3.1 Find the pattern TGGAC in the text AGCTTGGAC.

$$\Sigma = (A, G, C, T) \tag{3.16}$$

Action of the KMP flow chart (Table 3.2):

T:	A	G	C	T	T	G	G	A	C
	1	2	3	4	5	6	7	8	9

Algorithm 3.4 *Knuth-Morris-Pratt Scan Algorithm*

```
Input:    P, T, the pattern and text strings; m, the
length of P; fail the array of failure links setup in KMP
flowchart representation algorithm
   The length of P is determined when fail array is set up.
Output:    Return value is the index in T where a copy of P
begins, or -1 if a
   Match for P is found.
   Int kmpscan (Char []P, m Char[]T, int m, int [] fail)
```

KMP Cell Number	Text Index	Scanned Character	Success or Failure
1	1	A	F
0	2	G	Get next char
1	2	G	F
0	3	C	Get next char
1	3	C	F
0	4	C	Get next char
1	4	T	S
2	5	T	F
1	5	T	S
2	6	G	S
3	7	G	S
4	8	A	S
5	9	C	Stop

TABLE 3.2 KMP Cell Numbers 5, 6, 7, 8, 9 Indicate the Matched Pattern in the String in Example 3.1

```
Int Match
Int j, k
//j indexes text characters
//k indexes the pattern and fail array match = -1
j = 1; k = 1;
While [end text (T, j) = = false]
If (k > m)
  Match = j - m//match found                                  (3.17)
Break;
If (k = = 0)
j++:
k = 1//Start pattern over
          Else if (t_j = = pk)
j++
k++
          Else
//follow fail arrow
k = Fail(k)
//continue loop of return match
```

Algorithm 3.5 *Knuth-Morris-Pratt Flowchart*

```
Input:    P, a string of characters; m, the length of P
Output: Fail, the array of failure links: defined for
indexes, 1 - m. The array is passed in, and the algorithm
fills it.
Void KMP setup (Char []P, int M, int [] fail)
         Int k, s
         1.      fail [] = 0
         2.      for k = 2 , k ≤ m, k++
         3.      S = fail [k - 1]
         4.      While (s ≥ 1)
         5.      If P = = P_k-1                                 (3.18)
         6.      break;
         7.      S = fail[S]
         8.      Fail[k] = S + 1
```

The complexity is $O(m^2)$.

3.4.3 Boyer-Moore Algorithm

In the Boyer-Moore algorithm [13], text characters may be skipped over entirely. A good algorithm should be able to jump faster past places in the text where the pattern cannot appear. The Boyer-Moore (BM) algorithm always scans the pattern from right to left. It uses two heuristics to decide how far the pattern may be slid over the text string after a mismatch. Let P be the pattern of length m and T a text string of length n.

Che w C hew wc hewch ewchewe

The Chief Defect of Henry VIII was chewing little bits of chewing gum.

Since *w* and *C* do not match, there is now *w* in the first four letters and the *chew* is moved four matches. This is repeated a few times. Lining up the *e*, the pattern was found. The letter where there is a match is lined up after the text is moved. The number of comparisons required is less in this approach than in other methods.

Algorithm 3.5 *Boyer-Moore Algorithm*

```
Input:    Pattern String, P
          m - length of P
          alphabet size, α = |Σ|
Output:   Array charjump defined on indexes 0, 1, . . . , α - 1.
      The array is passed in and the algorithm fills it.
          Void Compute jumps (Char [], int m, int α, int [],
          charjump)
          Char Ch;
          Int k;
          For (ch = 0; Ch < α, int [], Charjump)
          Char Ch;
          Int k
          For (ch = 0; Ch < α_j ch++)                      (3.19)
          Char jump [ch] = m;
          For (k = 1; k < m; k++)
    Charjump [Pk] = m = k;
```

The time-taken complexity can be seen to be $\theta(m + |\Sigma|)$.

There is another algorithm to compute jumps based on partial matches. The pattern is studied to match up a substring. This can be achieved in $O(m)$ time. The behavior of the BM algorithm depends on the size of the alphabet and the repetition within the strings. In empirical studies using natural-language text and $m \geq 5$, the algorithm did only roughly 0.24–0.3 character comparisons per character in the text, up to the point of the match or the end of the text.

For binary strings, BM does not do quite as well; in another study, roughly 0.7 comparisons were done for each text character. In all cases, with $m \geq 5$, the average number of comparisons is bounded by Cn for a constant $C < 1$. If the pattern is quite small ($m \leq 3$), then the overhead of preprocessing the pattern is not worthwhile. BM does more comparisons than the straightforward approach. There are several improvements and modifications to the BM algorithm that make it run faster. Some of the problems at the end of this chapter discuss these improvements. Two extensions to the pattern-matching problem are often useful. Find all occurrences of the pattern in the text, and find any one of a finite set of patterns in the text.

The KMP and BM algorithms search for an exact copy of the pattern in the text. However, in many applications, an exact copy cannot be expected. A spelling corrector, for example, may search a dictionary for an entry that is similar to a given misspelled word. In speech recognition, samples may vary. Other applications in which close but not exact matches are sought range from identifying sequences of amino acids to recognizing bird songs. A dynamic programming solution to the problem of finding an approximate match for a pattern in a string has been developed. The approximate match problem means that the match between pattern and text has at most k differences. Differences tables can be constructed and the solution sought in $O(mn)$ time.

3.4.4 Finite Automaton

A finite automaton [14] is built during the execution of string-matching algorithms. The text string is scanned, and all occurrences of the pattern P are searched for. These string-matching automata are efficient. Each text character is examined exactly once, taking constant time per character. The matching time used after preprocessing the pattern to build the automaton is therefore $O(n)$. The time to build the automaton, however, can be large if Σ is large.

A finite automaton is defined. A special string-matching automaton is examined. An illustration is presented on how this can be used to detect a given pattern in a text. The method of how to construct a string-matching automaton for a given input pattern is given:

A *finite automaton* M is defined as a 5-tuple $(Q, q_0, A, \Sigma, \delta)$

where Q is a finite set of states

$q_0 \propto Q$ is the start state

$A \subseteq Q$ is a distinguished set of accepting states

Σ is a finite input alphabet

δ is a function from $Q \times \Sigma$ into Q called the *transition function* of M

The finite automaton begins in state q_0 and reads the characters of its input string one at a time. If the automaton is in state q and reads input character a, it moves from state q to state $\delta(q, a)$. Whenever its current state q is a member of A, the machine M is said to have *accepted* the string read so far. An input that is not accepted is said to be *rejected*. A finite automaton M induces a function ϕ called the *final state function* from Σ^+ to Q such that $\phi(w)$ is the state M ends up in after scanning the string w. Thus M accepts a string w if and only if $\phi(w) \propto A$. The function ϕ is defined by the recursive relation

$$\phi(\varepsilon) = q_0$$
$$\phi(wa) = \delta[\phi(w), a] \qquad \text{for } w \propto \Sigma^*, a \propto \Sigma \tag{3.20}$$

There is a string-matching automaton for every pattern P. This automaton must be constructed from the pattern in a preprocessing step before it can be used to search the text string. In order to specify the string-matching automaton corresponding to a given pattern $P(1, \ldots, m)$, an auxiliary function σ called the *suffix function* is defined corresponding to P. The function σ is a mapping from Σ^* to $(0, 1, \ldots, m)$ such that $\sigma(x)$ is the length of the longest prefix of P that is a suffix x:

$$\sigma(x) = \max(k: P_k] \, x) \tag{3.21}$$

The suffix function is well defined because the empty string $P_0 = \varepsilon$ is a suffix of every string. As examples, for the pattern $P = ab$, $\sigma(\varepsilon) = 0$, $\sigma(ccaca) = 1$, and $\sigma(ccab) = 2$. For a pattern P of length m, $\sigma(x) = m$ if and only if $P] \, x$. It follows from the definition of the suffix function that if $x] \, y$, then $\sigma(x) \le \sigma(y)$. The string-matching automaton that corresponds to a given pattern $P(1, \ldots, m)$ is as follows: The state set Q is $(0, 1, \ldots, m)$. The start state q_0 is state 0, and state m is the only accepting state. The transition function δ is defined by the following equation for any state q and character a:

$$\delta(q, a) = \sigma(P_q a)$$

Algorithm 3.5 *Finite-Automaton Matcher (T, δ, m)*

```
Length (T) = n
q = 0
   for i = 1, n
     do q = δ(q, T[i])                          (3.22)
       if q = m
       then print "Pattern occurs with shift" i - m
```

The matching time on a text string of length n is $O(n)$. This matching time does not include the preprocessing time required to compute the transition function δ.

3.5 Suffix Trees in String Algorithms

The suffix tree data structure has many applications in string algorithms. The suffixes of a given string are stored in the tree. All the possible substrings of the given sequence are represented by some unique path descending from the root. All the suffixes of a sequence can be encoded in linear space. A large amount of information can be retrieved from the index. It has been deployed intensively in pattern-matching problems on strings, matrices, and trees. One such exercise consists of locating all the occurrences of a given string called the pattern y as a substructure of another string called the text x. A procedure to speed up the linear time algorithms for string matching both in practice and on average using suffix trees was introduced by Crochemore [8]. The dynamic version of the static

Aho-Corasick dictionary automaton [15] was obtained by Amir and colleagues [16]. They used a dynamic set of strings to define the suffix tree. The Aho-Corasick algorithm searches for strings and is a kind of dictionary-matching algorithm. The elements of a finite set of strings (the *dictionary*) are located within an input text. All patterns are matched "at once." It can be completed in $O(m + n)$ time, where m and n are the lengths of the pattern and text, respectively. In some applications, the text is fixed and static, as in *Webster's English Dictionary* or in DNA sequences, and the string-matching query is repeated online for different patterns many times. Thus the suffix tree T is built on $x\$$. The assumption is that y occurs at least once in x. The completeness prefix property guaranteed that there is one-to-one correspondence between all occurrences of y in x and the leaves of T that are descending from the extended locus of y. The longest prefix of y occurring in x can be found in time proportional only to the length of such a prefix. Association of the number of descending leaves with each node of T can lead to knowledge of the frequency or number of occurrences of y in x without accessing all the leaves explicitly. Methods are being developed where the dynamic case can be handled without building the suffix tree T from scratch each time.

Suffix trees can be used to speed up the dynamic programming computation for solving approximate string matching. Landau and Vishkin [17] computed the longest common prefix of any two given suffixes of x using the suffix tree T in constant time. The *common prefix property* has locus in the least common ancestor (LCA) of the two corresponding leaves that can be computed in constant time after a linear time preprocessing to answer LCA queries. Chang and Lawler [9] used suffix trees on the pattern y to obtain matching statistics for text x in linear time. For each position j of x, find the longest prefix of $x[j{:}n]$ occurring as a substring of y and its corresponding extended locus in the suffix tree for y. An alternative solution for matching statistics is applying the external matching problem for file transmission by building one suffix tree at time on the string $y@w$, where @ is a separator and w is taken over $O(|x|/(|y|)$ overlapping substrings of x of size $2|y|$. The problem of finding the palindromes of maximal length in a string x can be solved in linear time with suffix trees for a constant-sized alphabet. First, a suffix tree T is built on the string $w = x@x^K\$$, where @ is a distinct separator not occurring elsewhere and x^R is the reversed string of x. The T can be preprocessed to answer LCA queries, and the technique mentioned can be applied for finding the longest common prefix of any two suffixes of w. For each position j of x, the maximal palindrome having center in j can be found, i.e., maximum k such that $x[j{:}j + k - 1] = x[j - k;j - 1)^R$. The former condition is for palindromes having center in j. This is for palindromes of even length. For those with odd length, $x[j + 1{:}j + k] = x[j - k;j - 1]^R$.

3.6 Look-up Tables

Software such as BLAST that gets used a lot for sequence queries and CAP3 for genome assembly uses a data structure called a *look-up table.* The data structure is simple in construct, where the positions of occurrences of subsequences of a certain length in a couple or more strings are recorded. Each entry in the look-up table points to a linked list of specific locations within the input set of strings where the substring corresponding to the index for the entry occurs. A look-up table for string S can be constructed in $O(|\Sigma|^w + n)$ time, where Σ is the alphabet of the string, w is the window size or prescribed length, and n is the length of the sequence S. $|\Sigma|^w$ is the number of possible substrings of length w.

Example 3.2 Construct the look-up table for the following sequence with $w = 2$:

S: cgtattctggcaggg

The mapping is as follows (Fig. 3.4): $c \rightarrow 0$, $a \rightarrow 1$, $t \rightarrow 2$, and $g \rightarrow 3$. The substring *gg* corresponds to the index $(33)_9 = 14$. The entry at index 14 indicates that the substring *gg* occurs in the sequence at positions 9, 13, and 14.

Construction of the look-up table consists of the following steps:

Step 1: Create and initialize a null list. This can be done in $O(|\Sigma|^w)$ time.

Step 2: Insert substrings one at a time. Compute index = $F(\cdot)$ in $O(w)$ time. Use the following identity for insertion:

$$F(s[k+1 \dots k+w+1]) = (F(s[k \dots k+w] - f(s[k])|\Sigma|^{w-1}) \times |\Sigma| + f(s[k+w+1]) \quad (3.23)$$

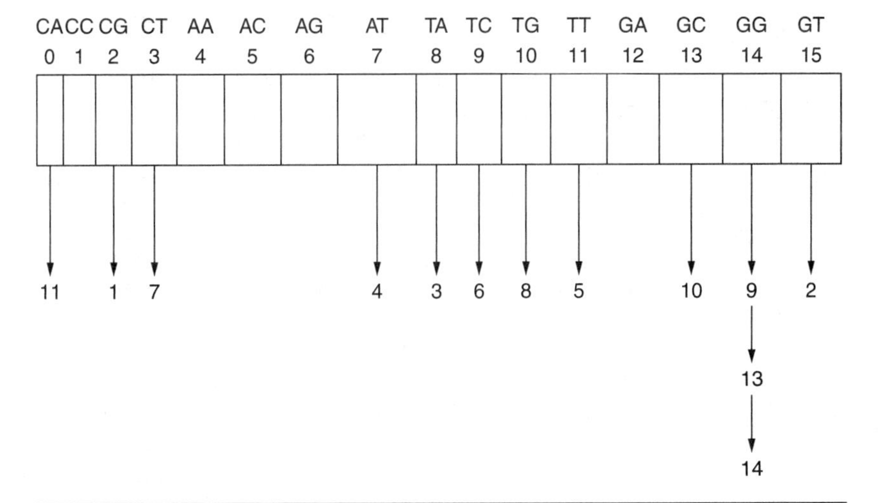

Figure 3.4 Look-up table for sequence *S*: cgtattctggcaggg.

This can be done in $O(1)$ time. The total size of linked lists is $O(n)$, and the size of the look-up table data structure is $O(|\Sigma|^w + n)$. This can be extended to more than one sequence or a set of strings. Given the look-up table for a database of strings available and a query string of length w, all occurrences of the query string in the database can be retrieved in $O(w + k)$ time, where k is the number of occurrences.

Summary

Suffix tree construction and representation of a sequence in a suffix tree are described. The generalized suffix tree can be used to represent a set of strings and stores all the suffixes of all the strings. The algorithm for suffix tree construction can be completed in $O(n)$ time and $O(n)$ space. Tandem repeats can be found in a sequence in $O[n \lg(n) + occ]$ time efficiency. Suffix trees can be used to obtain pairwise sequence alignment. One of the sequences is streamed against another sequence that is stored in a suffix tree. Where the query sequence branches off from the stored tree can be caught. In this way, all matches between the sequences can be determined.

String algorithms can be used to find patterns P in a text T. Nineteen such algorithms are discussed in this chapter. The Rabin-Karp algorithm can be executed in $\theta(m)$ preprocessing time and $O(n - m + 1)m$ matching time. The Knuth-Morris-Pratt algorithm can be completed in $\theta(m)$ preprocessing time and $O(n)$ matching time. The Boyer-Moore algorithm can be performed in $O(m + \sigma)$ preprocessing and $O(n)$ matching time, and the finite-automaton algorithm can be run in $O(m|\Sigma|)$ preprocessing time and $O(n)$ matching time. Suffix trees can be used in string matching. They can be tapped into to improve the speed in approximate string matching using dynamic programming. Look-up tables can be constructed in $O(|\Sigma|^w + n)$ time, where Σ and w are the alphabet and window sizes, respectively. The Raita algorithm, Shift algorithm, Simon algorithm, Colussi algorithm, Galil and Giancarlo algorithm, not-so-naïve algorithm, Horspool algorithm, quick-search algorithm, Berry-Ravindran algorithm, Smith algorithm, reverse-factor algorithm, turbo reverse-factor algorithm, forward DAWG matching algorithm, McCreights algorithm, construction of suffix trees, the Karkainnen and Sander algorithm, lazy suffix trees, exact string matching using suffix trees, suffix forests, hash tables, and finding the lowest common ancestor (LCA) are discussed in end-of-chapter exercises. CHAOS, LAGAN, MULTI-LAGAN, Shuffle-LAGAN, F index and pairwise alignment of sequences, GLASS, QUASAR, hash table–based tools, AVID, flat trees, and distributed suffix trees are also discussed in end-of-chapter exercises.

References

[1] E. Ukkonen, "Online construction of suffix trees," Algorithmica 14 (1995), 249–260.

[2] S. Aluru (ed.), *Handbook of Computational Molecular Biology*. New York: Chapman Hall/CRC Computer and Information Series, 2006.

[3] M. Hohl, S. Kurtz, and E. Ohlebusch, "Efficient multiple genome alignment," *Bioinformatics* 18 (2002), S312–320.

[4] L. Kaderali and A. Schliep, "Selecting signature oligonucleotides to organisms using DNA arrays,"*Bioinformatics* 18 (2002), 1340–1349.

[5] S. Kurtz and C. Schleiermacher, "REPuter: Fast computation of maximal repeats in complete genomes," *Bioinformatics* 15 (1999), 426–427.

[6] A. L. Delcher, S. Kasif, R. D. Fleischmann, et al., "Alignment of whole genomes," *Nucleic Acids Res.* 27 (1999), 2369–2376.

[7] S. B. Needleman and C. D. Wunsch, "A general method applicable to the search for similarities in the amino acid sequence of two proteins," *J. Mol. Biol.* 28 (1970), 443–453.

[8] M. Crochemore, "An optimal algorithm for computing the repetitions in a word," *Inform. Process. Lett.* 12 (1981), 244–250.

[9] W. I. Chang and E. L. Lawler, "Sublinear expected time approximate string matching and biological applications," *Algorithmica* 12 (1994), 327–344.

[10] D. Gusfield, *Algorithms on Strings, Trees and Sequences: Computer Science and Computational Biology*. Cambridge, UK: Cambridge University Press, 1997.

[11] R. M. Karp and M. O. Rabin, "Efficient randomized pattern-matching algorithms," *IBM J. Res. Dev.* 31 (1987), 249–260.

[12] D. E. Knuth, J. H. Morris, Jr., and V. R. Pratt, "Fast pattern matching in strings," *SIAM J. Comput.* 6 (1977), 323–350.

[13] R. S. Boyer and J. S. Moore, "A fast string searching algorithm," *Commun. ACM* 20 (1977), 762–772.

[14] T. H. Cormen, C. E. Leiserson, R. L. Rivest, and C. Stein, *Introduction to Algorithms*. Boston: MIT Press, 2001.

[15] A. V. Aho and M. J. Corasick, "Efficient string matching: An aid to bibliographic search," *Commun. ACM* 18 (1975), 333–340.

[16] A. Amir, G. Benson, and M. Farach, "Let sleeping files lie: Pattern matching Z-compressed file," *J. Syst. Sci.* 52 (1992), 879–884.

[17] G. M. Landau and U. Vishkin, "Fast string matching with k differences," *J. Comp. Sys. Sci.* 37 (1988), 63–78.

[18] T. Raita, "Turning the Boyer-Moore-Horspool string searching algorithm," *Software Pract. Exp.* 22 (1992), 879–884.

[19] R. Baeza-Yates, G. Navarro, and B. Ribeiro-Neto, "Indexing and searching," *Modern Information Retrieval*. Reading, MA: Addison-Wesley, 1999, Chap. 8, pp. 191–228.

[20] I. Simon, "String matching algorithms and automata," in *Results and Trends in Theoretical Computer Science*. Graz, Austria, 1994.

[21] L. Colussi, "Correctness and efficiency of the pattern matching algorithms," *Inform. Comput.* 95 (1994), 225–251.

[22] Z. Galil and R. Giancarlo, "On the exact complexity of string matching: upper bounds," *SIAM J. Comput.* 21 (1992), 407–437.

[23] A. Cardon and C. Charras, *Introduction a' l'algorithmique et a' la Programmation*. Paris: Ellipses, (1996), Chap. 9, p. 254.

[24] R. N. Horspool, "Practical fast searching in strings," *Software Pract. Exp.* 10 (1980), 501–506.

[25] D. M. Sunday, "A very fast substring search algorithm," *Commun. ACM* 33 (1990), 132–142.

[26] T. Berry and S. Ravindran, "A fast string matching algorithm and experimental results," in J. Holub, M. Simanek, et al. (eds.), *Proceedings of the Prague Stringology Club Workshop* '99. Collaborative Report DC-99-05. Prague: Czech Tech University, 1999, pp. 16–26.

[27] P. D. Smith, "Experiments with a very fast substring search algorithm," *Software Pract. Exp.* 21 (1991), 1065–1074.
[28] R. Baeza-Yates, G. Navarro, and B. Ribeiro-Neto, "Indexing and searching," in *Modern Information Retrieval.* Reading, MA: Addison-Wesley, 1999, Chap. 8, pp. 191–228.
[29] M. Crochemore, "Off-line serial exact string searching," in A. Apostolico and Z. Galil (eds.), *Pattern Matching Algorithms.* Oxford, UK: Oxford University Press, 1997, Chap. 1, pp. 1–53.
[30] M. Crochemore and W. Rytter, *Text Algorithms.* Oxford, UK: Oxford University Press, 1994.
[31] E. M. McCreight, "A space-economical suffix tree construction algorithm," *J. ACM* 23 (1973), 262–272.
[32] J. Karkkainen and P. Sanders, "Simpler linear work suffix array construction," in *Proceedings of the 30th International Colloquium on Automata, Languages and Programming.* Netherlands: Eindonven, 2003, pp. 943–955.
[33] R. Giegerich, S. Kurtz, and J. Stoye, "Efficient implementation of lazy suffix trees," in *Proceedings of the 3rd Workshop on Algorithm Engineering.* London, UK: Springer, 1999, pp. 30–42.
[34] K. R. Sharma, "Approximate solution to multiple sequence alignment problem using suffix forest method," in *CHEMCON 2005.* New Delhi, India, Indian Institute of Chemical Engineers, 2005.
[35] T. Kahveci, V. Ljosa, and A. K. Singh, "Speeding up whole genome alignment by indexing frequency vectors," *Bioinformatics* 20 (2004), 2122–2134.
[36] K. R. Sharma, "Binomial-tree representation of maximum increasing subsequence problem," in *41st Annual Convention of Chemists.* New Delhi, India, Indian Chemical Socierty, 2004.
[37] B. S. Baker and R. Giancarlo, "Sparse dynamic programming for longest common subsequence from fragments," *J. Algorithms* 42 (2002), 231–254.
[38] N. Bray, I. Dubchak and L. Pachter, "AVID: A Global Alignment Program," *Genome Research*, 13, 1, (2003), 97–102.

Exercises

1.0 Construct the look-up table for the sequence S given in Example 3.2 for $w = 3$.

2.0 Given the suffix tree shown in Fig. 3.5, deduce the sequence from which it was constructed.

3.0 Draw the suffix tree of the sequence S: GCGTACCGCGAA.

4.0 Find the pattern TATT in the text GCTTGCTATT using the KMP algorithm.

5.0 Using the suffix tree representation, approximately align the sequences S and T given below in linear time.

S: ACTGACGAGCATCATCGATGCAC
T: GAAGACATCGTCGAT

6.0 Align the –10 signal in *E. coli* promoter sequences TATAAT with the sequences GTTACGTAA. Use the scoring function 2 for a match, –2 for a mismatch, and –4 for gap. Does the complementary sequence of S match better. What is the time taken?

7.0 Using suffix tree representation and streaming of T over S in Exercise 6.0, obtain the global alignment in $O(n)$ time using the concept used in MUMer.

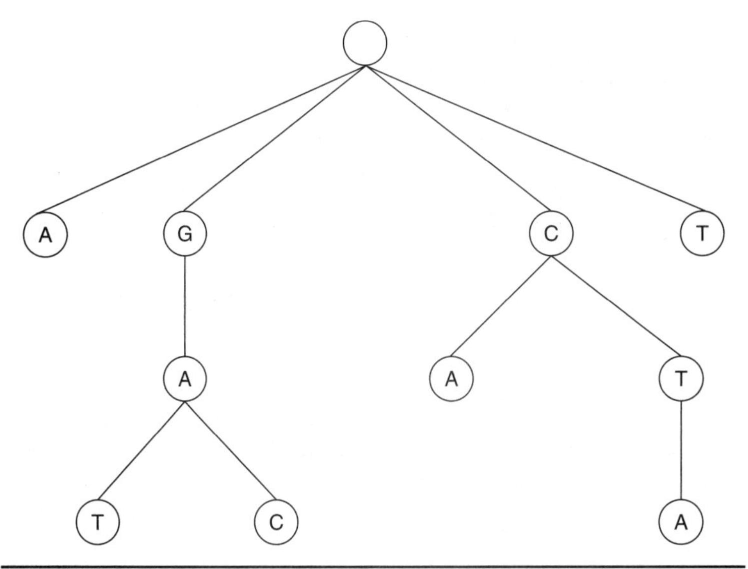

FIGURE 3.5 Suffix tree for Exercise 2.0.

8.0 Discuss the space requirements for a suffix tree representation of a sequence. How does it depend on whether it is a nucleotide sequence or an amino acid sequence?

9.0 *Raita algorithm.* An algorithm was proposed by Raita [18] that is called the *Raita algorithm.* Here, the last character of the pattern is compared first with the rightmost character of the text of the window. If they match, then the first character of the pattern is compared with the leftmost character of the text of the window. On obtaining a match, the middle character of the pattern is compared with the medley characters of the text of the window. Finally, if they match, the other characters from the second to the penultimate character of the pattern are compared again with the medley characters of text of the window. Show that the preprocessing phase of the Raita algorithm consists of computing the bad-character shift function and that it can be done in $O(m + \sigma)$ time and $O(\sigma)$ space complexity, where σ is the size of the alphabet Σ.

10.0 In Exercise 9.0, further prove that the searching phase of the Raita algorithm has an $O(mn)$ worst-case time complexity.

11.0 *Shift or algorithm.* Bitwise techniques were used to develop the Shift or algorithm [19]. It is advantageous to use the Shift or algorithm when the pattern size is no longer than the memory word size of the machine. Let A be a bit array size of length n. Vector A_j is the value of the array A after text character $y(i)$ has been processed. Information about all matches of prefixes of x that end at position j in the text for $0 < i < n - 1$ is contained in the array

$$A_j(i) = 0 \qquad \text{if } x(0, i) = y(j - i, j) \tag{3.24}$$

$$A_j(i) = 1 \text{ otherwise}$$

A_{j+1} can be obtained from A_j as follows: For each $A_j(i) = 0$:

$$A_{j+1}(i+1) = 0 \qquad \text{if } x(i+1) = y(j+1) \tag{3.25}$$
$$Aj_{+1}(i+1) = 1 \text{ otherwise}$$
$$A_{j+1}(0) = 0 \qquad \text{if } x(0) = y(j+1)$$
$$A_{j+1}(0) = 1 \text{ otherwise}$$

A complete match can be reported if $A_{j+1}(m-1) = 0$. The transition from A_j to A_{j+1} can be calculated as

$$A_{j+1} = \text{SHIFT}(A_j) \quad \text{or} \quad S_{y(j+1)}$$

where S_c is a bit array of size n for each c in alphabet Σ such that for $i < n-1$, $S_c(i) = 0$; iff $x(i) = c$. The positions of the character c in the pattern x are stored in array S_c. Assuming that the pattern length is no longer than the memory-word size of the machine, show that the space and time complexity of the preprocessing phase is $O(n + \sigma)$.

12.0 Show in Exercise 11.0 in the Shift or algorithm that the time taken for the searching phase to be completed is $O(m)$.

13.0 *Simon algorithm.* When constructing the finite automaton, the size of the automaton can be large, as discussed in Sec. 3.4.4. Simon [20] noted that there are only few significant edges in M, and the other edges lead to the initial state and hence can be deduced. The significant edges are the forward edges going from the prefix of x of length k to the prefix of length $k+1$ for $0 \leq k \leq m$ (m such edges) and the backward edges from the prefix of x of length k to a smaller non-zero-length prefix (bounded by m edges). The bound on the significant edges is $O(m)$. For each state of the automaton, now it is only necessary to store the list of its significant outgoing edges. A table L of size $m-2$ of linked lists is used. The list of the targets of the edges starting from state u is given by the element $L(i)$. During computation of the table, the integer l is computed such that $l+1$ is the length of the longest border of x. This obviates the need to store the list of the state $m-1$. Show that the preprocessing phase of the Simon algorithm can be completed in $O(m)$ space and time.

14.0 Show that in the Simon algorithm described in Exercise 13.0, the searching phase can be completed in $O(m + n)$ time. At most, $2n - 1$ text character comparisons are completed during the searching phase. Show that the maximal number of comparisons for a single text character called the delay is bounded by $\min[1 + \lg(m), \sigma]$.

15.0 *Colussi algorithm.* A refinement of the KMP algorithm discussed in the Sec. 3.4.2 was suggested by Colussi [21]. The set of patterns is divided into two disjoint subsets. Each attempt thereafter consists of two phases. In phase I, the comparisons are performed from left to right with text characters aligned with pattern positions called *noholes,* for which the value of the KMP NEXT function is greater than –1. The second phase consists of comparing the remaining positions called *holes* from right to left. The strategy offers two advantages: (1) when a mismatch occurs during the first phase after the appropriate shift, it is not necessary to compare the text characters aligned with noholes compared during the previous attempt, and (2) when a mismatch occurs during the second phase, it means that if a suffix of the pattern matches a factor of the text, and after a corresponding shift, a prefix of the pattern will still match a factor of the text, then it is not necessary to compare this factor again. Show that the space and time needed for the preprocessing phase are $O(m)$.

16.0 In the Colussi algorithm described in Exercise 15.0, show that the time taken for the searching phase is $O(n)$ and that in the worst-case scenario, $3n/2$ text comparisons are made.

17.0 *Galil and Giancarlo algorithm.* Another refinement of the KMP algorithm and variation of the Colussi algorithm is the Galil and Giancarlo algorithm [22]. For $x \neq c^m$, the searching phase is modified as follows: Let l be the last index in the pattern such that for $O \leq i \leq l$, $x(0) = x(i)$ and $x(0) \neq x(l+1)$. At the previous attempt, all the noholes were matched, and a suffix of the pattern was matched. So after the corresponding shift, a prefix of the pattern will start to match a part of the text. The window is positioned on the text factor $y(j, \ldots, j+m-1)$, and the portion $y(j, \ldots,$ last) matches $x(0, \ldots,$ last $j)$. During the next attempt the text character will be scanned beginning with $y(\text{last}+1)$ until either the end of the text is reached or a character $x(0) \neq y(j+k)$ is found. Two subcases can be identified: (1) $x(l+1) \neq y(j+k)$ and too little of $x(0)$ has been found ($k \leq l$); then the window is shifted and positioned on the text factor $y(k+1, \ldots, k+m)$, scanning of the text is resumed with the first nohole, and the memorized prefix of the pattern is the empty word. (2) $x(l+1) = y(j+k)$ and enough of $x(0)$ has been found ($k > l$); then the window is shifted and positioned on the text factor $y(k-l-1, \ldots, k-l+m-2)$, scanning of the text is resumed with the second nohole, and the memorized prefix of the pattern is $x(0, \ldots, l+1)$. Show that the preprocessing phase can be completed in $O(m)$ time and space.

18.0 Show that for the Galil and Giancarlo algorithm described in Exercise 17.0 the searching phase can be done in $O(n)$ time and that at most $4n/3$ text character comparisons are performed during the searching phase.

19.0 *Not-so-naïve algorithm* [23]. In the not-so-naïve algorithm, the character comparisons are made with the pattern positions in the following order: 1, 2, . . . , m −2, m − 1, 0. For each attempt where the window is positioned on the text factor $y(j, \ldots, j+m+1)$: If $x(0) = x(1)$ and $x(1) \neq y(j+1)$ if $x(0) \neq x(1)$ and $x(1) = y(j+1)$, the pattern is shifted by two positions at the end of the attempt and by one otherwise. Show that the preprocessing phase can be completed in constant time and constant space.

20.0 Prove that the searching phase of the not-so-naïve algorithm described in Exercise 19.0 has a worst-case time taken of $O(n^2)$ and can be completed in sublinear time in the average case.

21.0 *Horspool algorithm* [24]. For small alphabets, the Boyer-Moore algorithm is not very efficient because the bad-character shift is used.

Horspool proposed to use only the bad-character shift of the rightmost character of the window to compute the shifts in the Boyer-Moore algorithm. Show that the preprocessing phase can be completed in $O(m + \sigma)$ time and with $O(\sigma)$ space.

22.0 Show that the searching phase of the Horspool algorithm described in Exercise 21.0 can be completed in the worst case in $O(n^2)$ time. Prove that the average number of comparisons for text characters is between $1/\sigma$ and $2/(\sigma + 1)$.

23.0 *Quick-search algorithm* [25]. The bad-character shift table alone is used in this algorithm. The length of the shift is at least equal to 1 after an attempt where the window is positioned on the text factor $y(j, \ldots, j + k + 1)$. The character $y(j + m)$ can be used for the bad-character shift of the current attempt if the character is involved in the next attempt. For c in Σ,

$$qsBc\ (c) = \min(i{:}0 \leq i < m) \tag{3.26}$$

$x(m -1, i) = c$ if c occurs in x; otherwise $= m$

The comparisons between characters of pattern and text can be in any order during the searching phase. Show that the preprocessing phase can be completed in $O(m + \sigma)$ time and $O(\sigma)$ space.

24.0 Further show that the searching phase of the quick-search algorithm described in Exercise 23.0 in the worst case can be completed in $O(n^2)$ time.

25.0 *Berry-Ravindran algorithm* [26]. Shifts are performed by considering the bad-character shift table for the two consecutive text characters immediately to the right of the window. For each pair of characters (a, b) with a, b in Σ, the rightmost occurrence of ab in $a \times b$ is computed during the preprocessing phase.

$$\begin{array}{ll} 1 & \text{if } x(m-1) = a \\ brBc(a, b) = \min m - i + 1 & \text{if } x(i) \times (i+1) = ab \\ m + 1 & \text{if } x(0) = b \\ m + 2 & \text{otherwise} \end{array} \tag{3.27}$$

Show that the preprocessing phase can be completed in $O(m + \sigma^2)$ time and space.

26.0 In Exercise 25.0, a shift of length $brBc[y(i + m), y(j + m + 1)]$ is performed after an attempt where the window is positioned on the text factor $y(j, \ldots, j + m -1)$. In order to be able to compute the last shifts of the algorithm, $y(n + 1)$ is set to the null character, and the text character $y(n)$ is equal to the null character. The searching phase of the algorithm can be completed in $O(mn)$ time taken.

27.0 *Smith algorithm* [27]. The preprocessing phase of the algorithm comprises of computation of the bad-character shift function and the quick-search bad-character shift function. Shorter shifts can be achieved by computing the shift with the text character just next the rightmost text character of the window compared with using the rightmost text character of the window. Show that the preprocessing phase of the Smith algorithm can be completed in $O(m + \sigma)$ time taken and requiring $O(\sigma)$ space.

28.0 *Reverse-factor algorithm* [28]. This is an improvement on the Boyer-Moore algorithm. More prefixes are matched of the pattern by scanning the characters of the window from right to left, and then the lengths of the shifts are bettered. Use is made of the smallest suffix automaton of the reverse pattern. The suffix automaton is a directed acyclic word graph (DAWG).

$$S(w) = (Q, q_0, T, E)$$

$$L[S(w)] = (u \text{ in } \Sigma^*\text{: exists v in } \Sigma^* \text{ such that } w = vu)$$

During the preprocessing phase, the smallest suffix automaton is computed for the reverse pattern x^R. Show that this can be achieved in $O(m)$ time and $O(m)$ space, respectively.

29.0 In the reverse-factor algorithm described in Exercise 28.0, during the searching phase, the characters of the window are parsed from right to left with the automaton, $S(x^R)$ starting with state q_0. Stop where there is no more transition defined for the current character of the window from the current state of the automaton. The length of the longest prefix of the pattern that has been matched is now known. This is the length of the path taken in the suffix automaton from the start state to the final state. Then the right shift to perform is a trivial next step. Show that in the worst case the time taken for the searching phase by the algorithm is $O(n^2)$. On average, it is optimal. $O[n - \lg_\sigma(m)/m]$ inspections of text characters are performed on average. Show that this is the best bound that can be reached.

30.0 *Turbo reverse-factor algorithm* [29]. It is possible to complete the searching phase in linear time taken. In the reverse-factor algorithm, it is sufficient to save in memory the prefix u of x matched during the previous attempt. During the current attempt, on reaching the right end of u, it can be readily shown that it is sufficient to read again at most the rightmost half of u in the turbo reverse-factor algorithm. A Disp(z, w) is defined as the displacement of z in w to be the least integer $d > 0$ such that $w(m - d - [z] - 1, \ldots, m - d) = z$, where word z is a factor of word w. Generally, a prefix u is found in the text in the previous attempt, and in the present attempt, the factor v with length $m - [u]$ is matched in the text immediately to the right of u. When v is not a factor of x, then the shift is computed as in the reverse-factor algorithm. If v is a suffix of x, then the occurrence of x has been detected. If v is not a suffix and is a factor of x, then the min[Per(u), $[u]/2$] rightmost characters of u are scanned again. If u is periodic, let z be the suffix of y with length Per(u). z is now an acyclic word. Thus z can occur in u at distances multiple of Per(u), which implies that the smallest proper suffix of uv that is a prefix of x with length equal to $[uv]$ – disp(zv, x) = m – disp(zv, x). If u is not periodic, it is sufficient to scan the right part of u of length $[u]$ – Per(u) < $[u]/2$ to find a nondefined transition in the automaton. Function Disp is implemented in the automaton S directly without changing the complexity of its construction. Show that the preprocessing phase that consists of building the suffix automaton of x^R can be completed in $O(m)$ time.

31.0 Show that in the turbo reverse-factor algorithm described in Exercise 30.0 the searching phase can be completed in $O(n)$ time taken. At most, $2n$ inspections of text characters are performed in the algorithm. Show that this is optimal on average. $O[n - \lg_\sigma(m)/m]$ inspections of text in the average case are performed.

32.0 *Forward DAWG matching algorithm*[30]. The longest factor of the pattern ending at each position in the text is computed by the forward DAWG matching algorithm. DAWG uses the smallest suffix automaton of the pattern. The preprocessing phase of the forward DAWG algorithm consists of computing the smallest suffix automaton for the pattern x. Show that the time taken and space needed both can be completed in linear time. The searching phase consists of parsing the characters of the text from left to right with the automaton $S(x)$ starting with state q_0. Length (q) is the longest path from q_0 to p for each state q in $S(x)$. The notion of suffix links is used. $S(p)$ is denoted the suffix link for each state p. A transition defined for $y(j)$ for the first state of Path(p) for which such a transition is defined is taken for each text character $y(j)$ sequentially, where p is the current state. p then is updated with the target state of this transition or with the initial state q_0 if no transition exists labeled with $y(j)$ from a state of Path(p). An occurrence of x is found when length(p) $= m$. Exactly n text character inspections are performed. Show that the worst-case time taken is $O(n)$.

33.0 *McCreight's algorithm for construction of suffix trees* [31]. The suffixes are inserted in the order of $S_1, S_2, \ldots, S_n$. T_i is the tree after insertion of suffix S_i. The run time for insertion is $|S_i| = n - i + 1$. Show that the total run time would be $O(n^2)$. A linear time construct of the suffix tree can be achieved by using suffix links. Insertion of a suffix is speeded up. In order to insert S_i if the end of the path labeled β is found soon, comparison of characters in S_i can start beyond the prefix β. Show that the suffix tree construction can be completed in linear time in this fashion.

34.0 *Karkkainen and Sander's algorithm* [32]. Let S be a sequence of length n on a alphabet $\Sigma = \{1, 2, \ldots, n\}$, and assume that n is a multiple of 3. The algorithm consist of three steps: (1) $2n/3$ suffixes are sorted recursively, (2) $n/3$ suffixes are sorted using the result of step 1, and (3) the two sorted arrays are merged. Show that in this way the suffix arrays and hence suffix trees can be constructed in $O(n)$ time.

35.0 *MUMs during alignment of a pair of sequences.* Consider two sequences S and T:

S: t c g a t
T: a g g a t

Construct the generalized suffix tree GST of the two sequences S and T. Show that traversal of the tree is sufficient to identify internal nodes corresponding to MUMs.

36.0 Show that the space required for the algorithm described in Exercise 35.0 can be reduced by building the suffix tree of only one string S_1 and streaming the other string S_2 to identify the MUMs.

37.0 How many internal nodes are present in an n leaf suffix tree? What is the maximum number of children it can have?

38.0 What is the difference between a generalized suffix tree and suffix tree?

39.0 What is the difference between suffix tree and suffix array and suffix link?

40.0 What is a suffix forest?

41.0 *All strings that contain pattern P.* Use the suffix tree representation to find the query pattern P in a set of strings, $S_1\$$, $S_2\$$, . . . , $S_1\#$. Construct a GST generalized suffix tree. Enter two different digits $ and # that do not appear in any of the strings but are stored along with the suffixes. At each node, store a list of all strings S_i that are the start point of a suffix represented by an information node in the GST.

42.0 *Repeats.* Represent a sequence S using a suffix tree. Find the longest substring of S that appears at least m times, where $m > 1$.

43.0 Show that the query in Exercise 42.0 can be found in $O(n)$ time, where $n = |S|$. (*Hint:* Traverse the labeled suffix tree at the branch nodes with the sum of the label lengths from the root. Traverse the tree visiting branch nodes with information node count $\geq m$. Return to the visited branch node with longest label length.)

44.0 What is the difference between a tandem repeat and a tandem array?

45.0 What is the difference between a look-up table and a hash table?

46.0 What is the difference between a tuple and a hash table?

47.0 Describe the use of suffix tree in REPuter software development.

48.0 What is meant by a *seed* in a sequence?

49.0 At what length of the alphabet and for what sequence is a suffix tree not a profitable method of representation of a sequence?

50.0 Can a suffix tree be used to identify errors in a given sequence?

51.0 What is the difference between approximate alignment of a pair of sequences and optimal alignment of a pair of sequences?

52.0 What is a Multiple Genome Aligner (MGA)? How is a suffix tree used in MGA software?

53.0 What is the difference between the approached of Ukkonen [1] and that of McCreight [31] toward the construction of a suffix tree?

54.0 What is a palindromic repeat?

55.0 What is Multiple Exact Match (MEM)? How is a suffix tree used in MEM software?

56.0 How does ClustalW software use the suffix tree representation of biologic sequences?

57.0 Discuss the space and time savings when a query is streamed against a suffix tree when seeking an alignment.

58.0 *Lazy suffix trees.* Kurtz and colleagues [33] provided an efficient implementation of lazy suffix trees. A subtree is evaluated not before it is traversed for the first time. Write-only top-down construction can alleviate some of the concerns of using a suffix tree as a data structure. This can be seen in the word algorithm. Discuss how this can be more efficient in space and time than McCreight's algorithm [31].

59.0 *Exact string matching using suffix trees.* Given an input of pattern of length m and text of length n, prepare an output of all occurrences of P in T. Create a suffix tree for T. Maximally match P in the suffix tree. Show that this can be completed in $O(n)$ time and $O(n)$ space. Prepare output with all the leaf positions below the match point. Can this be completed in $O(m + k)$ time, where k is the number of matches?

60.0 *Set of patterns using suffix trees.* Given an input of a set of patterns (P_i) of total length m and text T of length n, prepare an output with the positions of all occurrences of each pattern P_i in T. Create a suffix tree T with a preprocessing time and space need of $O(n)$. Maximally match each P_i in the suffix tree. The output contains all leaf positions below the match point in $O(m + k)$ time, where k is the total number of matches.

61.0 Compare the Aho-Corasick approach [15] to building a keyword tree of a set of patterns P in $O(m)$ preprocessing time and $O(n + k)$ search time with the suffix tree approach. Show with matching statistics that the suffix tree approach for finding the set of patterns as discussed in Exercise 60.0 has a similar tradeoff as the Aho-Corasick approach.

62.0 *Lowest common ancestor.* Given an input of suffix tree T and two nodes v and w of T, prepare an output with the LCA of v, w in T. Can this be completed in linear time?

63.0 *Longest common extension.* Given two strings S_1 and S_2, find the length of the longest substring of S_1 beginning at i that matches substring S_2 beginning at j using suffix trees. Show that this can be completed in $O(n)$ time and $O(1)$ query time.

64.0 Discuss how the space requirements for constructing a suffix tree can be further reduced compared with the $O(n)$ discussed in Sec. 3.1.

65.0 *Suffix forest.* Sharma [34] showed that the suffix forest method can be used to approximately align multiple sequences. Compare this with construction of a generalized suffix tree, and discuss the pros and cons of each method.

66.0 *Hash tables.* A hash function can be used to convert a sequence of length n into a smaller number that serves as the digital fingerprint of the sequence. The Rabin-Karp string search algorithm, as discussed in Sec. 3.4.1, makes use of hashing to compare strings. Keys are associated with values in a hash table. The look-up operation is supported efficiently by use of a hash table. Discuss the potential of hash tables to represent sequences. Can they come in handy to align a pair of sequences?

67.0 Can the suffix tree representation of sequences be used to obtain the optimal global alignment of two sequences S and T in less time and space than the $O(n^2)$ used by the dynamic programming approach?

68.0 Can the suffix tree representation of sequences be used to obtain optimal local alignment of two sequences S and T in less time and space than the $O(n^2)$ used by the dynamic programming approach?

69.0 How can the affine gap penalty parameters be incorporated into the suffix tree construction used for obtaining local alignment with increased biologic significance?

70.0 *Rapid global alignment of human and mouse genomes.* Alignment between two sequences with lengths m and n would consume $O(mn)$ time using the dynamic programming approach discussed in Sec. 2.4. In related genomic sequences, such as the human and mouse genomes, parts of the sequences are largely conserved between the two species and align well, and these parts of the sequences are separated by longer sequences that are not well conserved and are difficult to align. Rather than seeking a global alignment, a pragmatic approach would be to find islands that align readily. This can be accomplished by performing local alignments in linear time, and then an optimal alignment can be attempted by concatenating the local alignments. Suffix trees can be used to obtain the local alignments between the segments of the two sequences. All suffixes of a sequence are stored in a suffix tree such that each path from the root to the leaf node corresponds to a suffix of the sequence are stored in the suffix tree. Every leaf corresponds to a different suffix. The number of leaves is the number of suffixes. Discuss the time and space needed for (1) constructing the suffix trees for sequences S and T, (2) obtaining local alignments between the segments of two sequences, and (3) chaining all the local alignments to obtain a global alignment. Is the resulting alignment optimal?

71.0 *CHAOS.* CHAOS is a pairwise local alignment-finding software developed at Stanford University. It is optimized for noncoding and other poorly conserved regions of the genome. Both exact matching and degenerate seeds are used. Homology in the presence of gaps is detected. Discuss the advantages of using CHAOS.

72.0 *LAGAN.* LAGAN is a parameterizable pairwise global alignment software developed at Stanford University. Local alignments generated by CHAOS are used as anchors, and the search area of the Needleman-Wunsch algorithm is limited to around these anchors. Discuss the speed and space savings of using this approach compared with the dynamic programming method.

73.0 *MULTI-LAGAN.* The approach discussed in Exercises 69.0 to 71.0 can be extended to multiple-sequence alignment by generalization. Progressive pairwise alignments are performed by a user-specified phylogenetic tree. Discuss the optimality and speed and space savings gained by using this method.

74.0 *SHUFFLE-LAGAN.* SHUFFLE-LAGAN is a novel global alignment algorithm developed by computer scientists at Stanford University. Rearrangements such as inversions, transpositions, and duplications are found in the framework of global alignment. Regions of conserved synteny are aligned using LAGAN, and a map of the rearrangements between sequences is built using CHAOS. Discuss the advantages of using this approach.

75.0 Discuss the hash table–based approach used in FASTA software for sequence queries.

76.0 *F-index and alignment of sequences S and T.* F-index is about 2 percent of the size of the sequence. A Boolean match table is constructed by partitioning sequence S into substrings. These substrings are searched in the F-index of sequence T. The columns of the match table correspond to substrings of sequence S, and the rows correspond to substrings of sequence T. Entries of true and false in the table are used to mark up corresponding substrings with similarities and dissimilarities, respectively. The match table is divided into slices and submitted for processing to a alignment tool such as BLAST. This technique is called *match table–based pruning* (MAP) [35]. Discuss the speed and space savings of this approach. What happened to the optimality of alignment.

77.0 Show that the $O(n^2)$ time taken by Needleman and Wunsch's dynamic programming method for obtaining global optimal alignment can be speeded up to $O(rn)$, where r is the amount of error allowed. This can be done by filling only the required part of the distance matrix.

78.0 *GLASS.* In the GLASS software, speedup of the dynamic programming solution for optimal global alignment for a pair of sequences is obtained. Exact matches of long substrings are found first. The extraction of k mers is required. Show that the space and time complexity are still high using this approach.

79.0 *QUASAR.* In this software tool, a suffix array is built on one of the sequences. The exactly matching seeds are counted using the suffix array. If the number of seeds for a region exceeds a selected threshold, the region is searched using BLAST. Discuss the time and space needed in this approach.

80.0 *Hash table–based tools.* Some of the hash table–based software tools developed include BLAST, MegaBLAST, BL2SEQ, WU-BLAST, SENSEI, FLASH, PipMaker, BLASTZ, PatternHunter, and BLAT. A hash table is constructed in all these tools on one of the sequences. All substrings of certain length l are inserted in the hash table. The length l varies for different applications. In BLAST, the values used are $l = 11$ for nucleotides and $l = 3$ for proteins. Exactly matching substrings called seeds of length l are found using the hash table. The seeds are extended in both directions during the second phase. Combinations are used if needed to seek better alignments. Discuss the time and space efficiency of this approach for (1) short queries and (2) long queries.

81.0 *AVID.* Pachter and colleagues [38] developed a global alignment method called AVID in which suffix trees and hash tables are used to represent sequences S and T prior to obtaining alignment. Knowledge of conserved regions in long genomes such as the human and mouse genomes is tapped into in this procedure. The input sequences S and T are preprocessed using the RepeatMasker program developed at the University of Washington. Masked and unmasked sequences are used during the alignment process. Matches are divided into repeat matches, clean matches, etc. Maximal unique matches (MUMs) are found by construction of suffix trees. The problem of finding all maximal matches is transformed into finding maximal repeated substrings in one string. The two sequences are concatenated, and a character is placed between them. A maximal repeat that crosses the boundary of sequence S represents a maximal match between sequences S and T. The anchoring and alignment of sequences are completed in a recursive fashion. An anchor is a set of nonoverlapping, noncrossing matches. *Noisy* matches are delineated. Matches then are ordered. The gap score used was zero, and the mismatch score used was infinity. Discuss the speed and space savings that can be expected with this approach.

82.0 Find the maximum decreasing subsequence from $\{21, 27, 15, 18, 16, 14, 17, 28, 13\}$. Discuss the time taken and space needed.

83.0 Compare the recursion solution obtained for Exercise 82.0 with that representing subsequence information in the form of a binomial heap [36]. [*Hint:* The time taken to identify the longest subsequence is almost $O(n)$ because it is the largest branch of the binomial heap. Space required in the worst case is $O(2n)$ and may be less in the average case.]

84.0 Show that there is an $n \lg n$ solution for the maximum increasing subsequence problem by keeping track of the indices of a sequence, predecessor, and lengths of subsequences.

85.0 *Sparse dynamic programming method for LCS from fragments.* Given a pair of sequences S and T of length n and m, respectively, and a set of M of matching substrings of S and T, find the LCS based only on the symbol correspondence induced by the substrings. Giancarlo and Baker [37] developed an algorithm that solves the problem in $O[\,|M|\lg(M)]$ using balanced trees. Show by an example that this is an improvement over the Hunt-Szymanski algorithm discussed in Chap. 2.

86.0 When Johnson's version of flat trees was used in Exercise 85.0, show that the solution can be obtained in $O[\,|M|\lg \lg \min(M, nm/M)]$.

87.0 Show that the algorithm discussed in Exercise 86.0 can be adapted to finding the LCS problem in $O[(m + n) \lg(\Sigma) + |M|\lg(M)]$ time using balanced trees.

88.0 Show that the algorithm discussed in Exercise 87.0 can be adapted to finding the LCS problem in $O[(m + n) \lg(\Sigma) + |M|\lg \lg \min(M, nm/M)]$ time using Johnson's version of flat trees.

89.0 What is the connection, if any, between suffix tree and DAWG?

90.0 How can a suffix tree representation be used to find approximate repeats in DNA sequences?

91.0 What is the difference between local and global alignment of sequences?

92.0 How are suffix trees used in fast look-ups in databases?

93.0 How are suffix trees used to calculate substring frequencies?

94.0 How are suffix trees used in motif- and pattern-finding algorithms?

95.0 How are suffix trees used in hybrid dynamic programming methods?

96.0 How suffix trees are used in oligo construction and microarray design algorithms?

97.0 What are affix trees?

98.0 How are suffix trees used in resequencing projects?

99.0 Why is use of a suffix tree an advantage when main memory is limited?

100.0 What is clustered storage, and how is it used in implementation of suffix trees?

101.0 What is a distributed suffix tree?

CHAPTER 4

Multiple-Sequence Alignment

Objectives

The objectives of this chapter are to

- Define multiple-sequence alignment and grading functions.
- Define sum-of-pairs' scores.
- Show that the optimal multiple-sequence alignment problem is NP complete.
- Introduce the center star algorithm to come within twice the optimality.
- Introduce iterative multiple-sequence alignment methods.
- Discuss suboptimal multiple-sequence alignment by the greedy method.

4.1 What Is Multiple-Sequence Alignment?

The theory of evolution states that a common ancestor exists for several known organisms. The protein sequences in different organisms that evolved from a common ancestor can be expected to be homologous but for a few misalignments. The homologous relationships can be captured by multiple-sequence alignment (MSA).

MSA methods are a topic of increasing interest. Both the development of grading functions and the hunt for an optimal alignment from all the possible alignments are of importance. The discussions apply to both protein and DNA alignments. Multiple alignments usually are found from primary sequences. Expert knowledge of protein sequence evolution can be used to produce high-quality multiple-sequence alignments. Important factors are specific sorts of columns in alignments, such as highly conserved residues or buried hydrophobic residues; the influence of secondary and tertiary structures, such as the alternation of hydrophobic and hydrophilic columns in an exposed beta sheet; and expected patterns of insertions and deletions and the tendency to alternate with blocks of conserved sequence. The changes that occur in columns and in the patterns of gaps are dictated by phylogenetic relationships between sequences. RNA alignments are constrained by a secondary-structure model. The secondary structure in many cases can be inferred from the primary-sequence information.

Databases of proteins usually feature protein families. Proteins with similar secondary structures are categorized as a *protein family.* They have similar function and evolutionary history. The structure, function, and evolutionary history of an identified protein sequence depends on the protein family of which the protein is a member. The connection between structure, function, and origin of the molecule and the protein sequence distribution is not strong.

Another motivation for seeking MSA arises in the analysis of repeats in sequences. Sections of the DNA primary-sequence distribution are replicative. These happen several times throughout the genome. Sometimes the repeats can be off by a minor number of insertions, deletions, and substitutions. For example, an ALU replication is roughly 250 bp in length and is found to occur more than 500,000 times in the human genome. Nearly 60 percent of the human genome can be attributed to these repetitions. There is no known biologic function that can be attributed to the repeats. Sixty percent of 3 billion base pairs is 1.8 billion base pairs. Thus only 1.2 billion base pairs can be attributed with all the known functions.

In MSA, homologous residues among a set of sequences are aligned together in columns. *Homologous* refers to both structural and evolutionary cohesivity. A column of aligned residues occupies similar three-dimensional structural positions, and all diverge from a common ancestral residue.

For example, Fig. 4.1 shows a multiple alignment of immunoglobins. A crystal structure is generated owing to one of the sequences. This sequence structure and alignments with other related sequences reveal conserved characteristics of the immunoglobin superfamily, including conserved beta strands and certain key residues in the sequences. With the exception of trivial cases of highly identical sequences, it is not possible to unambiguously identify structurally or evolutionary homologous positions and create a single, correct multiple alignment. Chothia and Lesk [1] examined pairwise

```
VTISCTGSSSNIGAG-NHVKWYQQLPG
VTISCTGTSSNIGS--ITVNWYQQLPG
LRLSCSSSGFIFSS--YAMYWVRQAPG
LSLTCTVSGTSFDD--YYSTWVRQPPG
PEVTCVVVDVSHEDPQVKFNWYVDG--ATLVCLISDFYPGA--VTVAWKADS--
AALGCLVKDYFPEP--VTVSWNSG---VSLTCLVKGFYPSD--IAVEWESNG—
```

Figure 4.1 Multiple sequence alignment in immunoglobins.

structural alignments in several different protein families and found that for a given pair of divergent but clearly homologous (30 percent identical) protein sequences, usually only about 50 percent of the individual residues were superimposable in the two structures. In principle, there is always strikingly correct evolutionary alignment even if the structures diverge. An evolutionary correct alignment may be more difficult to infer than a structural alignment. The structural alignment has a independent point of reference—the superimposition of crystal or nuclear magnetic resonance (NMR) structures. Sequence tends to diverge more rapidly than structure. Parts of a protein are not alignable by structure or not alignable by sequence.

4.2 Definitions of Multiple Global Alignment and Sum of Pairs

4.2.1 Multiple Global Alignment

A multiple global alignment maps the given sequences $S_1, S_2, \ldots, S_l$ to sequences $S_1', S_2', \ldots, S_l'$ that may contain spaces where

$$|S_1'| = |S_2'| = \cdots = |S_l'| \tag{4.1}$$

Removal of spaces from S_1' leaves S_i, for $1 \le i \le l$. In multiple alignment, there are various grading methods, and it is not clear which is the best. A function $d(x, y)$ that measures the distance between characters x and y is defined. This grading method is called the *sum of pairs.* The distance function assigns higher grades the more distant apart the two sequences are:

$$\Sigma\, d(S'[i], T'[j]) \qquad \text{where } l = |S'| = |T'| \tag{4.2}$$

4.2.2 Sum of Pairs

The sum-of-pairs (SP) grade for a multiple global alignment A of l sequences is the sum of the grades of all alignments induced by A. The grading function is assumed to be symmetric. The issue of a separate gap penalty is not discussed. The optimal SP global alignment of sequences $S_1, S_2, \ldots, S_l$ is an alignment that has the minimum possible SP grade for these l sequences.

4.3 Optimal MSA by Dynamic Programming

The dynamic programming methods described in Chap. 2 can be generalized for the problem of aligning l sequences each of length n. The dynamic programming table has l dimensions. The dimensions of the table are $(n + 1)^l$. Each entry depends on $2^l - 1$ adjacent entries.

The running-time complexity of the algorithm is $O(n)^l$. If n is around 460, such as the typical length of a protein, it would be feasible only for small values of l, perhaps 6 or 7. Typical protein families have 1005 members. In order to have an algorithm that works for l in the hundreds, the running time needs to be in polynomial time. As l appears as a power exponent such as n^l, when l is large it is no longer polynomial.

4.4 Theorem of Wang and Jiang [2]

The optimal MSA problem using dynamic programming is NP complete. NP stands for *nondeterministic polynomial*–bounded.

4.5 What Are NP Complete Problems?

Most computer algorithms have time efficiency of polynomial time on input sizes of n. Not all problems can be solved in polynomial time. Turing's halting problem cannot be solved by any computer no matter how much time is provided. As a matter of general rule, problems that can be solved within polynomial time are considered tractable, and problems that require superpolynomial time are considered intractable. No polynomial time solution is possible for NP complete problems.

A problem has a polynomial time solution *if and only if* there is some algorithm that solves it in $O(n^c)$ time, where c is a constant and n is the size of the input. For example, for the 2-sequence optimal global alignment problem, the time complexity is $O(n^2)$; for the 2-sequence alignment with arbitrary gap penalty function, the time complexity is $O(n^3)$; and for the 100-sequence alignment problem, the time complexity is $O(n^{100})$, comparison sorting in $O(n \lg n)$, and counting sort in $O(n)$. The $O(n^{100})$ is a polynomial time solution, but it is impractical.

NP complete problems are equivalent in the sense that if any one of them has a polynomial time solution, then all of them do. In 1971, Cook defined the notion of NP completeness. He defined NP complete problems to be problems that have a property that can be verified in polynomial time whether or not a supplied solution is correct. Karp [3] showed that a diverse array of problems is NP complete. Many problems in graph theory, combinatorial optimization and scheduling, and symbolic computation have been proven NP complete. There are some methods of dealing with NP complete problems in bioinformatics:

1. Consider only small inputs using a non–polynomial time search algorithm.
2. For inputs that are nonpolynomial on worst-case inputs, consider average inputs.
3. Give up guaranteed optimality of solutions by settling for an approximate algorithm.

4. Heuristics: Genetic algorithms can be used to seek approximate solutions. Rigorous analysis of heuristic algorithms is generally unavailable.
5. Problems to be solved in practice may be more specialized than the general one that was proved NP complete.

4.6 Center-Star-Alignment Algorithm [4]

MSA can be performed in polynomial time using the center-star-alignment algorithm. The SP grades are less than twice those of the optimal solutions. The distance function has the following properties:

1. $d(x, x) = 0$
2. Triangle inequality: $d(x, z) \leq d(x, y) + d(y, z)$ for all characters of x, y, z (4.3)

Distance along one edge of a triangle is at most the sum of the distances along the other two edges.

Algorithm 4.1 *Center-Star-Alignment Algorithm*
d is defined as the grade of the minimum global alignment distance of sequences S and T.

```
Input: Set of T of l sequences
       S1 is found that minimizes
                 d = ΣD(S1, S)                    (4.4)
```

The dynamic programming algorithm is run on each of ${}^{l}C_2$ pairs of sequences in T. The remaining sequences in T are called $S_2, \ldots, S_l$. Addition of these sequences consecutively to a multiple alignment that initially contains only S_1 is as follows: Suppose that $S_1, S_2, \ldots, S_{c-1}$ is already aligned as $S_1', S_2', \ldots, S_{i-1}'$. S_i'' and S_i' are produced by executing the dynamic programming algorithm on S_1'. S_1 is added. S_i'' is obtained from S_i' by adding spaces to those columns where spaces were added. S_i' is replaced by S_i''.

4.6.1 Time Analysis

Theorem The center-star-alignment approximation algorithm runs in time $O(l^2n^2)$ when given sequences each of length at most n.

Each of the ${}^{l}C_2$ grades of $D(D, T)$ can be computed in $O(n^2)$ time. The total time taken is $O(l^2n^2)$. After adding S_i to the multiple alignment, the length of S_1' is at most n, so the time to add all the n sequences to the MSA is

$$\sum_{1}^{l-1} O[(in)n] = O(l^2n^2) \tag{4.5}$$

All that remains to be shown is that a solution that is less than a factor of 2 worse than the optimal solution can be produced using the algorithm. Let M be the alignment produced by this algorithm, and let $d'(i, j)$ be the distance M induced on the pair $S_{i'}$ $S_{j'}$ and let

$$g(M) = \sum_{i=1}^{l} \sum_{j=1}^{l} d(i, j) \tag{4.6}$$

$g(M)$ is exactly twice the SP score of M^* because every pair of sequences is counted twice. Then for all i, $d'(1, k) = D(s_{1'}, S_k)$. This is so because the algorithm used an optimal alignment of S_1'_ and $S_k d(\_, \_) = 0$. Let M^* be the optimal alignment, $d'(i, j)$ be the distance M^* induces on the pair $S_{i'}$ S_j and

$$g(M^*) = \sum_{i=1}^{l} \sum_{j=1}^{l} d^*(i, j) \tag{4.7}$$

Theorem SP grade less than twice that of the optimal SP alignment is produced by the center-star-alignment algorithm.

$$\frac{g(M)}{g(M^*)} \leq \frac{2(l-1)}{l} < 2 \tag{4.8}$$

Proof Obtain an upper bound on $v(M)$ and a lower bound on $v(M^*)$ and then take their quotient.

$$g(M) = \sum_{i=1}^{l} \sum_{j=1}^{l} d'(i, j) \tag{4.9}$$

$$\leq \sum_{i=1}^{l} \sum_{j=1}^{l} [d'(i,k) + d'(k, j)] \tag{4.10}$$

$$\text{Triangle inequality} = 2(l-1) \sum_{l=2}^{l} d(k,1) = 2(l-1) \sum_{l=2}^{l} D(S_1, S_l) \tag{4.11}$$

Equation (4.11) follows because each $d(l, 1) = d(1, l)$ occurs in $2(l - 1)$ terms.

$$g(M^*) = \sum_{i=1}^{l} \sum_{j=1}^{l} d^*(i, j) \tag{4.12}$$

$$\leq \sum_{i=1}^{l} \sum_{j=1}^{l} D(S_i, S_j) \tag{4.13}$$

$$\geq \sum_{i=2}^{l} \sum_{j=2}^{l} D(S_i, S_j) \tag{4.14}$$

Combining these inequalities,

$$\frac{g(M)}{g(M^*)} \leq \frac{2(l-1)}{l} < 2 \tag{4.15}$$

4.7 Progressive Alignment Methods

The solution method in progressive alignment is by constructing a succession of pairwise alignments. Initially, two sequences are chosen and aligned by standard pairwise alignment. Then a third sequence is chosen and aligned to the first alignment, and this process is iterated until all sequences have been aligned. This method was suggested by Feng and Doolittle [5], among others. Different algorithms differ in (1) the way that they choose the order to do the alignment, (2) whether the progression involves only alignment of sequences to a single growing alignment, and (3) in the procedure used to align and score sequences or alignments against existing alignments. These alignments are heuristic in nature. A *guide tree* is usually built. This is a binary tree whose leaves represent sequences and whose interior nodes represent alignments. The root node represents a complete multiple alignment. The nodes furthest from the root represent the most similar pairs.

Algorithm 4.2 *Feng-Doolittle Progressive MSA*
Calculate a diagonal matrix of $N(N-1)/2$ distances between all pairs of N sequences by standard pairwise alignment, covering raw alignment scores to approximate pairwise distances. Construct a guide tree from the distance matrix using the clustering algorithm of Fitch and Margoloash [6].

Starting from the first node added to the tree, align the child nodes (which may be two sequences, a sequence and an alignment, or two alignments). Repeat for all other nodes in the order in which they were added to the tree until all sequences have been aligned.

The distance D is calculated as

$$D = -\log S_{\text{eff}} = -\log \frac{(S_{\text{obs}} - S_{\text{rand}})}{(S_{\text{max}} - S_{\text{rand}})} \tag{4.16}$$

where S_{obs} is the observed pairwise alignment score, S_{max} is the maximum score, the average of the score of aligning either sequence to itself, and S_{rand} is the expected score for aligning two random sequences of the same length and residue composition.

The chosen center sequence is always attempted to be aligned with the unaligned sequences in the center-star algorithm discussed in the preceding section. However, there might be cases in which *clusters* are formed because some of the sequences are very "close." A technical hurdle is how to define *close* and *cluster.* The cluster of sequences may have to be merged after alignment of sequences in the same cluster first.

MAFFT software was developed [15] with increased speed to obtain MSA. Fast Fourier transformation (FFT) is used in the procedure. FFT is a interesting method especially for obtaining periodicity in real systems. Although similarities is the object of study, very little has been done using FFT. Homologous segments can be detected rapidly by reading of the peaks in the frequency spectrum. A scoring system is also introduced that is designed for sequences with large insertions that are distantly related with similar sequence width. The correlation between two amino acid sequences can be calculated. The homologous segments can be found, and a homology matrix is divided. The procedure is extended to group alignments. A suitable similarity matrix is defined with appropriate gap penalty. The CPU implementation was found to be speedier than T-COFFEE, CLUSTALW, DILAIGN, BALIBASE, etc.

A variation of the progressive alignment strategy is called *iterative pairwise alignment.* For example, a sequence that is not aligned is selected and aligned to the previously obtained aligned sequences. Optimal pairwise alignments between individual sequences in the MSA, without regard to spaces inserted, are used to identify the "nearest" sequence. All that remains is to show how to seek an alignment of a sequence with a group of sequences. The method that was used to add S_i to the center-star alignment can be set as a macro and run.

4.8 The Consensus Sequence

Given an MSA, it is sometimes useful to derive from it a *consensus sequence* that can be used to represent the entire set of sequences in the alignment.

Definition Given a multiple alignment of M sequences $S_1, S_2, \ldots, S_l$, the consensus character of column i of M is the character C_i that minimizes the sum of distances to it from all the characters in column i.

$$\min imize \sum_{j=1}^{l} d(S_j^{'}[i], C_i) \quad (4.17)$$

Let $d(i)$ be the minimum sum. The consensus sequence is the concatenation $C_1, C_2, \ldots, C_i$ of all the consensus characters, where $l = |S_1'| = |S_k'|$. The alignment error of M then is defined to be

$$\sum_{i=1}^{l} d'(i) \quad (4.18)$$

4.9 Greedy Method

A substantial reduction in the volume of computations to minimize Eq. (4.18) can be achieved using the greedy method to construct multiple alignments from pairwise alignments. The simplest such method fixes the alignment of the pair of sequences i, j with minimum distance. Of the remaining pairs, the minimum distance pairwise alignment is fixed. If each member of the pair is already in a fixed alignment, then the new fixed alignment joins those two aligned groups. The resulting multiple alignment is an upper bound. It is seldom optimal.

4.10 Geometry of Multiple Sequences

The geometries of multiple sequences are referred to as *line geometries* because any two points (sequences) can be joined by a straight line in the metric space. This geometry has some highly non-Euclidean properties that are not well understood. In the geometry of geodiscs, spaces are referred to as *straight*. The problem of aligning several sequences can be studied using this technique. If the k sequences are related by a binary tree, they can be aligned in $O(rn^2)$ time by a heuristic method naturally suggested by the geometry. If the original sequences are formed out of an alphabet Σ, define a weighted-average sequence to be a finite sequence $S = S_1, S_2, \ldots, S_n$ where each S_i has the form $S_i = (p_0, p_1, \ldots)$, where $p_i \geq 0$ and

$$\Sigma p_i = 1 \tag{4.19}$$

If p_i corresponds to the proportion of the ith element of A and b_0 corresponds to the proportion of deletions, -, it is then easy to convert a usual sequence into a weighted-average sequence by taking a statistical summary of the letters aligned to a given position. The letter, -, is thought of as a space indicating a deletion in the sequence in which it appears as an insertion m.

$$d(a,b) = \left(\sum_i w_i \left| p_i - q_i \right|^{\alpha} \right)^{1/\alpha} \tag{4.20}$$

where w_i is the weighting factor and $\alpha \geq 1$ is a constant. In order to compute the global distance $D(S, T)$ between two weighted sequences, the usual dynamic programming algorithm is employed. Here

$$S = S_1, S_2, \ldots, S_n$$

$$T = T_1, T_2, \ldots, T_m \tag{4.21}$$

$$D_{ij} = D(a_1, \ldots, a_i, b, \ldots, b_j) \tag{4.22}$$

$$D_{0j} = D(\_, b_1, \ldots, b_j) \tag{4.23}$$

$$D_{i0} = D(a_1, \ldots, a_i, \_) \tag{4.24}$$

$$D_{00} = 0 \tag{4.25}$$

Then

$$D_{ij} = \min[D(i-1), + d(a_i, \_), D(i-1, j-1) + d(a_i, b_j), D(I, j-1) + d(\_, b_j)]$$

$$D(n, m) = D(S, T)$$

For an optimal alignment of S and T, define

$$g(\lambda) = \lambda S \oplus (1-\lambda)T \tag{4.26}$$

where

$$g_i(\lambda) = \lambda S_i + (1-\lambda)T_i^* \tag{4.27}$$

and the last + sign is simple vector addition. In the case $\lambda = 1/2$, $U(1/2)$ is an equal weighting of S_i' and T_i^* from an optimal alignment of S and T, and more can be shown in that direction. The following theorem states that the resulting metric space in a line geometry.

Theorem Let $g(\lambda) = \lambda S \oplus (1-\lambda)T$ (4.28)

Then

$$D(S, T) = D[S, V(\lambda)] + D[b, V(\lambda)] \tag{4.29}$$

and

$$D[S, U(\lambda)] = (1-\lambda)D(S, T) \tag{4.30}$$

This theorem can be proved using the triangular inequality. As a corollary to this theorem,

$$D[U(\lambda_1), U(\lambda_2)] = |\lambda_1 - \lambda_2| D(S, T) \tag{4.31}$$

The theorem implies that a weighed-average sequence can be found to represent any point on the line between two sequences.

Theorem If U satisfies $D(S, U) + D(U, T) = D(S, T)$, then each $g_i = \lambda_i S_i + (1-\lambda_i)T_i^*$ for some optimal alignment of S and T.

The proof of this theorem can be found in Waterman [7]. It may be conjectured that the geometry for more than two sequences immediately follows. Unfortunately, the geometric properties of even three sequences are far from simple. The problem of aligning r sequences when a binary tree relating the sequences is assumed does have a practical heuristic solution.

Suppose that two sets of sequences $S_1, S_2, S_3, \ldots, S_n$ and $T_1, T_2, \ldots, T_m$ have been aligned by some method. Each such alignment can be easily made into weighted-average sequence S^* and T^*. The metric $D(\_, \ldots)$ can be applied to align these alignments. Note that $\lambda S^* \oplus (1-\lambda)T^*$ can be formed from any alignment that gives $D(S^*, T^*)$ but that the number of sequences involved m and n do not contribute to the complexity of computing $D(S^*, T^*)$.

Consider three sequences S_1, S_2, and S_3. Let them be related by a tree. S_1 and S_2 are nearest neighbors. Thus $e_2 = \frac{1}{2}S_1 \oplus \frac{1}{2}S_2$ occupies the midpoint of a line between S_1 and S_2. If all distances had the properties of Euclidean geometry, the center of gravity would be a point on a line from the midpoint e_2 to S_3, two-thirds of the length from S_3 and one-third from C_2. Therefore, the desired sequence is $C_3 = \frac{1}{2}S_3 \oplus \frac{2}{3}[e_2]$. This algorithm generalizes to r sequences, and other weightings can be used.

Summary

Multiple sequence alignment involves lining up more than two sequences and finding matches among them. The dynamic programming methods discussed in Chap. 2 for pairwise sequence alignment can be extended to multiple sequences. For k sequences, the size of the dynamic programming table would be $(n + 1)^k$. Running time needed would be $O(n)^k$. For sequence length greater than 30, this would be infeasible when k is greater than 4. When k is large, the time taken cannot be represented by a polynomial. The optimal sum-of-pairs alignment is NP complete, non–deterministic polynomial bounded.

Center-star alignment can be used to obtain multiple sequence alignment in polynomial time with a grade of alignment within twice the optimal solution. Progressive alignment methods of MSA are discussed. Variations of this approach include the iterative alignment method. Consensus sequence and the greedy method for MSA also are discussed. End-of-chapter exercises include analysis of COSA, CLUSTALW, T-COFFEE, suffix forest, DIALIGN, MUSCLE, MAFFT, PSI-BLAST, STAMP, JalView, etc.

References

[1] C. Chothia and A. M. Lesk, "The relation between the divergence of sequence and structure in proteins," *EMBO J.* 5 (1986), 823–826.

[2] L. Wang and T. Jiang, "On the complexity of multiple sequence alignment," *J. Comput. Biol.* 9 (1994), 337–349.

[3] R. M. Karp, "Reducibility among combinatorial problems." In R. E. Miller and J. W. Thatcher (eds.), *Complexity of Computer Applications*. New York: Plenum Press, (1972), pp. 85–104.

[4] M. Tompa's course notes, Computational Biology, CSE 527, University of Washington, Seattle, WA, Winter 2000, DBI-9601046 and NSF Grant, DBI-997 4498.

[5] D. F. Feng and R. F. Doolittle, "Progressive sequence alignment as a prerequisite to correct phylogenetic tree," *J. Mol. Evol.* 25 (1987), 351–360.

[6] W. M. Fitch and E. Margoliash, "Construction of phylogenetic trees," *Science* 155 (1967), 279–284.

[7] M. S. Waterman, *Introduction to Computational Biology*. New York: Chapman and Hall, 1995.

[8] E. Althaus, A. Caprara, H. P. Lenhof, and K. Reinert, "Multiple sequence alignment with arbitrary gap costs: Computing an optimal solution using polyhedral combinatorics," *Bioinformatics*. 18 (2002), S4–S16.

[9] P. H. A. Sneath and R. P. Sokal, *Numerical Taxonomy.* San Francisco: Freeman, 1973.
[10] R. Chenna, H. Sugawara, T. Koike, et al., "Multiple sequence alignment with the clustal series of programs," *Nucleic Acids Res.* 31 (2003), 3497–3500.
[11] C. Notredame, D. G. Higgins, and J. Heringa, "T-Coffee: A novel method for fast and accurate multiple sequence alignment," *J. Mol. Biol.* 302 (2000), 205–217.
[12] K. R. Sharma, "On the use of suffix trees in multiple sequence alignment," 230th ACS National Meeting, Washington, DC, August–September 2005.
[13] A. R. Subramanian, J. Weyer-Menkhoff, M. Kaufmann, and B. Morgenstern, "DIALIGN-T: An improved algorithm for segment-based multiple sequence alignment," *Bioinformatics.* 6 (2005), 66.
[14] R. C. Edgar, "MUSCLE: Multiple sequence alignment with high accuracy and high throughput," *Nucleic Acids Res.* 32 (2004), 1792–1797.
[15] K. Katoh, K. Misawa, K. Kuma, and T. Miyata, "MAFFT: A novel method for rapid multiple sequence alignment based on fast Fourier transform," *Nucleic Acids Res.* 30 (2002), 3059–3066.
[16] S. F. Altschul, T. L. Madden, A. A. Schaffer, et al., "Gapped Blast and PSI-Blast: A new generation of protein database search programs," *Nucleic Acids Res.* 25 (1997), 3389–3402.
[17] R. B. Russell and G. J. Barton, "Multiple protein sequence alignment from tertiary structure comparison: Assignment of global and residue confidence levels," *Proteins.* 14 (1992), 309–323.
[18] M. Clamp, J. Cuff, S. M. Searle, and G. J. Barton, "The Jalview Java alignment editor," *Bioinformatics.* 20 (2004), 426–427.

Exercises

1.0 What are NP complete problems?

2.0 What is the Turing's halting problem?

3.0 Name two applications of MSA?

4.0 How is MSA needed in the alignment in finding the common ancestor among several organisms?

5.0 How is MSA used in finding repetitive sequences?

6.0 Why is MSA performed on protein sequences?

7.0 What is meant by the distance between two sequences?

8.0 Discuss the proof of triangle inequality?

9.0 How is a smaller input going to help in dealing with NP complete problems?

10.0 How is using a average input going to help in dealing with NP complete problems?

11.0 How is solving a specialized problem compared with the general problem going to help in dealing with NP complete problems?

12.0 How are genetic algorithms used in obtaining solutions to NP complete problems?

13.0 How does giving up optimality help in obtaining solutions to NP complete problems?

14.0 What are progressive alignment methods?

15.0 What is the difference between the progressive alignment method and the iterative alignment method?

16.0 Can the order of selection of sequences make a difference in the results of the progressive alignment method?

17.0 What is a guide tree?

18.0 What is meant by a clustering algorithm?

19.0 What is meant by a consensus sequence?

20.0 How is the greedy approach applied to MSA?

21.0 *COSA* [8]. COSA is an integer linear programming (ILP) method. It can be used instead of the multidimensional dynamic programming method to obtain MSA. An objective function is maximized subject to some constraints. The similarity grade is maximized—$\Sigma x.w_x$. Four constraints are required for optimal alignment. A cutting-plane algorithm is adapted. Show that solving an ILP is NP complete.

22.0 *UPGMA* [9]. The unweighted pair group method with arithmetic mean is a bottom-up data-clustering method. Two groups of sequences or alignments can be aligned to form a single alignment. All the possible $^{N}C_2$ pairwise alignments among N sequences are calculated, and the distance matrix is obtained. A guide tree can be constructed from the matrix. Then groups of sequences are aligned progressively following the branching order in the tree. UPGMA is such a distance-matrix method. Show that the accuracy of alignment is not superior using this method but that it can generate large alignments rapidly.

23.0 *CLUSTALW* [10]. ClustalW is a popular computer software using the progressive alignment methods described in Sec. 4.7. There are three main steps: (1) obtain a pairwise alignment, (2) construct a phylogenetic tree, and (3) obtain the multiple sequence alignment. Pairwise alignments are computed for all sequences, and similarities are stored in a matrix. This is then converted into a distance matrix, where the distance measures reflect the evolutionary distance between each pair of sequences. From this distance matrix, a guide tree, or phylogenetic tree, for the order in which pairs of sequences are to be aligned and combined with previous alignments is constructed using a neighbor-joining clustering algorithm. Sequences are aligned progressively at each branch point starting from the least distant pair of sequences. Discuss the time-taken and space-needed efficiency in this approach. What is the degree of optimality?

24.0 *T-COFFEE* [11]. Tree-based consistency objective function for alignment evaluation (T-COFFEE) is an MSA software using a progressive approach. It generates a library of pairwise alignments to guide the multiple-sequence alignment. Discuss the degree optimality and time-taken and space efficiency of this approach?

25.0 What are the advantages of using multiple-sequence alignments of genomic DNA sequences and a multiple-sequence alignment of a group of homologous proteins?

26.0 What are the advantages of using multiple-sequence alignment instead of pairwise-sequence alignment?

27.0 *Suffix Forest* [12]. Given k sequences, construct k suffix trees. How will you obtain a approximate multiple-sequence alignment using suffix forest? How close is it to optimality? What is the time-taken and space efficiency of this approach?

28.0 *DIALIGN* [13]. Segment-based multiple-sequence alignment is used in this approach. It is an implantation of an improved algorithm. Show that the time-taken efficiency would be $O(kn^2)$, where k is the length of the fragment size. A greedy strategy is employed. The weights for each fragment are recalculated. Is it NP complete?

29.0 *MUSCLE* [14]. This is used for creating multiple alignments of protein sequences. Elements of the algorithm include fast distance estimation using $k$mer counting, progressive alignment using a new profile function called the *log-expectation score,* and refinement using tree-dependent restricted partitioning. Discuss the degree of optimality and time-taken and space efficiency of the method used in the MUSCLE software.

30.0 *MAFFT* [15]. MSA can be performed using MAFFT software. The fast search for anchor points is obtained by the fast Fourier transform (FFT) method. The guide tree is constructed rapidly. Accurate alignments also can be constructed rapidly. An initial alignment is obtained using the progressive method twice. A roughly estimated guide tree is used to align sequences in the first phase. Show that the time taken would be $O(n^2l)$. The guide tree is constructed in a similar fashion to the UPGMA method. The progressive method is used in the second phase. FFT preprocessing is used. Discuss the degree of optimality and time-taken and space efficiency of this approach.

31.0 *PSI-BLAST* [16]. This is a profile-based methods. A database is searched with a single sequence for any high-scoring sequences that are found. These are built into a multiple alignment. This multiple alignment is used to derive a search "profile" for a subsequent search of the database. This process is repeated until no new sequences are found or after a prespecified number of iterations. Discuss the optimality of alignment and time-taken and space efficiency of this approach.

32.0 *STAMP* [17]. Two or more structures can be aligned using STAMP simultaneously. Multiple alignments are sought using hierarchical methods. Structures are superimposed, assuming that the alignment is correct. The structural similarity is provided in a matrix of grades for all possible pairs of residues. A dynamic programming algorithm is used to obtain the best grade and an alignment of the sequences. The process is repeated until convergence. Discuss the degree of optimality and time-taken and space efficiency of this approach.

33.0 *JalView* [18]. Automatic multiple-sequence alignments can be improved by manual editing. JalView is a Java alignment editor. It has a number of core alignment viewing and editing options. Principal components analysis (PCA) can be performed. How is this tool going to help improve the accuracy of the alignment? What is the additional time taken and space needed?

34.0 What are the differences between structural and evolutionary alignments?

35.0 What is the importance of the quality of alignment in MSA?

36.0 What is a subalignment during MSA?

37.0 What is meant by automatic alignment?

38.0 For the methods of MSA discussed, should the genomes be linear or circular?

39.0 Where is MSA used in finding the protein secondary structure?

40.0 What are the requirements on the quality of alignment in MSA in order to obtain the protein secondary structure?

41.0 Where does the affine gap penalty figure during MSA?

42.0 Given k sequences, does there exist one unique set of sequences for all the possible cell values in a k-dimensional dynamic programming table?

43.0 What are the considerations of stability of alignment in the dynamic programming method of MSA?

44.0 What are the considerations of stability of alignment in the center-star alignment method of MSA?

45.0 What are the considerations of stability of alignment in the progressive alignment method of MSA?

46.0 What are the considerations of stability of alignment in iterative alignment method of MSA?

47.0 What are the considerations of stability of alignment in the greedy method of MSA?

48.0 Consider a set of k sequences that differ by a few errors. Is obtaining the multiple-sequence alignment of these sequences NP complete? Why?

49.0 Can the banded diagonal approach discussed in Chap. 2 for pairwise alignment be extended to MSA?

50.0 Can the inverse dynamic programming method be applied to multiple sequences?

51.0 Consider a sparse k-dimensional dynamic programming table. Is this problem NP complete? Why?

52.0 Can the dynamic array method of Hirschberg used for pairwise alignment be extended to MSA?

53.0 Is there a tradeoff between time efficiency and degree of optimality during MSA? How can this be tapped into?

PART 2

Probability Models

CHAPTER 5

Hidden Markov Models and Applications

Objectives

The objectives of this chapter are to

- Construct zeroth-order, first-order, second-order, and *k*th-order Hidden Markov models (HMMs).
- Represent DNA sequences using the HMM.
- Characterize the HMM.
- Learn the forward, backward, and Viterbi algorithms.
- Apply probability models to
  - Phylogenetic tree construction.
  - Evolution.
  - The proteome.
- Seek pairwise and multiple alignment using the HMM.
- Accomplish protein family characterization.
- Model periodicity in DNA by wheel HMMs.
- Understand the Chargaff's parity rule.
- Accomplish signal peptide and signal anchor prediction.

5.1 Introduction

Hidden Markov models (HHMs) are constructed by using concepts such as conditional probability. They are used in a variety of applications in bioinformatics. They are classified under a useful class of *probabilistic models.* HMMs are a special case of neural networks, stochastic networks, and Bayesean networks. Sequence consensus,

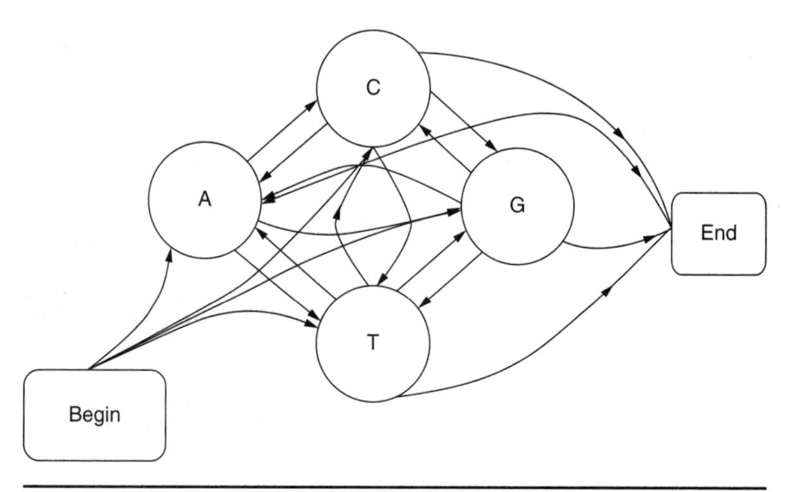

FIGURE 5.1 Markov chain model for DNA sequence with a begin and end state.

profiles, flexible patterns, and blocks can be special cases of the HMM approach. A DNA sequence can be represented using an HMM. Such an example is shown in Fig. 5.1. In the early 1990s, Krogh and colleagues [1] at the University of California at Santa Cruz described preliminary results on modeling protein sequence multiple alignments with probabilistic HMMs. Information available in biologic sequences can be captured using Markov models and heuristics.

Two HMM software packages for sequence analysis were developed and made available free of charge. There is a lot of interest centering around HMMs in the literature. They are still viewed as black boxes instead of natural models of sequence alignment problems. Many of the key papers where HMMs are described are in the field of speech recognition and therefore not readily accessible to the bioinformatics community. HMMs can be applied to a lot of problems, such as protein structure modeling, gene finding, phylogenetic analysis, modeling time series, speech recognition, modeling coding and noncoding regions of DNA, protein subfamilies, and machine learning techniques, among others.

5.2 *k*th-order Markov Chain

A *Markov chain* is a sequence of random variables whose probabilities at a time interval depend on the value of the number at the previous time or times. The controlling parameter in a Markov chain is the transition probability. This is a conditional probability for the system to go to a particular new state given the current state of the system. In a kth-order Markov chain, the distribution of X_t depends on the k values immediately preceding it.

$$\text{Transition probability of } X_t = P(X_t = X / X_{t-k}, X_{t-k-1}, \ldots, X_{t-1}) \quad (5.1)$$

The transition probabilities in a first-order Markov model for X_t would depend on only one previous value, X_{t-1}. Dyad dependencies can be modeled using a first-order model. The transition probabilities in a zeroth-order Markov model for X_t would not depend on the previous values and would be independent of them. Similarly, in a second-order Markov model, the transition probabilities for X_t would depend on two previous values, X_{t-1} and X_{t-2}. A kth-order Markov chain is said to be stationary for all t and u:

$$P(X_t = X/X_{t-k}, X_{t-k+1}, \ldots, X_{t-1}) = P(X_u/X_{u-k}, X_{u-k+1}, \ldots, X_{u-1}) \quad (5.2)$$

That is, for a stationary Markov chain, the distribution of X_t is independent of the value of t and depends only on the previous k variables. The transition probabilies for a first-order Markov model to represent the primary sequence structure of DNA with the beginning and ending base pair can be represented in the form of a diagram similar to the one shown in Fig. 5.1. That diagram is a directed graph with nonzero t_{ij} connections and can be called the *architecture of the Markov chain*. The arrows point to the next occurrence of the base pair.

5.3 DNA Sequence and Geometric Distribution [2–4]

The chain sequence length distribution of DNA can be represented using the geometric distribution. The mechanism of formation of the polynucleotide may have a role in the parameter of the geometric distribution. For instance, a terpolymer formed by free-radical polymerization can be modeled with respect to the sequence distribution as follows: When three termonomers enter a long copolymer chain at M_1, M_2, and M_3 concentration with reactivity ratios r_{12}, r_{21}, r_{23}, r_{32}, r_{13}, and r_{31},

$$P_{22} = \frac{1}{\left[\dfrac{M_1}{r_{21}M_2} + \dfrac{M_3}{r_{23}M_2}\right]} \quad (5.3)$$

Let

$$\beta = \frac{M_3}{r_{23}M_2};\ \gamma = \frac{M_3}{r_{23}M_2} \quad (5.4)$$

Then the probability of an M_2M_2 dyad is

$$P_{22} = \frac{1}{(1+\beta+\gamma)} \quad (5.5)$$

The sequence length of the repeats of monomer 2 in the chain is given by

$$N_{2x} = \frac{1}{[1+\beta+\gamma]^{x-1}}\left[1-\frac{1}{(1+\beta+\gamma)}\right] \tag{5.6}$$

The mean of the distribution is given by

$$\lambda = \frac{(\beta+\gamma)}{(1+\beta+\gamma)} \tag{5.7}$$

The variance σ^2 of the distribution can be written as

$$\sigma^2 = \frac{1}{(\beta+\gamma)^2} + \frac{1}{(\beta+\gamma)} \tag{5.8}$$

For a tetrapolymer with four monomers, such as the case for DNA polynucleotides,

$$P_{22} = \frac{1}{(1+\beta+\gamma+\delta)} \tag{5.9}$$

where $$\delta = \frac{M_4}{r_{24}M_2}$$

The sequence length of a single base in the polynucleotide chain can be given by a geometric distribution:

$$N_{2x} = \frac{1}{[1+\beta+\gamma+\delta]^{x-1}}\left[1-\frac{1}{(1+\beta+\gamma+\delta)}\right] \tag{5.10}$$

The mean and variance of the distribution can be written as

$$\lambda = \frac{(\beta+\gamma+\delta)}{(1+\beta+\gamma+\delta)}; \qquad \sigma^2 = \frac{1}{(\beta+\gamma+\delta)^2} + \frac{1}{(\beta+\gamma+\delta)} \tag{5.11}$$

The polymer compositions can be related to the monomer compositions by simple relations. Thus the run lengths of A, AA, AAA, AAAA, etc. for each of the four bases can be calculated. Modifications to $(\beta+\gamma+\delta)$ can be made depending on the mechanism of formation of the polynucleotide chain. The assumption that the composition of adenine, guanine, cytosine, and thymine occurs in equal proportions in the polymer chain can be used to simplify the

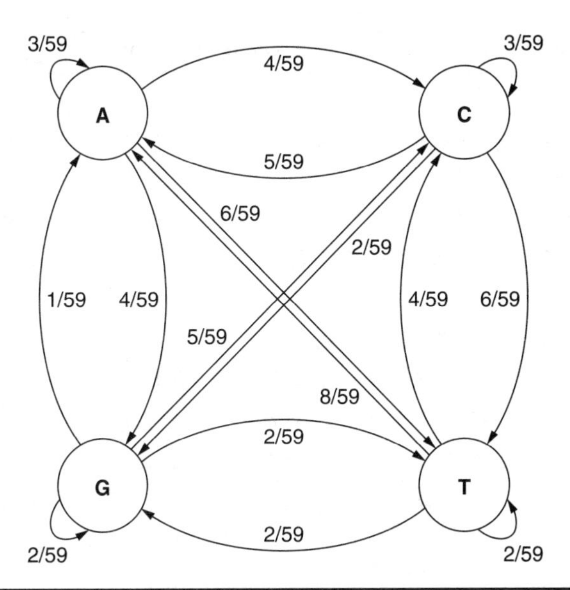

Figure 5.2 660 base pairs of DNA in *Homo sapiens* [5].

terms. The triad and tetrad probabilities, such as AGC or AAC, AGG, etc., can be computed via the dyad probabilities.

Example 5.1 Chaves and colleagues [5] submitted the DNA sequence with 660 bases in *Homo sapiens* shown in Fig. 5.2 to the National Center for Biotechnology Information (NCBI). Develop a Markov model of the third order to represent this information. Calculate the transition probabilities, and represent the information in the form of a suitable table.

$$\text{Number of triads that need to be studied} = 4^3 = 64 \tag{5.12}$$

$$\text{Alphabet} = \{A, C, G, T\} \tag{5.13}$$

$$\text{Number of transition probabilities that need to be calculated} = 4 \times 64 = 256 \tag{5.14}$$

The 256 transition probabilities $P(A/AAA)$, $P(G/AAA)$, . . . are calculated from the information provided in Fig. 5.2 and presented in the Table 5.1. Columns 3–6 are conditional probability values for the base pair shown at the top of the column given the preceding triad that occurred in the sequence in column 2. A triad number is also given to the 64 possible triads for DNA.

Example 5.2 Develop a first-order Markov model for the DNA sequence given in Example 5.1 to represent the first 60 base pairs. Calculate the transition probabilities, and represent the information in the form of a suitable diagram.

ctatatatcttaatggcacatgcagcgcaagtaggtctacaagacgctacttcccctatc

$$\text{Alphabet} = \{A, C, G, T\} \tag{5.15}$$

$$\text{Number of transition probabilities that need to be calculated} = 4 \times 4 = 16 \tag{5.16}$$

Triad No.	Triad	P(A/Triad)	P(G/Triad)	P(C/Triad)	P(T/Triad)
1	AAA	3/657	0	6/657	3/657
2	AAC	5/657	1/657	6/657	4/657
3	AAG	3/657	0	1/657	1/657
4	AAT	3/657	3/657	4/657	5/657
5	ACC	5/657	4/657	0	3/657
6	AGG	1	0	2/657	2/657
7	ACG	4/657	2/657	4/657	3/657
8	ATG	2/657	2/657	5/657	0
9	ATC	7/657	2/657	5/657	5/657
10	AGC	1/657	1/657	1/657	2/657
11	ATT	4/657	2/657	4/657	1/657
12	AGT	4/657	0	2/657	1/657
13	ACT	7/657	1/657	5/657	3/657
14	ATA	4/657	2/657	3/657	2/657
15	AGA	4/657	1/657	3/657	1/657
16	ACA	4/657	3/657	3/657	6/657
17	GGA	1/657	1/657	2/657	0
18	GGC	1/657	2/657	1/657	1/657
19	GGT	2/657	0	3/657	0
20	GCA	3/657	1/657	2/657	1/657
21	GCG	2/657	1/657	1/657	0
22	GCC	2/657	0	6/657	0
23	GCT	2/657	1/657	1/657	3/657
24	GTA	0	2/657	3/657	3/657
25	GTG	0	1/657	0	0
26	GTC	2/657	0	4/657	4/657
27	GTT	1/657	1/657	0	1/657
28	GAA	3/657	2/657	3/657	1/657
29	GAG	0	1/657	3/657	2/657
30	GAC	1/657	6/657	2/657	3/657
31	GAT	0	1/657	2/657	1/657
32	GGG	0	2/657	1/657	1/657

Table 5.1 Transition Probabilities in the Third-Order Markov Model to Represent the DNA Sequence from *Homo sapiens*

Triad No.	Triad	*P*(A/Triad)	*P*(G/Triad)	*P*(C/Triad)	*P*(T/Triad)
33	CCA	2/657	1/657	4/657	5/657
34	CCG	2/657	1/657	2/657	1/657
35	CCC	5/657	2/657	7/657	8/657
36	CCT	9/657	3/657	4/657	6/657
37	CAA	4/657	3/657	3/657	5/657
38	CAG	2/657	2/657	1/657	1/657
39	CAC	4/657	2/657	4/657	3/657
40	CAT	3/657	4/657	8/657	4/657
41	CGA	0	3/657	4/657	2/657
42	CGG	1/657	1/657	1/657	1/657
43	CGC	2/657	0	2/657	3/657
44	CGT	1/657	0	4/657	0
45	CTA	5/657	4/657	8/657	3/657
46	CTG	3/657	0	3/657	3/657
47	CTC	6/657	0	5/657	0
48	CTT	5/657	1/657	4/657	4/657
49	TTA	1/657	3/657	3/657	3/657
50	TTG	2/657	1/657	1/657	0
51	TTC	4/657	1/657	7/657	0
52	TTT	1/657	0	5/657	2/657
53	TAA	2/657	0	4/657	6/657
54	TAG	4/657	2/657	0	3/657
55	TAC	6/657	4/657	1/657	6/657
56	TAT	4/657	1/657	5/657	1/657
57	TCA	5/657	1/657	4/657	7/657
58	TCG	1/657	0	0	2/657
59	TCC	0	0	10	11/657
60	TCT	2/657	4/657	2/657	3/657
61	TGA	4/657	1/657	1/657	1/657
62	TGG	1/657	0	2/657	1/657
63	TGC	3/657	1/657	3/657	1/657
64	TGT	1/657	1/657	1/657	1/657

TABLE 5.1 (*Continued*)

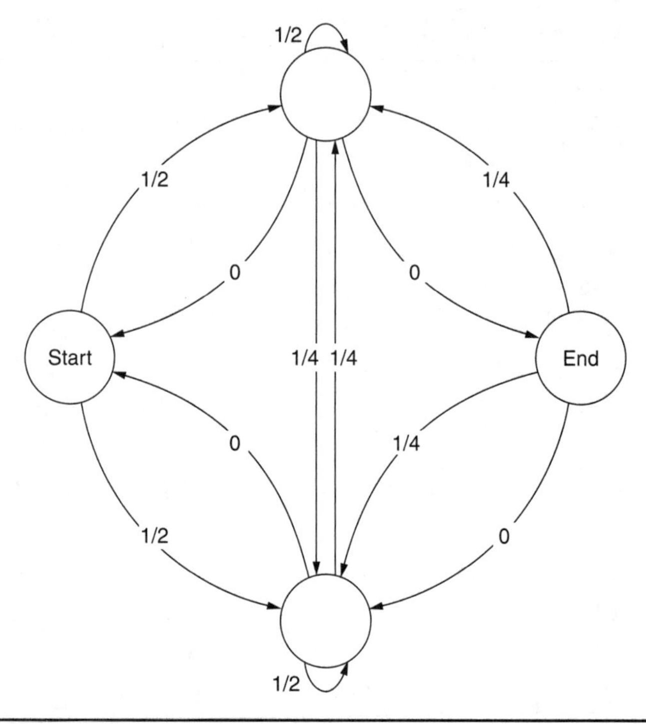

FIGURE 5.3 First-order Markov model with 16 transition probabilities to represent 60 base pairs in *Homo sapiens* [5].

The 16 transition probabilities $P(\mathrm{A}/\mathrm{A})$, $P(\mathrm{G}/\mathrm{A})$, . . . are calculated (Fig. 5.3) from the information provided in Table 5.2. Columns 3–6 are conditional probability values for the base pair shown at the top of the column given the preceding base pair that occurred in the sequence in column 2. A base pair number is also given to the four possible base pairs, adenine, guanine, cytosine, and thymine.

DNA strings can be generated from a four-letter alphabet {A, C, G, T}. A simple sequence model can be developed by assuming that the sequences have been obtained by independent tosses of a four-sided

No.	Given Base Pair	P(A/#)	P(G/#)	P(C/#)	P(T/#)
1	A	3/59	4/59	4/59	6/59
2	G	1/59	2/59	5/59	2/59
3	C	5/59	2/59	3/59	6/59
4	T	8/59	2/59	4/59	2/59

TABLE 5.2 Transition Probabilities in the First-Order Markov Model to Represent the DNA Sequence from *Homo sapiens*

die. Let the data be represented by D and the model by M. The model M has four parameters, namely, P_A, P_C, P_G, and P_T for the probabilities of the bases adenine, cyotosine, guanine, and thymine. Thus

$$P_A + P_B + P_G + P_T = 1 \tag{5.17}$$

Equation (5.17) is written based on the simple surmise that each sequence member has to be among the four nucleotide base pairs. Further,

$$P(D/M) = P_A^{\,na} P_C^{\,nc} P_G^{\,ng} P_T^{\,nt} \tag{5.18}$$

where na, nc, ng, and nt are the number of times the letters A, C, G, or T, respectively, appear in the sequence O.

$$P(D/M) = \Pi P_x^{\,nx} \qquad x \in A \tag{5.19}$$

where N: $D = \{0\}$, with $O = x^1 \ldots x^N$, where $x_i \in A$.

The negative logarithm of Eq. (5.19) yields

$$-\log[P(M/D)] = \Sigma n_x \log P_x \qquad x \in A \tag{5.20}$$

Functional regions can be identified from biologic sequence data. This includes the problem of how to identify relatively long functional regions such as genes. A site is a short sequence that contains some signal that is often recognized by some enzyme. Examples of nucleotide sequence sites include the origins of replication, the sites where DNA polymerase binds, transcription start and stop sites, ribosome binding in prokaryotes, promoters or transcription factor binding sites, and intron splicing sites.

Consider a large sample A of length n sites and a large sample B of length n nonsites. Given a sequence $S = S_1, S_2, \ldots, S_n$ of length n, is S more likely to be a site or a nonsite? Once this can be determined, then the entire genome can be screened, testing every length n sequence and thereby generate a complete list of candidate sites. For example, the cyclic AMP receptor CRP is a transcription factor in *Escherichia coli.* Its binding sites are DNA sequences of length approximately 22". Stormo and Hertzel [6] identified 23 bonafide CRP binding sites from unaligned DNA fragments. Positions 3–9 of 22 sequence positions are shown in Table 5.3.

The most relevant information from these 23 sites needs to be identified. To do this, a profile is constructed. A probability profile shows the distribution of residues in each of the n positions. For instance, the profile for the information provided in Table 5.3 is a 4 × 7 matrix (Table 5.4). The elements of the matrix comprise of A_{rj}, the fraction of sequences in A_{rj} that have a residue r in position j.

T	T	G	T	G	G	C
T	T	T	T	G	A	T
A	T	T	T	G	C	A
C	T	G	T	G	A	G
A	T	G	C	A	A	A
G	T	G	T	T	A	A
A	T	T	T	G	A	A
T	T	G	T	G	A	T
A	T	T	T	A	T	T
A	C	G	T	G	A	T
A	T	G	T	G	A	G
C	T	G	T	A	A	C
C	T	G	T	G	A	A
G	C	C	T	G	A	C
T	T	G	T	G	A	T
T	T	G	T	G	A	T
G	T	G	T	G	A	A
C	T	G	T	G	A	C
A	T	G	A	G	A	C
T	T	G	T	G	A	G

TABLE 5.3 Positions 3–9 from 23 CRP Binding Sites [6]

A	0.35	0.04	0	0.043	0.13	0.83	0.26
C	0.17	0.087	0.043	0.043	0	0.043	0.3
G	0.13	0	0.78	0	0.83	0.043	0.17
T	0.35	0.87	0.17	0.91	0.043	0.087	0.26

TABLE 5.4 Transition Probabilities for CRP Binding Sites

A_{rj} can be thought of in terms of probability. Let $t = t_1 t_2 \ldots t_n$ be chosen randomly and uniformly from A. Then

$$A_{rj} = P(t_j = r / t \propto A) \tag{5.21}$$

A_{rj} is the probability that the j^{tj} residue of t is the residue r, given that t is chosen randomly from A. For example, $A_{T2} = 0.87$. It is assumed that these events are independent. Residue that occurs at position *j* is independent of the residues occurring at other positions. Residues of any two different positions are uncorrelated.

The probability of two independent events are multiplied together to calculate the probability that they both occur. The probability that a randomly chosen site has a specified sequence $r_1, r_2, \ldots, r_n$ is determined as follows:

$$P(t = r_1, r_2, \ldots, r_n / t \text{ is a site}) = P(t_1 = r_1 \text{ and } t_2 = r_2 \ldots)$$
$$= \Pi P(t_j = r_j / t \text{ is a site}) = \Pi A_{rj} \quad (5.22)$$

For example, the probability that a randomly chosen CRP binding sites will be CTGTGAC is given by

$$P(t = \text{CTGTGAC}/t = \text{site})$$
$$= 0.17 \times 0.87 \times 0.78 \times 0.91 \times 0.83 \times 0.83 \times 0.3$$
$$= 0.0447 \quad (5.23)$$

The sequences corresponding to this value are

AT G T GA C
T T G T GA C

Using the profiles *A* and *B*, the question of whether a given sequence *S* is more likely to be a site or a nonsite remains. A likelihood ratio is defined as follows: Given the sequence $S = S_1, S_2, \ldots, S_n$, the likelihood ratio denoted by $LR(A, B, S)$ is defined to be

$$\frac{P(t = S / t \rightarrow \text{site})}{P(t = S / t \rightarrow \text{nonsite})} = \frac{\Pi A_{sij}}{\Pi B_{sij}} = \Pi \frac{A_{sij}}{B_{sij}} \quad (5.24)$$

5.4 Three Questions in the HMM

The hidden Markov model (HMM) is a finite set of states, each of which is associated with a probability distribution. Transition probabilities are used to govern the transitions among the states. Some states are hidden from the external observer. These are the hidden states. They are used to generate the desired output from the given input. For instance, an outcome can be generated given the associated probability distribution. In Sec. 5.2, the geometric distribution was identified as one that can describe DNA polynucleotide sequence distribution. Complete description of the HMM requires the following [7]:

Number of states N

Number of observation symbols σ in the alphabet Σ

A set of transition probabilities, namely,

$$A_{ij} = P(q_{t+1} = j / q_t = i) \qquad 1 < i, j \leq \underline{N} \tag{5.25}$$

where q_t denotes the current state. The normal stochastic constraints are met by the transition probabilities such that

$$A_{ij} \geq 0, \quad 1 < i, j \leq N, \quad \text{and} \quad \Sigma^N A_{ij} = 1 \tag{5.26}$$

A probability distribution in each of the states B

Initial state distribution $\pi = (\pi_i)$

Given a sequence $S =$ ATCCTTTTTTTCA, three questions arise in an HMM. These questions are as follows:

1. *Evaluation question:* How likely is this sequence for a particular HMM?

$$P(O/\lambda) \tag{5.27}$$

2. *Decoding question:* What is the most probable sequence of transitions and emissions through the HMM underlying the production of this particular HMM?
3. *Learning question:* How should the transition and emission parameters be revised in light of the observed sequence?

$$\text{Maximize}[P(O/\lambda)] \tag{5.28}$$

For sequences that appear frequently in bioinformatics, the main alphabets of the HMM are the 20 different amino acids for proteins and the 4-letter nucleotide base pairs set for DNA/RNA. Selection of the architecture of the HMM depends on the problem at hand. The directed graph associated with nonzero A_{ij} connections is called the *architecture of the HMM.* The hidden states and interconnections are examples of the structural parameters of the architecture. Depending on the task at hand, a 64-letter alphabet of triplets of codons can be used or a three-symbol set (α, β, and γ) for the secondary structure of proteins and other alphabets such as the hydrophobic alphabet, the charge alphabet, the functional alphabet, the chemical alphabet, the structural alphabet, and the hydrogen-bonding alphabet can be used.

More complex HMM architectures than the one with two hidden states may be considered. The linear aspects of sequences can be captured by left-right architectures. An architecture is left-right if it prevents returning to any state once a transition from that state to any other state has occurred. In the standard HMM architecture, in addition to the start and end, there are other classes of states—main states, delete states, and insert states. Assuming that the emissions

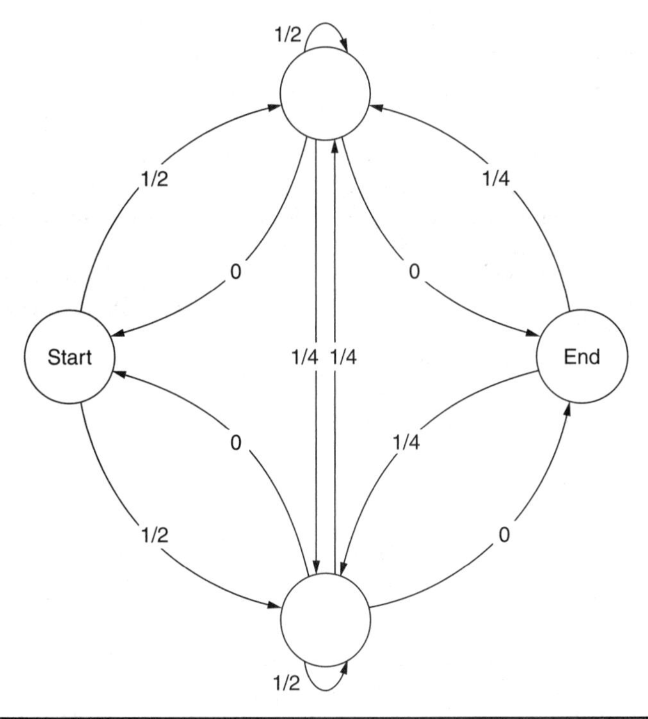

Figure 5.4 HMM with four states for sequence S = ATCCTTTTTTTCA.

and transitions depend on the current state only and not on the past, two special states, i.e., the start state and the end state, are chosen. The transition and emission probabilities are the parameters of the model. The sequence S = ATCCTTTTTTTCA is represented by an HMM with two states in addition to the start and end states (Fig. 5.4).

For the sake of mathematical and computational tractability, the following assumptions are made in the theory of HMMs:

1. *Markov assumption:* Transition probabilities are defined in Eq. (5.28). The next state depends only on the current state. The resulting model is a first-order HMM. Higher-order HMMs with greater complexity can be used.
2. *Stationarity assumption:* State transition probabilities are independent of the real time at which the transitions take place. A_{ij} remains the same in Eq. (5.28) regardless of the t_1 or t_2 considered for q.
3. *Output independence assumption:* The current observation is independent of previous observations. For an HMM, λ that describes sequence $O = o_1, o_2, \ldots, o_T$.

$$P(O/q_1, q_2, \ldots, q_T, \lambda) = \prod P(O_t/q_t, \lambda) \tag{5.29}$$

5.5 Evaluation Problem and Forward Algorithm

Given a sequence O and the HMM, for λ that is used to represent the sequence O, find $P(O/\lambda)$. When calculated from simple probabilistic arguments, the number of operations needed is on the order of N^T, where N is the number of states of the HMM and T is the length of the sequence. This is large even for a moderate length of sequence. To save time, an auxiliary variable called the *forward variable* $\alpha_t(i)$ is defined. The forward variable is defined as the probability of the partial observation sequence O when it terminates at state i. That is,

$$\alpha_t(i) = P(o_1, o_2, \ldots, o_t, q_t = i/\lambda) \tag{5.30}$$

It can be seen that the following recursive relationship is valid:

$$\alpha_{t+}(j) = b_j[o_{t+1}\Sigma^N\alpha_t(i)A_{ij}] \qquad 1 \le j \le N, 1 \le t \le T-1 \qquad i = 1 \tag{5.31}$$

where $$\alpha_1(j) = \pi_j b_j(o_1) \qquad 1 \le j \le N \tag{5.32}$$

and $$\mathrm{P(O}/\lambda) = \Sigma^N\alpha_T(i) \qquad i = 1 \tag{5.33}$$

The time taken to complete the task is $O(N^2T)$. This is less than $O(N^T)$, especially for long sequences [8]. It is linear with respect to the length of the sequence. The backward variable can be defined in a similar fashion.

5.6 Decoding Problem and Viterbi Algorithm

The problem is to find the most likely state sequence for a given sequence of observations O and an HMM λ. The solution depends on the definition of "most likely state sequence." One approach is to find the most likely state q_t at $t = t$ and to concatenate all such q_t's. Some of the time, the solution from this method is not physically meaningful. Another method has been developed called the *Viterbi algorithm* [9]. Here, the whole state sequence with the maximum likelihood is found. An auxiliary variable is defined as

$$\delta_t(i) = \max[P(q_1, q_2, \ldots, q_{t-1}), q_t = i, o_1, o_2, \ldots, o_{t-1}/\lambda] \tag{5.34}$$

This auxiliary variable denotes the highest probability that the partial observation sequence and state sequence up to t can have when the current state is i. It can be seen that the following recursive relationship will hold:

$$\delta_{t+1}(i) = b_j(o_{t+1})\{\max[\delta_t(i)A_{ij}]\} \qquad 1 \le i \le N, 1 \le t \le T-1 \tag{5.35}$$

where $$\delta_1(i) = \pi_j b_j(o_1) \qquad 1 \le j \le N \tag{5.36}$$

A pointer to the winning state is kept throughout the recursion process. The state j^* is finally found where it is the arg max$[\delta_T(j)]$. Starting from this state, the sequence of states is backtracked as the pointer in each state indicates. This gives the required set of states. This algorithm is like a search graph whose nodes are formed in the states of the HMM in each of the time instants t in the closed interval of $(1, T)$.

The learning problem generally is how to adjust the HMM parameters so that the given set of observations called the *training set* is represented by the model in the best way for the intended application. Quantity for optimization changes with the application. Some examples of optimization criteria are maximun likelihood (ML) and maximum mutual information (MMI).

5.7 Relative Entropy

Given the sequence $S = S_1, S_2, \ldots, S_n$, the log-likelihood (LLR) (A, B, S) is defined by

$$\log_2 \text{LR}(A, B, S) = \log_2 \prod A_{sj,j} / B_{sj,j} \tag{5.37}$$

S is more likely to be a site if LLR $(A, B, S) \geq \log_2 L$. To test for sites, a scoring matrix W is defined whose entries are the log-likelihood ratios:

$$W_{r,j} = \log_2(A_{rj}/B_{rj}) \tag{5.38}$$

The weight matrix for the example of CRP samples A and B is shown in Table 5.5.

To compute LLR (A, B, S) from the preceding definition, the corresponding scores from W: LLR $(A, B, S) = W_{sj,j}$ need to be added. When the entry $A_{rj} = 0$, a problem arises because the entry becomes $-\infty$ if the residue r cannot occur in position j of any site for biologic reasons. Often this is a result of having too small a sample A of sites. In this case, there are various "small sample correction" formulas that replace A_{rj} with a small positive number.

The log-likelihood matrix shown in Table 5.5 is an example of a weight matrix. A score is assigned to each sequence $S = S_1, S_2, \ldots, S_n$ according to the formula $\sum W_{sj,j}$ in weight matrix A that is $C \times n$.

A	0.48	−2.5	−∞	−2.5	−0.94	1.7	0.061
C	−0.52	−1.5	−2.5	−2.5	−∞	−2.5	0.28
G	−0.94	−∞	1.6	−∞	1.7	−2.5	−0.52
T	0.48	1.8	−0.52	1.9	−2.5	−1.5	0.061

Table 5.5 Log-Likelihood Weight Matrix for CRP Binding Sites

A large portion of the genome is taken to be the background distribution when computing log-likelihood ratios. $B_{rj'}$, the background distribution of residue r in the entire genome, is the frequency with which residue r appears in the genome as a whole:

$$B_{rj} = B_{rj}^{1} \qquad \text{for all } j \text{ and } j^{1} \tag{5.39}$$

A uniform distribution is a fair estimate for the nucleotide composition of *E. coli,* and $B_{rj} = 0.25$. This is not a fair estimate for other organisms. For instance, the nucleotide composition for the archaeon *Methanococcus Jannaschii* is approximately

$$\begin{aligned} B_{A1,j} &= BT_{1j} = 0.34 \\ B_{C1,j} &= BG_{1j} = 0.16 \end{aligned} \tag{5.40}$$

A sample space is the set of all possible values of some random variable S. A probability distribution P for a sample space S assigns a probability $P(S)$ to every $s \in S$ satisfying

$$P \leq P(S) \leq 1 \tag{5.41}$$

$$\Sigma P(s) = 1 \tag{5.42}$$

The sample space is a set of length n sequences. The site profile A induces a probability distribution on this sample space according to the definition, as does the nonsite profile B.

Definition Let P and Q be probability distributions on the sample space S. The relative entropy, information content, or Kullback-Leibler measure of P [10] with respect to Q is denoted $D_b(P[]Q)$ and is defined as follows:

$$D_b(P[]Q) = \Sigma P(S) \log_b[P(S)/Q(S)] \tag{5.43}$$

By convention, define $P(s) \log_b[P(S)/Q(S)]$ to be O whenever $P(S) = 0$, in agreement with the fact from calculus that limit as $X \rightarrow 0$,

$$x \log x = 0 \tag{5.44}$$

Since $\log[P(S)/Q(S)]$ is the log-likelihood ratio, $D_b(P[]Q)$ is a weighted average of the log-likelihood ratio with the weights $P(S)$.

Definition The expected value of a function $f(S)$ with respect to probability distribution P on sample space S is

$$E[f(s)] = \Sigma p(s) f(s) \tag{5.45}$$

In these terms, the relative entropy is the expected value of LLR (P, Q, S) when S is picked randomly according to $P(S)$. That is, it is the expected log-likelihood score of a randomly chosen site.

Now, when P and Q are the sample distribution, the relative entropy will be zero. The relative entropy measures how different the distributions P and Q are. The relative entropy needs to be large to be able to distinguish between sites and nonsites. The relative entropy is the measure of how informative the log-likelihood ratio test is. When the sample space is all length n sequences and independence of the n positions is assumed, it can be proved that the relative entropy satisfies

$$D_b(P[]Q) = \Sigma D_b(P_j[]Q_j) \tag{5.46}$$

where P_j is the distribution P imposed on the jth position. When $b = 2$, the relative entropy is measured in bits. Unless specified otherwise, this will be the usual case.

Theorem For any probability distribution P and Q over a sample space D_b, $(P[]Q) \geq 0$, with equality if and only if P and Q are identical.

$\ln(x) \leq x - 1$ for all real numbers x, with equality if and only if $x = 1$

The reason is that the curve $y = \ln(x)$ is concave downward, and its tangent $x = 1$ is the straight line $y = x$. Thus

$$\ln(1/x) = -\ln(x) \geq 1 - x$$

The inequality with $x = Q(s)/P(S)$ is used up below:

$$D_b(P[]Q) = \Sigma P(s)\lg_b \frac{P(S)}{Q(S)} \tag{5.47}$$

$$\geq \frac{1}{\ln(b)} \Sigma[P(s) - Q(s)] \tag{5.48}$$

$$= 1/\ln(b)\Sigma[P(s) - \Sigma Q(s)] \tag{5.49}$$

because $\Sigma P(S) = \Sigma Q(S) = 1$.

Note that the relative entropy is equal to 0 if and only if $x = Q(s)/P(s) = 1$ for all s. P and Q are identical probability distributions.

5.8 Probabilistic Approach to Phylogeny

A phylogenetic tree is one in which the evolutionary relationships among various species that are believed to have a common ancestor are shown. The conditional probability P(Data/tree) is the likelihood of the sequence occurring given the tree and the posterior probability. P(Tree/data) is a way of constructing the tree given the data. Each node with descendants represents the recent common ancestor, with edge lengths corresponding to time estimates. $P(x^*/T, t_0)$ is the probability of a set of data that can be defined and calculated given a tree. A model of evolution is needed and selection of events that change sequences along the edges of a tree.

During the course of evolution, residues are substituted by others, deletions and insertions occur among groups of residues, and more complex constraints are imposed by the structures of nucleic acids and proteins. Models for deletions and insertions can be sought. Let $P(b/a, t)$ denote the probability of a residue a having been substituted by a residue b over an edge length t. For two aligned, gapless sequences x and y, $P(x/y, t) = \prod P(x_u/y_u, t)$, where u represents sites in the alignment. All possible forms for the substitution probabilities $P(b/a, t)$ for residues a and b are examined. Given a residue alphabet of size K, these can be written as a $K \times K$ matrix that depends on t, which is denoted by $S(t)$:

$$S(t) = \begin{matrix} P(A_1/A_1, t) & P(A_2/A_1, t) & \cdots & P(A_k/A_1, t) \\ P(A_1/A_2, t) & P(A_2/A_2, t) & \cdots & P(A_k/A_2, t) \\ \cdots & \cdots & \cdots & \cdots \\ P(A_1/A_{K,t}) & P(A_2/A_{K,t}) & \cdots & P(A_K/A_{K,t}) \end{matrix} \tag{5.50}$$

For several important families of substitution matrices, the family is multiplicative, that is,

$$S(t)S(s) = S(t + s) \tag{5.51}$$

for all values of lengths s and t. The probabilities should satisfy

$$\Sigma P(a/b, t)P(b/c, s) = P(a/c, s + t) \tag{5.52}$$

for all a, c, s, and t. The substitution process is Markovian and stationary, and the probabilities are multiplicative. Jukes and Cantor [11, 12] proposed a model for DNA sequences. This assumes that a matrix R of rates of substitution takes the form shown in Fig. 5.5.

The nucleotides undergo transitions at the same rate α. The substitution matrix for a short time $S(\epsilon)$ is approximately given by $S(\epsilon) \approx (I + R\epsilon)$, where I is the identity matrix with ones down the diagonal and zeros elsewhere.

$(I + R\epsilon)$ becomes by multiplicativity

$$S(t + \epsilon) = S(t)S(\epsilon) \approx S(t)(I + R\epsilon)$$

Figure 5.5 Substitution matrix R in Jukes and Cantor model [12].

	A	C	G	T
A	-3α	α	α	α
C	α	-3α	α	α
G	α	α	-3α	α
T	α	α	α	-3α

In the limit of small $\in$,

$$[S(t+\in) - S(t)]/\in \approx S(t)R \tag{5.53}$$

$$dS(t)/dt = S(t)^*R \tag{5.54}$$

Substituting for $S(t)$ in Eq. (5-60) gives

$$dr/dt = -3\alpha r + 3\alpha s \tag{5.55}$$

$$ds/dt = -\alpha s + \alpha r \tag{5.56}$$

These equations can be solved and the solutions written as

$$r_t = ¼[1 + \exp(-4\alpha t)] \tag{5.57}$$

$$s_t = ¼[1 - \exp(-4\alpha t)] \tag{5.58}$$

The matrix given by Eq. (5.58) constitutes the Jukes-Cantor model. At infinite time, the nucleotide equilibrium frequencies can be seen to be ¼. The Jukes and Cantor model does not capture some important features of nucleotide substitution. For instance, transitions, namely, purine to purine or pyrimidine to pyrimidine substitutions, are common. Transversions, where the nucleotide type is changed, is less common. Kimura [13] proposed a model with the rate matrix shown in Fig. 5.6.

The matrix can be solved to give

$$s_t = ¼[1 - \exp(-4\beta t)] \tag{5.59}$$

$$y_t = ¼\{1 + \exp(-4\beta t) - 2\exp[-2(\alpha + \beta)t]\} \tag{5.60}$$

$$r_t = 1 - 2s_t - u_t \tag{5.61}$$

FIGURE 5.6 Rate matrix in Kimura's model [13].

$$\begin{pmatrix} -2\beta-\alpha & \beta & \alpha & \beta \\ \beta & -2\beta-\alpha & \beta & \alpha \\ \alpha & \beta & -2\beta-\alpha & \beta \\ \beta & \alpha & \beta & -2\beta-\alpha \end{pmatrix}$$

5.9 Sequence Alignment Using HMMs

The similarity between two sequences can be scored using probabilistic models. The gapped alignment process can be converted into HMMs. The reliability of the alignment can be explored. A finite-state automaton with three states can be used to represent pairwise alignment with affine gap penalties. The match is given by state *M*, and the insert is given by the *X* and *Y* states (Fig. 5.7). The recurrence relation for updating the values in the dynamic programming matrix was given in Chap. 2. This is used for global alignment of sequences. Suitable changes can be included for local alignment. The HMM is derived from the machine diagram shown in Fig. 5.8. The symbols that derive from the states are assigned probabilities, and transition values are provided between states. For example, state *M* has probability distribution

FIGURE 5.7 Finite state machine diagram for affine gap alignment.

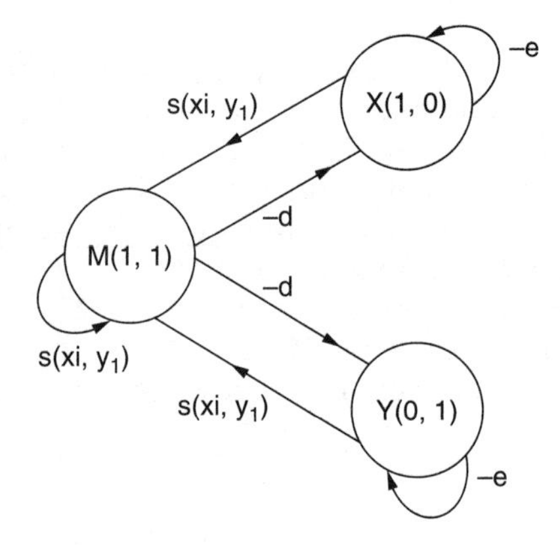

FIGURE 5.8 Probabilistic model for affine gap alignment.

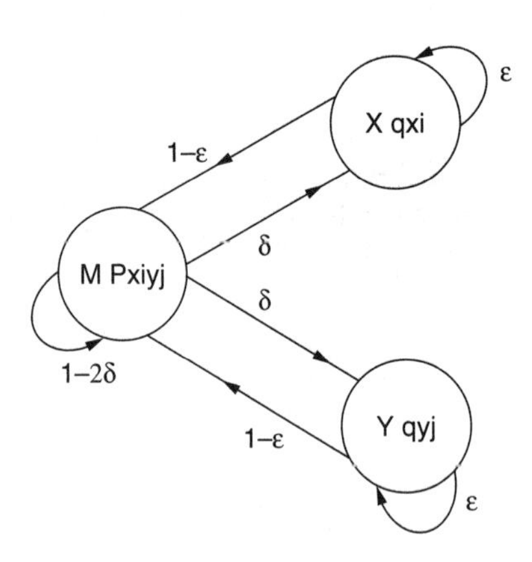

P_{ab} for emitting an aligned pair *a:b,* and states X and Y will have distributions q_a for emitting symbol a against a gap. q_{xi} represents state X, and state X emits symbol x_i from sequence x. Transition probabilities are specified between states. The parameters of the model are indentified and shown in Fig. 5.8 [14]. The transition from M to an insert state is given by δ, and the probability of staying in an insert state is given by ε. A begin and end state may be added to Fig. 5.8. The addition of an end state may introduce another model parameter τ. This is to represent the probability of transition into the end state. Thus the HMM *emits a pairwise alignment.* Most discussions of HMMs can be extended to pair HMMs. They need an extra dimension of search space to store the extra emitted sequence. A pair HMM can be used to generate an aligned pair of sequences.

5.10 Protein Families

HMMs have been applied with success to many protein families [15], such as globins, immunoglobins, kinases, and G protein–coupled receptors (GPCRs). They have been used to model the secondary structures of proteins such as α-helices, β-sheets, and γ-coil structures, as well as the consensus patterns of protein superfamilies. FORESST, the database containing protein family secondary structures, and Pfam, the database containing protein families, were available in 1997. In 1997, Pfam contained 527 manually verified families, consisting of 39,113 sequence alignments and 6.8 million residues in the full alignments, and they are available for browsing and online searching via the World Wide Web. Pfam was developed to use HMM profile analysis to complement BLAST analysis in the *Caenor habditis elegans* genome project. Protein family databases typically are based on multiple sequence alignments of known family members. The main distinction between Pfam and most other protein family databases is that for all of Pfam, both the family definition and the search method span entire domains, including not only conserved motifs but also less conserved regions, insertions, and deletions. HMM profile methods allow variable conservation and insertions/deletions to be dealt with in a fairly robust way. Modeling of complete domains should facilitate more biologically meaningful sequence annotation and in some cases more sensitive detection.

For each protein domain family in Pfam, there are three important files. The seed alignment is a manually verified multiple alignment of representative sets of sequences. An HMM profile is built from the seed alignment for database searching and alignment purposes. A full alignment is generated automatically from the HMM seed profile by searching Swissprot for all detectable members and aligning them with the HMM profile. Most Pfam families are based on and cross-referenced to corresponding PROSITE entries. For comprehensiveness, all Swissprot sequences not in Pfam are clustered automatically by the Program Domainer (Pro Dom), which also constructs multiple alignments automatically and is the basis for the Pro Dom protein family database. The quality of these alignments tends to be low. These are made available as Pfam B. Pfam B contains 13,289 clusters, 62,611 subsequences, and 8.2 million residues. On average, alignments are 146 residues wide and contain 5 members. Fifty-eight percent of the sequences and 32 percent of the residues in Swissprot 34 are included in annotated Pfam alignments.

GPCRs are a family of transmembrane proteins capable of transducing a variety of extracellular signals carried out by hormones, neurotransmitters, odorants, and light. A total of 142 GPCR sequences extracted form PROSITE database were used to train an HMM architecture of length $N = 430$, the average length of the training sequences, using online Viterbi learning during 12 iterations through the entire training set. The entropy of the emission distribution of the main states of the model is derived. The amplitude profile of the entropy contains seven major oscillations directly related to the seven transmembrane domains. The structural feature was discovered by the HMM without any prior knowledge of α-helices or hydrophobicity. To test the discriminative abilities of the model, 1600 random sequences were generated with the same average composition as GPCRs in the training set with lengths 300, 350, 400, 450, 500, 550, 600, 650, 700, 750, 800, 1000, 1500, and 2000. For any sequence, random or otherwise, its raw score according to the model is calculated. The raw sequence is the negative log likelihood of the corresponding Viterbi path.

Random sequences with similar average composition are discriminated using the model from that of GPCRs. The scores of random sequences and the Swissprot sequences cluster among two similar lines. On average, the clustering along a line indicates that the cost of adding one amino acid is roughly constant. For very short sequences, the linearity is not preserved. These can have irregular Viterbi paths. The linearity becomes increasingly precise for very long sequences. Viterbi paths of very long sequences with a fixed average composition must rely on insert states and, in fact, are forced to loop many times in a particular insert state that becomes predominant as the length goes to infinity. The cost-effectiveness of an insert state k depends equally on two factors—its self-transition probability t_{kk} and the cross-entropy between its emission probability vector e_{kk} and the fixed probability distribution associated with the sequences under construction. Long random sequences generated using a fixed some $p - p_x$ as a function of sequence length on examination scores cluster along a regression line with slope

$$\min(-\log t_{kk} - \Sigma p_x \log e_{kx}) \qquad (5.62)$$

Furthermore, for a large fixed length l, the scores are approximately normally distributed according to the central limit theorem with variance

$$l[E_p \log^2 e_{hk} - E_p \log(e_{hk})] = \text{var}\, p(\log e_{hx}) \tag{5.63}$$

In particular, the standard deviation of the scores increases as the square root of length l.

Discrimination tests can be developed to decide whether a sequence belongs to the GPCR family or not. The scores produced by the model need to discriminate between GPCR and non-GPCR sequences. In the case of the HMM library, a fixed set of randomly generated sequences with the same average composition as Swissprot can be used across different models. In the GPCR example, for any sequence O of length l, the normalized score $Es(0)$ was used based on the residual with respect to the empirical regression line of the random sequences of similar average composition divided by the approximate standard deviation:

$$Es(0) = \frac{3.0381 + 122.11 - E(0)}{0.66(l)^{1/2}} \tag{5.64}$$

where $E(0)$ is the negative log likelihood of the Viterbi path. Setting of the detection threshold is an issue. The smallest score here on the training set is 16.03 for the sequence labeled UK33-HCMVA. This low score is isolated because there are no other scores smaller than 18. The threshold can be 16 or higher. The search algorithm presents no false negatives and two false positives. This is accomplished by removing very long sequences exceeding the maximal GPCR length, as well as sequences containing ambiguous amino acids. At short lengths, below the length of the model, Eq. (5.64) is not a reasonable approximation. It may be wise to try a mixed scheme where a normalization factor is calculated empirically at short lengths, $l < N$, and Eq. (5.64) is used for larger lengths, $l > N$. Thresholds may be set from the fact that the extreme score of a set of random sequences of fixed length follows an extreme value distribution.

By construction of a *hydropathy plot*, it should be possible to detect easily whether a given sequence belongs to the class of GPCRs. The hydropathy scales are used. Hydropathy plots (Fig. 5.9) of a number of sequences were constructed using a 20-amino-acid window. Examples of plots obtained for these sequences are shown in Fig 5.10. As can be seen, the data can be noisy and ambiguous. The vertical axis represents free energy for transferring a hypothetical α-helix of length 20 at the corresponding location from the membrane interior to water. A peak of 20 kcal/mol or more usually signals the possible presence of a transmembrane α-helix.

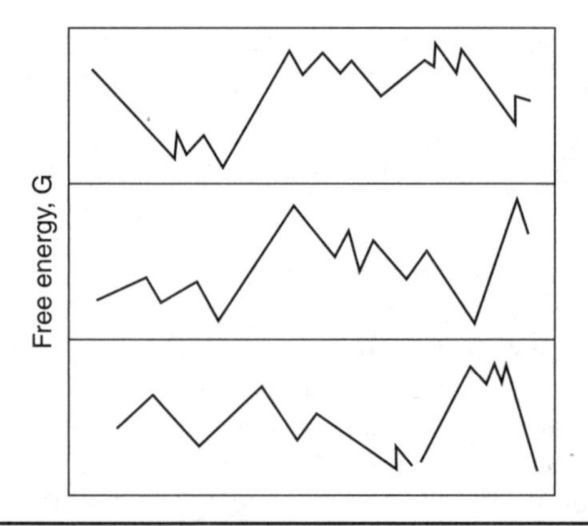

FIGURE 5.9 Hydropathy plots for three GPCRS of length 1000.

Detection from hydropathy plots alone cannot be relied on completely. Consensus-pattern hydropathy plots and HMMs should be considered complementary techniques. A hydropathy plot can be constructed from the HMM probabilities. This would display the expected hydropathy at each position rather than the hydropathy observed in any individual sequences. Signal amplification is effected, and the seven transmembrane regions are clearly identifiable.

5.11 Wheel HMMs to Model Periodicity in DNA

Periodic patterns in exons and introns can be indentified by using novel HMM architectures such as loop HMM and wheel HMM.Wheel HMMs are designed with a better ability to reveal periodic patterns in the presence of noise. The conventional left-right architecture is not ideal to represent exons owing to the large length variation. A different sort of loop model was trained on both exon and intron sequences. The HMM architecture was in the form of a wheel with the given number of main states, without flanking states, arranged linearly or any distinction between main states and insert states. Sequences can enter the wheel at any point. The point of entry can be determined using dynamic programming. The most likely periodicity can be revealed by using wheels with different numbers of states and comparing the negative log likelihood of the training set. Should the wheels of 9 states perform better than wheels of 10 states, the periodicity can be assumed to be related to the triplet reading frame rather than to structural aspects of the DNA. The wheel model architecture is displayed in Fig. 5.10.

Lengths of 10 nucleotides with sequences can enter the wheel at any point. The thickness of the arrows from outside represents the probability of starting from the corresponding state. A periodic pattern was inferred after training the emission parameters in the

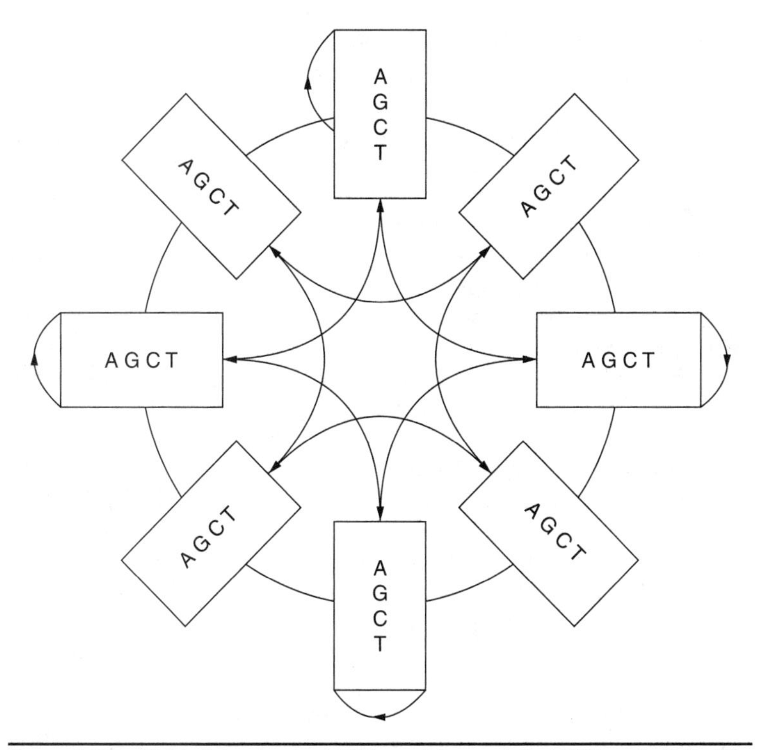

FIGURE 5.10 Eight-state circular HMM used for modeling DNA periodicity [8].

wheel model. By training wheels of many different lengths, it was found that models of length 10 yielded the best fit. This is confirmed by recognizing that the skip probabilities are not strong in these models. If the data were nine-periodic, a wheel model with a loop of length 10 should be able to fit the data by heavy use of the possibility of skipping a state in the wheel. State repeating in a 9-state wheel is nonequivalent to state skipping in a 10-state wheel. These wheel models do not contain independent insert states (as the left-right HMM architectures). A repeat of the same state does not give the same freedom in terms of likelihood as if independent inserts were allowed. HMM training procedure uses a regularization term favoring main states over skip states. All the experiments were repeated using several subsets of exons starting in one of the three codon positions in the reading frame without any significant change in the observed patterns of emission probabilities.

5.12 Generalized HMM (GHMM)

Genie is based on a generalized hidden Markov model (GHMM) that describes the grammar of a legal parse of a multiexon gene in a DNA sequence. Reese and colleagues [16] proposed an improved splice-site predictor for the gene-finding progam Genie. In Genie, probabilities are estimated for gene features by using dynamic programming to combine information from different sources. One of the toughest problems in gene finding is to determine the complete gene structure correctly. Two novel neural networks based on dinucleotide frequencies are used to overcome this. Significant improvements in the sensitivity and specificity of gene structure identification are achieved. Experimental results using a standard set of annotated genes show that Genie identified 82 percent of coding nucleotides correctly with a specificity of 81 percent versus 74 and 81 percent in the older system.

Gene-finding systems such as FGENEH, GenLang, and GenMark use known, recognized techniques in concert. The GRAIL Gene Parser combines mutiple statistical measures with database homology searching to identify gene features. The design in Genie is similar to that in the Gene Parser. Genie is a implementation of the GHMM whose states are arbitrary submodels emitting variable-length sequences rather than signal letters (as in HMM). A GHMM is defined in Fig. 5.11 with a simple gene structure syntax as an example.

A GHMM is an enhancement of the standard HMM often used for pattern recognition and time series in computational biology. A GHMM describes a more general model in which each state can emit one or more symbols according to an arbitrary distribution. Each state represents an independent submodel that may itself be an HMM or any statistical model. A simple GHMM that models eukaryotic gene structure is shown in Fig. 5.12. The arcs represent states that emit strings of bases, and the nodes represent transitions between states.

The GHMM is represented as a graph. Nodes in the graph represent transitions between states. This is different from typical graphic representations of regular HMMs. Each state corresponds to a submodel of an abstract gene feature such as an internal exon (E) or intron (I). For any sequence of bases x and state q, the submodel associated with the state q defines a likelihood for the sequence x. This likelihood is denoted by $P(x/q)$. When the GHMM is viewed as a generative statistical model, this is the probability of the sequence emitted when the Markov process is in state q. These likelihood

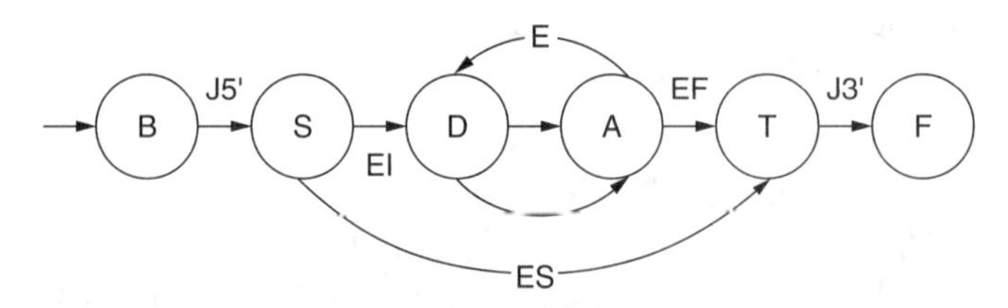

FIGURE 5.11 Simple GHMM for a sequence with multiple-exon genes.

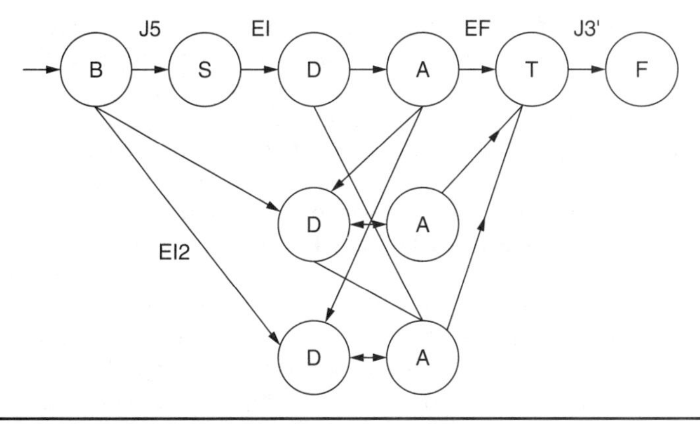

FIGURE 5.12 A GHMM including frame constraints.

functions, one for each state, are part of the definition of the GHMM. The graph of the GHMM has a unique source model B for begin and a unique single node f for final. The process of generating a string from a GHMM can be q, and the mode that the arc for state q leads to is denoted node(q). Once in this node, a next state is chosen at random from among the outgoing arcs from this node, independent of any previous choices made. The probability of choosing the next state r is denoted $P[r/\text{node}(q)]$.

Define a parse ϕ of the sequence X to be a pair consisting of a sequence of state $q_{1,\ldots,}\ q_k$ and a corresponding sequence of substrings $x_{1,\ldots,}\ x_{k'}$ where $X = x_{1,\ldots,}\ x_{k'}\ q_1$ is a state coming out of a unique source mode (B), and q_k is a state leading to the unique sink node (f). The GHMM defines a joint likelihood of the sequence $X = x_{1,\ldots,}\ x_k$ and the parse $\phi = q_{1,\ldots,}\ q_{k'};\ x_{1,\ldots,}\ x_{k'}$ according to the generative model described earlier. It is the joint independent probability of the subsequences given the corresponding states and the probability of the transitions between states, that is,

$$P(x, \phi) = P(q_1/B)\Pi P(x_i/q_i)\Pi P[q_{i+1}/\text{node}(q_i)] \tag{5.65}$$

Given only the observed sequence X, using a variant of the Viterbi algorithm, the parse ϕ can be calculated that maximizes Eq. (5.65). In a GHMM that represents gene structure, such as the one in Fig. 5.12, this parse probably represents the model prediction of the most likely gene structure within the sequence x. This variant of the Viterbi algorithm is used to find the most likely parse in a dynamic programming algorithm. The GHMM in Fig. 5.12 represents only the basic ordering of gene feature and does not fully capture the syntactic restrictions of a "legal gene parse." In an ideal DNA sequence, the parse is *frame consistent;* i.e., the total number of coding nucleotides is a multiple of three, and the reading frame is consistent from exon to exon. Additional states can be added to the model graph that only allow consistent parses. The model graph representing the resulting frame-consistent GHMM is shown in Fig. 5.12. The additional acceptor and donor transition nodes ensure that only syntactically correct parses are considered. The three levels represent the three frames. Exon lengths can be restricted in the likelihood functions $P(X/Q)$ to equal 0, 1, or 2 for the various exon states in this GHMM in such a way as to enforce frame consistency. This more complex state structure is used by Genie. Further extensions to the GHMM graph also can be added to make the model more realistic. For example, an arc leading back from node T to node S labeled with a state that generates noncoding bases between genes would allow the GHMM to model sequences that have multiple genes within them.

5.13 Database Mining

Given a trained model, the likelihood of any given sequence can be computed. These scores can be used in discrimination tests and in database searches to separate sequences associated with the training family from the rest. This is applicable to both complete sequences and fragments. Such scores can be calibrated as a function of sequence length. HMMs also can be used in classification problems, e.g., across protein families or across subfamilies of a single protein family. This can be done by training a model for each class, if class-specific training sets are available. This approach was used to build two HMMs that can reliably discriminate between tyrosine and serine–threonine kinase subfamilies. Otherwise unsupervised algorithms related to clustering can be used in combination with HMMs to represent the total number of protein superfamilies. The number of protein superfamilies is relatively small on the order of 1000. A global protein classification system with roughly 1 HMM per family is becoming a feasible goal from both an algorithmic and a computational standpoint. Global classification projects of this sort are currently under way and should become auxiliary tools in a number of tasks, such as gene finding, protein classification, and structure/function prediction.

5.14 Multiple Alignments

Multiple alignments can be derived by aligning the Viterbi paths with each other. Training a model can be done offline. The multiple alignment of K sequences after the training phase is completed requires the computation of K Viterbi paths and scales of $O(KN^2)$. All Viterbi paths consists only of main-state emissions or gaps with respect to main states. Multiple alignments derived by an HMM with both insert and delete states are potentially richer and, in fact, should be plotted in three dimensions rather than the two used by conventional multiple alignments. The insert and delete states of an

HMM represent formal operations on sequences. These need to be related to evolutionary events. When a single HMM is used as a basis, they correspond only to the first step of a full Bayesean treatment. HMMs also can be used in conjunction with substitution matrices. HMM emission distributions can be used to calculate substitution matrices, and substitution matrices can be used to influence HMMs during or after training. In the case of large training sets, most substitution information is already present in the data itself, and no major gains would be derived from an external infusion of such knowledge.

5.15 Classification Using HMMs

The organization of families of sequences into subclasses is called *classification.* It is used, for example, in phylogenetic reconstruction. There are two different ways of classification using HMMs. These are

1. Training several models in parallel and using some form of competitive learning (Fig. 5.13).
2. Looking at how likelihoods and paths cluster within a single model.

The number of sequences for some receptor classes is too small to train using the parallel approach. In the second approach, it is clear from visual inspection of the multiple alignment that there are clusterings and interesting relationships among the Viterbi paths corresponding to different receptor subgroups. The clustering of all the sequences in a given receptor subclass around a particular distance is striking. Olfactory receptors are the closest to being

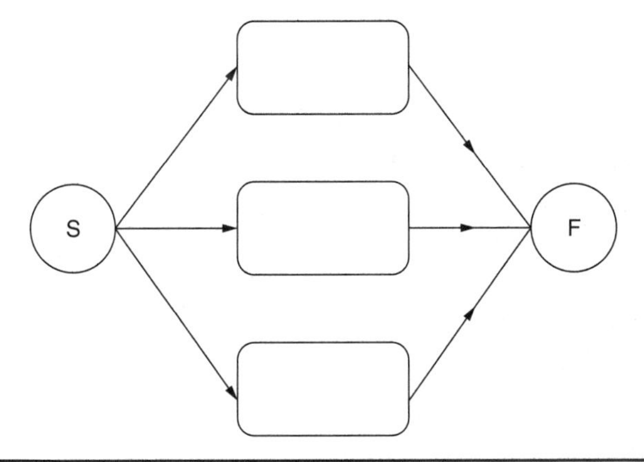

FIGURE 5.13 Representation of multiple HMM architecture for detecting subfamilies within a protein family [15].

random. Adrenergic receptors are the most distant from the random regression line and hence appear to be the most constrained.

There are also apparent differences in the standard deviation of each class. For instance, the angiotensin receptors occupy a narrow band, and only one angiotensin receptor type is known, whereas the opsin receptors are more spread out. Most classes have a bell-shaped distribution. There are exceptions. The opsins appear to have a bimodal distribution. This can be the result of the existence of subclasses within the opsins. The second peak corresponds mostly to rhodopsin (OPSD) sequences and a few red-sensitive opsins (OPSR). The presence of two peaks does not seem to result from differences between vertebrate and invertebrate opsins. With future database releases, it is possible to improve the resolution and reduce sampling effects. These results suggest a string relationship between the score assigned to a sequence by the HMM model and the sequence's membership in a given receptor class.

5.16 Signal Peptide and Signal Anchor Prediction by HMMs

Nielsen and Krogh [17] constructed an HMM designed both to discriminate between signals peptides and nonsignal peptides and to locate the cleavage site. The HMM was designed so that it took known signal peptide features into account. A prediction tool can be developed that can discriminate between signal peptides and anchors. The signal peptide model is shown in Fig. 5.14. An explicit modeling

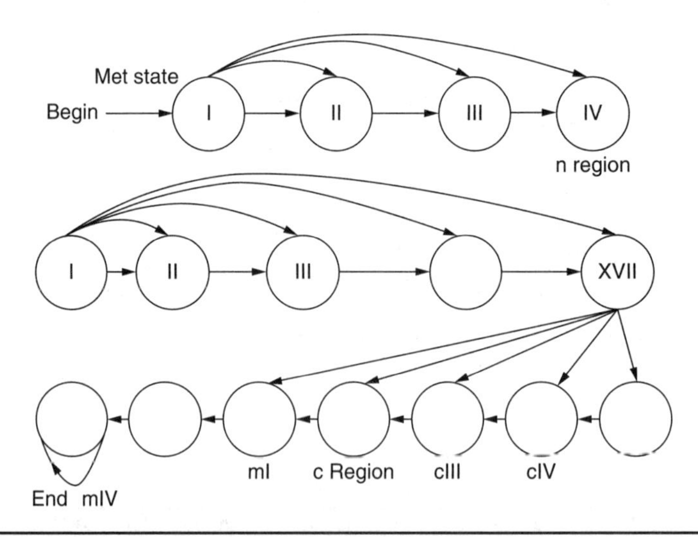

FIGURE 5.14 HMM used for signal peptide discrimination.

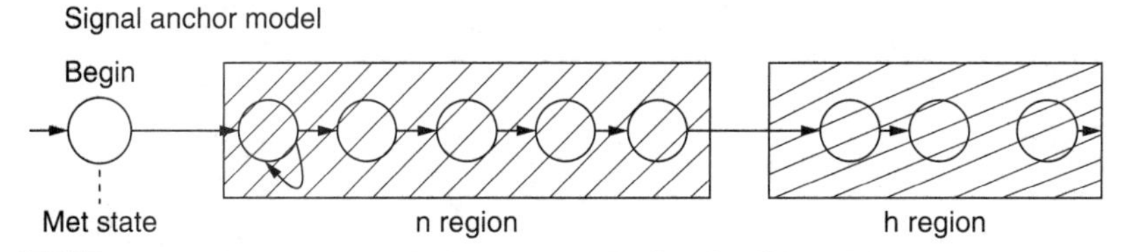

FIGURE 5.15 HMM design for delineating signal peptides and signal anchors.

of length distribution is implemented in the various regions using tied states that have the same amino acid distribution in the emission and transition probabilities associated with them. To discriminate among signal peptides, signal anchors, and soluble nonsecondary proteins, the model was augmented by a model of anchors. This is shown in Fig. 5.15. The entire model was trained using all types of sequences, including signal peptides, anchor sequences, and cytoplasmic and nuclear sequences. The prediction of which of three classes the protein belongs to is given by the most likely path taken throught the combined model. In terms of predictive performance, the combination of *C* score and *S* score networks had a discrimination level comparable with the HMM. The HMM was found to be better at recognizing signal anchors and therefore at detecting this type of membrane-associated protein.

5.17 Markov Model and Chargaff's Parity Rules

The Chargaff's first parity rule [18] states that in a place of double-helical DNA, the number of A's is equal to the number of T's, and the number of C's is equal to the number of G's. The Chargaff's second parity rule states that the same relation holds good for a piece of single-stranded DNA of reasonable size. The validity of Chargaff's second parity rule can be studied across different organisms and different coding and noncoding DNA at different length scales. For instance, genomic DNA in yeast was considered [8]. Symmetry was observed for both the strands of DNA, as shown in Table 5.6. The compositions were found to be stable.

It was found to be roughly 30 percent for A and T and 20 percent for C and G. In mitochondrial DNA, the same symmetry also was observed. To study the symmetries of double-stranded DNA, how often each nucleotide occurs on each strand over a length is counted. These frequencies correspond to a probabilistic Markov model of order 1.

For dinucleotides, a second-order Markov model may be used for determining whether Chargaff's second parity rule holds. A DNA Markov model of order *N* has 4*N* parameters associated with the

	A	C	G	T
Chr 1	30.3	19.4	19.9	30.4
Chr 2	30.7	19.4	19.0	31.0
Chr 3	31.1	19.7	18.9	30.3
Chr 4	31.1	18.9	19.0	31.0
Chr 5	30.6	19.0	19.5	31.0
Chr 6	30.7	19.3	19.4	30.6
Chr 7	31.0	19.0	19.0	31.0
Chr 8	30.9	19.4	19.5	30.6
Chr 9	30.5	19.4	18.9	30.6
Chr 10	31.0	19.2	18.9	31.0
Chr 11	30.9	19.2	18.9	31.0
Chr 12	30.7	19.3	19.2	30.9
Chr 13	31.0	19.1	19.1	30.8
Chr 14	30.8	19.3	19.3	30.6
Chr 15	31.1	19.2	19.0	30.9
Chr 16	31.0	19.0	19.0	30.9
Chr.mt	42.2	8.0	9.1	40.7
16 nucl chr.	30.9	19.2	19.1	30.8
All chr.	31.0	19.1	19.1	30.9

TABLE 5.6 Percent Values of the Nucleotide Bases

transition probabilities $P(X_N/X_1, \ldots, X_{N-1})$, also denoted $P(X_1, \ldots, X_{N-1} X_N)$, for all possible $X_1, \ldots, X_N$ in the alphabet, together with a starting distribution of the form $\pi (X_1, \ldots, X_{N-1})$.

Since the number of parameters grows exponentially, only models up to a certain order can be determined from a finite data set. A DNA Markov model of order 5, for instance, has 1024 parameters, and a DNA Markov model of order 10 has over 1 million parameters. Conversely, the higher the order, the larger is the data set needed to properly fit the model.

Summary

Markov models are explained in detail. A genome sequence from NCBI is obtained and modeled using geometric distribution and a Markov model. The *k*th-order Markov model is defined. Worked examples in the construction of zeroth-, first-, second-, and third-order Markov models are illustrated. The potential for the use of geometric distribution to model DNA sequences is explored. Rabiner's tutorial on HMM is referred to. The three questions in HMM, i.e., evaluation, decoding, and learning, are reviewed. The Markov, stationarity, and output independence assumptions are introduced to keep the problems mathematically tractable. The HMM is characterized completely. The number of operations needed to determine the sequence given the HMM, i.e., the evaluation problem, which usually takes time $O(N^T)$, where T is the length of the sequence and N is the number of states, can be completed in $O(N^2T)$ time using the forward algorithm. The Viterbi algorithm with optimal path is discussed. HMM applications such as construction of a phylogenetic tree, protein families, wheel HMMs to predict periodicity in DNA, the generalized HMM, database mining, multiple alignments, classication using HMMs, signal peptide and signal anchor prediction by HMMs, and Chargaff's parity rule predictions are discussed. Commercial software such as SAM, HMMER, HMMPRO, MetaMeme, PSI-BLAST, and PFAM are discussed and alanyzed as end-of-chapter exercises.

References

[1] A. Krogh, M. Brown, I. S. Mian, et al., "Hidden Markov models in computational biology: Applications to protein modeling," *J. Mol. Biol.* 235 (1994), 1501–1531.

[2] K. R. Sharma, "Hidden Markov model of order *n*," 230th ACS National Meeting, Washington, DC, 2005.

[3] K. R. Sharma, "Geometric distribution representation of DNA sequences," 230th ACS National Meeting, Washington, DC, 2005.

[4] K. R. Sharma, "Geometric distribution effects in quality of continuous copolymerization of alpha-methylstyrene acrylonitrile," 91st AIChE Annual Meeting, Dallas, TX, 1999.

[5] P. B. Chaves, M. F. Paes, S. L. Mendes, et al., "Noninvasive genetic sampling of endangered *Muriqui* (primates, *atelidae*): Efficiency of fecal DNA extraction," *Genet. Mol. Biol.* 29 (2006), 750–754.

[6] G. D. Stormo and G. W. Hertzel, "Identifying protein-binding sites from unaligned DNA fragments," *Proc. Natl. Acad. Sci. U.S.A.* 8 (1989), 1183–1187.

[7] L. R. Rabiner, "A tutorial on hidden Markov models and selected applications in speech recognition," *Proc. IEEE* 77. (1989), 257–286.

[8] P. Baldi and S. Brunak, *Bioinformatics: The Machine Learning Approach.* Boston: MIT Press, 2001.

[9] A. J. Viterbi, "Error bounds for convolutional codes and an asymptotically optimum decoding algorithm," *IEEE Trans. Inform. Theory.* 13 (1967), 260–269.

[10] S. Kullback and R. A. Leibler, "On information and sufficiency," *Ann. Math. Statistics.* 22 (1951), 79–86.

[11] R. Durbin, S. Eddy, A. Krogh, and G. Michison, *Biological Sequence Analysis: Probabilistic Models of Proteins and Nucleic Acids.* Cambridge, UK: Cambridge University Press, 1998.

[12] T. H. Jukes and C. R. Cantor, "Evolution of protein molecules." In H. N. Munro (ed.), *Mammalian Protein Metabolism.* New York: Academic Press, 1969.

[13] M. Kimura, "A simple method for estimating evolutionary rate of base substitution through comparative studies of nucleotide sequences," *J. Mol. Evolution.* 16 (1980), 111–120.

[14] R. Hughey and A. Krogh, "Hidden Markov models for sequence analysis: Extension and analysis of the basic method," *Comput. Appl. Biosci.* 12 (1996), 95–107.
[15] A.Krogh, M. Brown, I. S. Mian, et al., "Hidden Markov models in computational biology: Applications to protein modeling," *J. Mol. Biol.* 235 (1994), 1501–1531.
[16] M. G. Reese, D. Kulp, H. Tammana, and D. Haussler, "Genie: Gene finding in *Drosophila melanogaster*," *Genome Res.* 10 (2000), 529–538.
[17] H. Nielsen and A. Krogh, "Prediction of Signal Peptides and Signal Anchors by a Hidden Markov Model", *Proc. of the 6th International Conference on Intelligent Systems in Molecular Biology*, (1998), 122–130.
[18] E. Chargaff, "Some recent studies on the composition and structure of nucleic acids," *J. Cell Phys. Suppl.* 38 (1951), 41–59.

Exercises

1.0 Given the sequence

S: TATATGCGTAACCGGTT

construct a first-order HMM to represent the information in sequence *S*. Show the transition probabilities in Fig. 5.16.

2.0 Construct a second-order HMM to represent the information in sequence *T*.

T: ACGTTGACTGACTGTATACTGGTTAGTGT

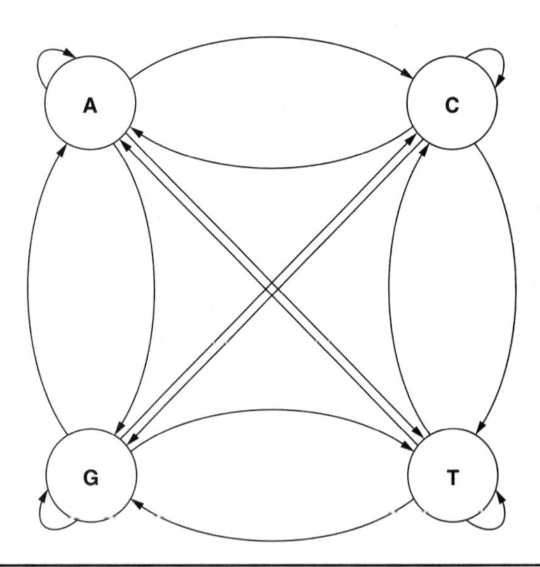

Figure 5.16 HMM of the first order to represent sequence *S*: TATATGCGTAACCGGTT.

3.0 Show by schematic the construction of a first-order HMM to represent the following sequence information:

GCCGCGCTTG

GCTTGGTGGC

TGGCCGTTGC

4.0 Chaves and colleagues submitted the DNA sequence with 660 bases in *Homo sapiens* shown in Fig. 5.2 to the NCBI. Develop a Markov model of the second order to represent this information. Calculate the 64 transition probabilities, and represent the information in the form of a suitable table such as Table 5.7. The 64 transition probabilities $P(A/AA)$, $P(G/AA)$, . . . can be calculated from the information provided in Fig. 5.2 and presented in a tabular form as shown in the table below. Columns 3–6 are conditional probability values for the base pair shown at the top of the column given the preceding dyad that occurred in the sequence in column 2. A dyad number is also given to the 16 possible dyads for DNA.

Dyad No.	Dyad	*P* (A/Dyad#)	*P* (C/Dyad#)	*P* (G/Dyad#)	*P* (T/Dyad#)
1	AA				
2	AC				
3	AG				
4	AT				
5	CA				
6	CC				
7	CG				
8	CT				
9	GA				
10	GC				
11	GG				
12	GT				
13	TA				
14	TC				
15	TG				
16	TT				

Table 5.7 Transition Probabilities in the Second-Order Markov Model to Represent 660 Base Pairs of a DNA Sequence from *Homo sapiens*

5.0 Eight hypothetical translation start sites are shown below:

ATG

ATG

ATG

ATG

ATG

GTG

GTG

TTG

Show the (1) site-profile matrix and (2) the log-likelihood ratio of the weight matrix.

6.0 The effect of background distribution that is nonuniform is studied. Consider the eight translation start sites of Exercise 5.0, but change the background distribution to $B_{Aj} = B_{Tj} = 0.375$, $B_{ij} = B_{gj} = 0.125$. The site-profile matrix remains unchanged. Find the weight matrix and relative entropies. Interpret the results using the Kullback-Leibler measure.

7.0 Verify the first-order HMM for the 23 CRP binding sites given in Table 5.3 (see Fig. 5.17).

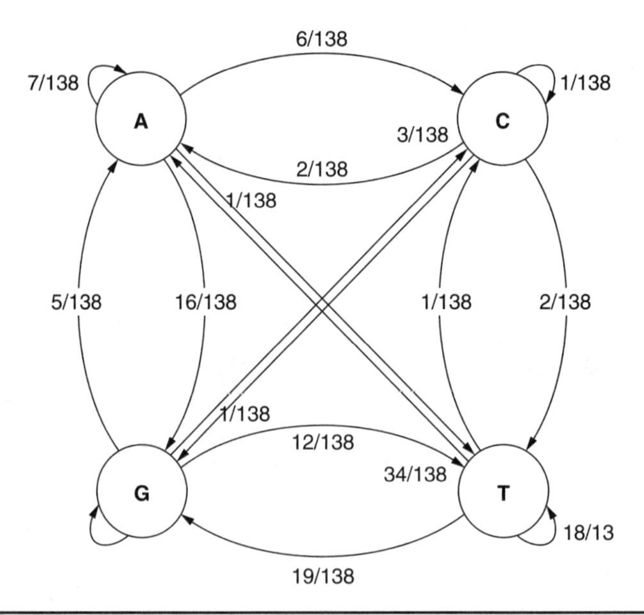

Figure 5.17 First-order HMM for the 23 CRP binding sites.

8.0 Develop a zeroth-order HMM model to represent the information in Fig. 5.2.

9.0 Find the probability of the sequence ATGTGAC using the HMM of the first order in Exercise 7.0.

10.0 Develop an HMM of the second order to represent the information in Table 5.3.

11.0 Given the following five sequences:

ACAATG

TCAACTATG

ACACATC

AGAATC

ACCGATC

construct an HMM of zeroth order to represent the information.

12.0 Construct an HMM of the first order to represent the five sequences in Exercise 11.0.

13.0 Show that

$$\partial e_{ix}/\partial W_{ix} = e_{ix}(1 - e_{ix}) \qquad \text{and} \qquad \partial e_{ix}/\partial W_{ix} = -e_{ix}e_{iy}$$

14.0 Give a generalized expression for k random variables for joint probability, Bayes' rule, and applications to HMM.

15.0 As an example of Markov chain model application, consider the C_pG islands. CG nucleoides are rarer in eukaryotic genomes than expected given the marginal probabilities C and G. But the regions upstream of genes are richer in CG dinucleotides. These are referred to as *C_pG islands*. Markov chains can be used to predict the C_pG islands. Given the set of sequences from C_pG islands, how can the probability parameters of the model be determined? Use the maximum likelihood estimation. Given a set of data D, a set of parameters θ is obtained to maximize $P(D/\theta)$. The sequences given are

ACCGCGCTTA

GCTTAGTGAC

TAGCCGTTAC

16.0 Given the sequences

GCCGCGCTTG

GCTTGGTGGC

TGGCCGTTGC

calculate the maximum likelihood estimates of A, C, G, and T.

17.0 Construct a first-order HMM model to represent the information provided in Exercise 16.0.

18.0 Verify Chargaff's parity rule for the 660 base pairs of DNA sequence given in Example 5.1.

19.0 Construct a wheel HMM for introns.

20.0 Distinguish protein quarternary structure from tertiary structure with an example.

21.0 Find the likely ancestor of the following five proteins

V	A	G	H	L	Cy	GL	Ser	His	Leu
V	A	G	–	L	–	GL	–	His	–
V	A	G	–	–	–	GL	Ser	His	–
V	–	–	–	L	Cy	–	Ser	Hia	Leu
V	A	G	H	L	–	–	–	His	Leu

22.0 Develop a statistical model for the following 12 related proteins.

PVAGTL

PCHSVL

PCHVTL

PCHGTL

PAHGPL

PAHGPL

PGGTPP

XGSLAA

STVTGG

YLLLTV

YLTTLL

23.0 Construct an HMM for the protein sequence of insulin. What order would you choose?

VAGHLCYG

24.0 Show the matrix for the Viterbi algorithm for PVAGHLCyG.

25.0 Estimate the database size requirements for a string HMM to represent 1000 protein families.

26.0 Distinguish left-right HMM models from ergodic models. Why is left-right preferred in bioinformatics?

27.0 Develop a loop HMM for the following sequence and indentify the periodicity.

AAGGCCCCAAGGCTGCAAGG

28.0 Rederive the Felsentein algorithm for variable rates of a substitution process.

29.0 Rederive expressions for α, β, and γ for the protein alphabet with 20 different amino acids. Obtain the expression for sequence fidelity. Draw neatly the Markov model for sequence evolution and the corresponding finite-state automaton for sequence alignment.

30.0 Discuss the utility and implications of a GHMM for a human proteome.

31.0 Why can't the training of the model for multiple alignments be conducted offline? Discuss with proof.

32.0 Can misclassification of protein families occur during HMM construction? If so, what are the remedies?

33.0 Discuss the propagation of noise when clustering and HMM are combined to generate classification during database mining.

34.0 Given Pfam and FORESST, what are the motivations for a database of protein tertiary structure? Discuss size requirements and issues.

35.0 What are some of the issues in constructing Viterbi paths for very long sequences.

36.0 Discuss the utility of a hydropathy plot for nucleic acids with alphabet (A, C, G, T).

37.0 Provide a fit with generalized normal distribution of opsins:

$$f(z) = A \exp[-(Az + bz^2 + cz^3 + dz^4)]$$

Point out the saddle points owing to rhodopsin sequences.

38.0 Why is a multiple-HMM architecture needed for detecting protein subfamilies.

39.0 Discuss the utility of an HMM in annotation of protein sequences, especially the regions whose functions are not known.

40.0 Can you expect an analogous relationship to the Chargaff's parity rule in a human proteome? How about other organisms?

41.0 Discuss two extensions of HMMs and their applications.

42.0 Discuss the limitations of HMMs and there advantages.

43.0 Distinsuigh the heuristic Markov model from shotgun sequencing.

44.0 *Maximum likelihood criterion (ML).* In the Viterbi algorithm, one way to settle for the optimization criteria is ML, the probability of a given sequence of observations O^w belonging to a class w given the HMM λ_w of the class w with respect to the parameters of the model λ_w. This probability of the total likelihood of the observations can be expressed as

$$L_{\text{toto}} = P(O^w/\lambda_w)$$

When only one class w at a time is considered, the subscript and superscript w can be dropped, and the ML criterion can be written as

$$L_{\text{toto}} = P(O/\lambda)$$

Discuss why there is no analytical solution for HMM λ that maximize L_{toto}.

45.0 *Baum-Welch algorithm.* This is an iterative method to obtain the solution set out in Exercise 44.0 and Sec. 5.6. This method can be derived using simple "occurrence counting" arguments or using calculus to maximize the auxiliary quantity:

$$Q(\lambda, \underline{\lambda}) = \Sigma P(q/O,\lambda)\ \lg[P(O,q,\lambda)]$$

Show that the convergence is gaurenteed in this method. Two more variables can be defined in addition to the forward and backward variables. Describe the Baum-Welch learning process where the parameters of the HMM are updated in such a way as to maximize the quantity $P(O/\lambda)$. With a initial guess of λ, the four variables are calculated recursively, and the HMM parameters are updated appropriately.

46.0 *Gradient-based method.* In continuation of the pursuits in Exercise 44.0 and Sec. 5.6, in the gradient-based method, any parameter θ of the HMM λ is updated by minimization of a certain J equivalent to the maximization of L_{toto}. J can be related to the model parameters via L_{toto}. Show that the gradient $\partial J/\partial\theta$ can be found from the two main parameter sets in the HMM, i.e., the transition probabilities A_{ij} and observation probabilities B_{ij}.

47.0 *Maximum mutual information (MMI) criterion.* In ML, the HMM is optimized only one class at a time, and the HMMs are not touched for other classes at that time. This procedure does not involve the concept of "discrimination," which is of great interest in pattern recognition. Thus the ML learning procedure gives a poor discrimination ability to the HMM system, especially when the estimated parameters (in the training phase) of the HMM system do not match with the speech inputs used in the recognition phase. This type of mismatch can arise for two reasons. One is that the training and recognition data have considerably different statistical properties, and the other is the difficulties of obtaining reliable parameter estimates in the training. Show that in the MMI criterion, on the other hand, HMMs of all the classes are considered simultaneously during training. Parameters of the correct model are updated to enhance its contribution to the observations, whereas parameters of the alternative models are updated to reduce their contributions. Show that this procedure gives a high discriminative ability to the system and thus that MMI belongs to the so-called discriminative training category.

48.0 *Alternating-sequence distribution.* Consider the sequence S: AUAUAUAUAUAUAUAUAUAU with the alternating-sequence-distribution microstructure. Construct a zeroth-order HMM to represent the sequence S. Show that $P(\text{A}) = P(\text{U}) = 0.5$.

49.0 Construct a first-order HMM to represent the sequence S in Exercise 48.0. Represent the information in the form of a schematic similar to Fig. 5.1. Show that $P(\text{A}/\text{A}) = P(\text{U}/\text{U}) = 0$, $P(\text{A}/\text{U}) = 9/19$, and $P(\text{U}/\text{A}) = 10/19$.

50.0 Construct a second-order HMM to represent the sequence S. Show that $P(\text{A}/\text{AU}) = 5/18 = P(\text{U}/\text{UA})$ and that the rest of the probabilities $P(\text{A}/\text{UU}) = P(\text{U}/\text{UU}) = P(\text{A}/\text{AA}) = P(\text{U}/\text{AA}) = 0 = P(\text{U}/\text{AU}) = P(\text{A}/\text{UA}) = 0$.

51.0 Given the results from Exercises 48.0, 49.0, and 50.0, what is the best HMM to represent the sequence S in Exercise 48.0 that has the alternating-sequence distribution.

52.0 *Block-sequence distribution.* Consider a sequence S: AAAAAAAAAAUUUUUUUUUU with the block-sequence-distribution microstructure. Construct a zeroth-order HMM to represent the sequence S. Show that $P(\text{A}) = P(\text{U}) = 0.5$.

53.0 How do the results of Exercise 52.0 compare with the results of Exercise 48.0? Should they be the same? What is missing in the zeroth-order HMM analysis?

54.0 Construct the sequence S' given the zeroth-order HMM results from Exercise 52.0. Is there a unique result?

55.0 Construct a first-order HMM to represent the sequence S in Exercise 53.0. Show that $P(\text{A}/\text{A}) = 9/19$, $P(\text{U}/\text{U}) = 9/19$, $P(\text{U}/\text{A}) = 1/19$, and $P(\text{A}/\text{U}) = 0$. Do the results of Exercise 54.0 differ from the results of Exercise 49.0. What is being done right during the construction of the first-order HMM?

56.0 Construct the sequence S' given the results of the first-order HMM in Exercise 55.0. Is the sequence given in Exercise 52.0 obtained?

57.0 Construct a second-order HMM to represent the sequence given in Exercise 52.0. Show that $P(\text{A}/\text{AA}) = 4/9 = P(\text{U}/\text{UU})$, $P(\text{U}/\text{AA}) = 1/18$, $P(\text{U}/\text{AU}) = 1/18$, and $P(\text{A}/\text{UU}) = P(\text{A}/\text{UA}) = P(\text{A}/\text{AU}) = P(\text{U}/\text{UA}) = 0$.

58.0 Construct the sequence S' given the results of the second-order HMM from Exercise 57.0. Is the sequence given in Exercise 52.0 arrived at?

59.0 Can you conclude based on the results from Exercises 48.0 through 58.0 that the order of the HMM selected to represent a given sequence depends on the microstructure of the sequence and only beyond a certain order do the results have one-to-one correspondence between the sequence and the model.

60.0 *Random-sequence distribution.* Consider the sequence S: UUUAUAUAAUUAAUAUAAAU with the random-sequence-distribution microstructure. Construct a zeroth-order HMM to represent sequence S. Show that $P(\text{U}) = P(\text{A}) = 0.5$.

61.0 Given the results of the zeroth-order HMM in Exercise 60.0, reconstruct the sequence S'. Is it different from the sequences given in Exercises 48.0, 52.0, and 60.0?

62.0 Construct a first-order HMM to represent the sequence given in Exercise 60.0. Show that $P(\text{A/A}) = 4/19$, $P(\text{U/U}) = 3/19$, $P(\text{A/U}) = 6/19$, and $P(\text{U/A}) = 6/19$.

63.0 Given the results of the first-order HMM in Exercise 62.0, reconstruct the sequence S'. Do you obtain the sequence you started out with in Exercise 60.0?

64.0 Construct a second-order HMM to represent the sequence given in Exercise 60.0. Show that $P(\text{U/UU}) = 1/18$, $P(\text{A/AA}) = 1/18$, $P(\text{U/UA}) = 1/6$, $P(\text{A/UA}) = 1/6$, $P(\text{A/UU}) = 1/9$, $P(\text{U/AA}) = 1/6$, $P(\text{A/AU}) = 2/9$, and $P(\text{U/AU}) = 1/18$.

65.0 Reconstruct the sequence S' given the results obtained from the first-order HMM in Exercise 64.0. Is there a one-to-one correspondence between the sequence and the model? Use the information given in Exercise 60.0 if necessary.

66.0 Construct a third-order HMM to represent the sequence S given in Exercise 60.0. Show that

$P(\text{A/AAA}) = 0$	$P(\text{A/AUA}) = 2/17$	$P(\text{A/UUA}) = 1/17$	$P(\text{A/AUU}) = 1/17$
$P(\text{U/AAA}) = 1/17$	$P(\text{U/AUA}) = 2/17$	$P(\text{U/UUA}) = 1/17$	$P(\text{U/AUU}) = 0$
$P(\text{A/UUU}) = 1/17$	$P(\text{A/UAA}) = 1/17$	$P(\text{A/UAU}) = 3/17$	$P(\text{A/AAU}) = 1/17$
$P(\text{U/UUU}) = 0$	$P(\text{U/UAA}) = 2/17$	$P(\text{U/UAU}) = 0$	$P(\text{U/AAU}) = 1/17$

67.0 Reconstruct the sequence S' given the results of the third-order HMM in Exercise 66.0. Do you get the same sequence given in Exercise 60.0?

68.0 Can the third-order HMM for sequence S given in Exercise 62.0 be constructed given the results in Exercise 64.0. Why?

69.0 Can the third-order HMM for sequence S given in Exercise 62.0 be constructed given the results in Exercise 62.0. Why?

70.0 *SAM.* The sequence-alignment modeling system (SAM) is a collection of software tools used for creating, refining, and using linear HMMs for biologic sequence analysis. The sequence of columns in a multiple-sequence alignment are represented by model states with provisions for arbitrary position-dependent insertions and deletions in each sequence. An expectation-maximization algorithm is used to train the models on a family of protein or nucleic acid sequences. The algorithms and methods in SAM can be accesed via the hotlink www.cse.ucsc.edu/research/compbio/sam.html. Discuss the advantages of using the HMM for seeking alignment compared with other database search strategies.

71.0 *HMMER.* HMMER is an implementation of profile HMM methods for sensitive database searches using multiple-sequence alignments as queries. An HMM is built based on the multiple-sequence alignment as input. Nine programs are supported in the HMMER@ package. These are Hmmalign, hmmbuild, hmmcalibrate. hmmconvert_hmmer, hmmemit, hmmfetch, hmmindex, hmmpfam, and hmmsearch. A number of utility programs that are not HMMs are also offered that may be useful. These are aftech, alistat, seqstat, sfetch, shuffle, and sreformat. Discuss the advantages of using HMMs when seeking multiple-sequence alignment. How close to the optimal can you come?

72.0 *HMMPRO.* HMMPRO is used for biologic sequence simulations. Models of protein families or DNA functional elements can be interactively built and analyzed using a graphic user interface (GUI). These models then can be used for multiple-sequence alignment, pattern discovery, and sensitive data mining. HMMpro 2.2 is available from www.netid.com. Some of the highlights of the software include support for editing individual emission and transition weights, the ability to fix emission or transition weights on a node-by-node basis during training, and support for importing and exporting HMMER 2.x models. Given the NP complete nature of the multiple-sequence-alignment problem, how close to optimality can one get? What are the advantages of using HMMPRO in seeking multiple-sequence alignment?

73.0 *Meta-MEME.* A motiff-based hidden Markov model (Meta-MEME) of biologic sequences is a software toolkit for building and using motif-based HMMs of DNA and proteins. Input is a set of protein sequences and motif models discovered by MEME. These models are combined in Meta-MEME, and the model is used to search a sequence database for homologues. Discuss the advantages of using Meta-MEME and the degree of optimality of multiple-sequence alignments.

74.0 *PSI-BLAST.* A position-specific scoring matrix (PSSM) is used and is a particular feature of BLAST 2.0. PSSM is constructed from a multiple alignment of the highest-scoring hits in an initial BLAST search. PSSM is generated by calculation of position-specific scores for each position in the alignment. High scores are awarded to highly conserved regions, and near-zero scores are awarded to weakly conserved positions. A second BLAST seach is performed, and the results from the iteration are used to refine the model. How many alignments are needed before the problem is considered NP complete? How close to the optimality would this procedure come? What are the advantages of using the PSI-BLAST in terms of biologic significance, increased sensitivity, etc.?

75.0 *PFAM.* PFAM is a large collection of multiple-sequence alignments and HMMs that includes many common protein domains and families. Multiple alignments can be looked up, protein domain architectures can be viewed, links to other databases can be followed, and protein structures can be viewed. What are the storage issues involved in PFAM?

76.0 *Profile HMMs.* A multiple-sequence alignment is converted to a position-specific scoring system by profile HMMs. This is suitable for searching protein sequences with weak homology. Compare profile HMM methods with pairwise sequence-comparison methods such as those of Smith and Waterman for global and Needleman and Wunsch for local alignments.

77.0 What are linear left-right models?

78.0 How is the affine gap penalty handled in profile HMMs?

79.0 The probability parameters in a profile HMM are usually converted to additive log-odds scores before aligning and scoring a query sequence [14]. The score for aligning a residue to a profile match state emitting residue x is P_x, the expected background frequency of residue x in the sequence database is f_x, and the score for residue c at this match state is $\log(P_x)/f_x$. Show that this gives rise to nontrivial optima.

80.0 What are the differences between the Jukes and Cantor substitution matrix and Kimura matrix in the model for evolution?

81.0 What are the three questions in HMM?

82.0 What are the three assumptions that are needed for the tractability of HMMs?

83.0 Discuss the time efficiency gained in the forward algorithm.

84.0 What is a Viterbi path?

85.0 What is meant by relative entropy of a sequence?

86.0 Can HMMs be used to obtain pairwise multiple-sequence alignment?

87.0 In the model of evolution, can the solution to the model equations exhibit subcritical damped oscillations?

88.0 What makes the system of equations in the model of evolution stable?

89.0 What is the difference between a GHMM and a loop HMM?

90.0 What are the limitations of a wheel HMM?

91.0 Can an HMM be constructed to obtain a semiglobal alignment?

92.0 Can an HMM be constructed to obtain a glocal alignment?

93.0 How would the design of an HMM differ when seeking a global alignment and when seeking a local alignment?

94.0 What are the pros and cons of representing a DNA sequence using a suffix tree versus using an HMM?

95.0 What are the pros and cons of representing a protein sequence using a suffix tree versus using an HMM?

96.0 Can the banded diagonal algorithms using the greedy strategy to obtain global alignment in lesser time be implemented using pairwise HMMs?

97.0 Can an HMM be used to represent a tRNA sequence? What would be different in the design?

98.0 Would the alignment of protein sequences that are encoded by the nucleotide sequence using an HMM result in a better alignment compared with aligning DNA sequences?

99.0 Given two sequences S and T, design an HMM to generate the supersequence S'.

100.0 Given two suffix trees of sequences S and T, design an HMM to construct a generalized suffix tree.

101.0 Construct an HMM to find tandem repeats in a DNA sequence.

102.0 What are the storage requirements to obtain a pairwise alignment using HMM?

103.0 Given an alignment of two sequences S' and T', construct an HMM to deduce the sequences S and T.

104.0 For very similar DNA sequences, what would be different in the construction of an HMM that is designed to obtain pairwise global alignment?

105.0 Given two sequences S and T and the alignments S' and T', what are the issues involved in constructing an HMM to obtain the scoring scheme and affine gap penalty parameters?

106.0 What would be the modifications to the HMM necessary to seek a more biologically meaningful alignment once more is known about the substitution and mutation rates in organisms whose sequences are being studied?

107.0 What is the biologic significance of the eigenvalues in the substitution matrix of Jukes and Cantor taking on imaginary values?

108.0 What is the biologic significance of the eigenvalues in the substitution matrix of Kimura taking on imaginary values?

CHAPTER 6

Gene Finding, Protein Secondary Structure

Objectives

The objectives of this chapter are to

- Learn the greedy algorithm for relative site-selection problems.
- Use binomial heap to obtain the maximum increasing subsequence.
- Learn the interpolated Markov Model (IMM) to find out its use in GLIMMER.
- Propose a solution to the Shine Dalgarno (SD) site-selection problem.
- Annotate genes using a dictionary.
- Devise GPHMM for cross-species gene finding.
- Be familiar with Steiner trees, the spliced alignment problem, and the fragment-matching problem.
- Be familiar with protein secondary structure and neural networks.
- Learn the Profilenetwork HeiDelberg (PHD) architecture of Rost and Sander and DAG-RNNS.
- Use hidden Markov models (HMMs) to obtain protein secondary structure.

6.1 Introduction

Gene finding, simply stated, refers to methods of finding regions in sequences of DNA that are functional. The explosive growth of biologic data resulting from the completion of various genome projects led to blossoming of the field of *genomics.* The sequencing of proteins and the relation between the protein and signals that govern the functions of the organism is the field of *proteomics*. These projects are not complete unless the genomes are functionally annotated. Functional genomics, or *metabolomics*, is the area that pertains to the mapping of every function of the organism to its originating gene. This continues to be a challenge. The accurate annotation of sequenced genomic data is a key technical hurdle. This is a fertile area of research. The coding regions, exons and introns of the genes, need to be identified. Very large databases of proteins, Expressed Sequence Tags (ESTs), and smaller databases of annotated genes are available to complete this task.

6.2 Relative Entropy Site-Selection Problem

The relative entropy site-selection problem was shown by Akutsu and colleagues [1] to be NP complete. Provided that the optimality constraint is relaxed, "good" solutions are plausible. Relative entropy is a function of $P(S)$, the fraction of sites containing each residue S and not the absolute number of sites. Increasing the length n of each site does increase the number of sites and will not increase the relative entropy. The relative entropy is a function of $P(S)$, the fraction of sites containing each residue S, and not the absolute number of sites. For example, a conserved protein has $P(S) = 1$ regardless of the number of sites present. It measures the degree of conservation. However, with more instances of a conserved residue, this measure needs to be increased. Increasing the length n of each site does increase the relative entropy because it is additive and always nonnegative. Normalization can be used when comparing relative entropies of different length sites by dividing by the length n of the site.

6.2.1 Greedy Approach

Hertz and Stormo [2] presented a "greedy" approach to develop an efficient algorithm for the relative entropy site-selection problem. Best choices at a local level without regard for ramifications on subsequent choices are picked using the greedy algorithm. The greedy method will result in solutions that are far from optimal for some input instances. The user specifies the length n of sites. The user also specifies a maximum number d of profiles to retain at each step. Profiles with lower relative entropy scores than the top d will be discarded. This is the greedy aspect of the algorithm.

Algorithm 6.1 *Hertz and Stromo Algorithm*
Input: Sequences $S_1, S_2, \ldots, S_k$ and d, the background distribution.

1. Create a singleton set, i.e., only one member for each possible length n substring of each of the k input sequences.

2. For each set S retained so far, add each possible length n substring from an input sequence S_i not yet represented in S. Compute the profile and relative entropy with respect to the background for each new set. Retain the d sets with the highest relative entropy.
3. Repeat step 2 until each set has k members.

In order to avoid exponential possible sets, the number of sets is pruned to d. The greedy nature of this pruning biases the selection from the remaining input sequences. The remaining sequences may not contain high scoring profiles chosen from the first few sequences. Superior scores would result if that were the case. The single-set-per-sequence assumption may be relaxed, and multiple substrings may be permitted to be chosen from the same sequence. A different stopping condition is needed. The procedure was applied to Cyclic AMP Receptor Proteins (CRP) binding sites by Hertz and Stormo [2]. Their best solution contained 19 correct sites plus 3 more from overlapping correct sites from 18 genes containing 24 known CRP binding sites.

6.2.2 Gibbs Sampler

An iterative approach is used in the Gibbs sampling method for solution to the relative entropy site-selection problem. By trial and error, one of a set of k starting strings is removed at random and then replaced with another one at random with probability proportional to its score. An improved score may result. The assumption made is one site per input sequence.

The stopping condition is sewn into the routine to let the iteration continue as desired. A fixed number of iterations or relative stability of the score could be the stopping condition. The calculations return a best solution T. Some degree of greediness was retained by the Gibbs sampler. The principle of it is to enable a strong signal in only a few sequences to outweigh a weaker signal in all the sequences.

Algorithe 6.2 *Gibbs Sampling Algorithm*

Input: Sequences $S_1, S_2, \ldots, S_k$, n and k background distribution

Algorithm: Initialize set T to contain substrings $t_1, t_2, \ldots, t_k$, where t_i is a substring of S_i chosen randomly and uniformly. Now perform a series of iterations, each of which consists of the following steps:

1. Choose one randomly and uniformly from $\{1, 2, \ldots, k\}$ and remove it from T.
2. For every j in $\{1, 2, \ldots, s - n + 1\}$,
   a. Let t_{ij} be the length n substring of Si that starts at position j.
   b. Compute Dj, the relative entropy of $T \cup t_{ij}$ with respect to the background.
   c. Let $pj = Dj/\Sigma hDh$.
3. Randomly choose ti to be tij with probability P_j, and add t_i to T.

Motifs were found in the protein families by Lawrence et. al. [3]. A helix-turn-helix motif and motifs in lipocalins and prenyltransferases were discovered.

6.3 Maximum-Subsequence Problem

A corollary to the problem of finding the coding regions in DNA is the maximum-subsequence problem. Given a sequence $X_1, X_2, \ldots, X_n$ of real numbers, where X_i corresponds to the score of the ith element of the sequence, the problem is to find a contiguous subsequence X_i, $X_{i+1}, \ldots, X_j$ that maximizes $X_i, X_{i+1}, \ldots, X_j$.

6.3.1 Bates and Constable Algorithm

The following algorithm for finding a maximum subsequence was given by Bates and Constable [4]. They use the principle of recursion. Suppose that the maximum subsequence of B, of $X_1, X_2, \ldots, X_k$, has score b and is known, how can the maximum subsequence of $X_1, X_2, \ldots, X_k, X_{k+1}$ be found ? If $X_{k+1} > 0$, add X_{k+1} to B, and if not, leave B unchanged. But what if X_{k+1} is not included in B? In this case, in addition to B, we will have to keep track of the score of the maximum suffix F of $X_1, X_2, \ldots, X_k$; F is the suffix $X_5, X_{5+1}, \ldots, X_k$ that maximizes $F = X_5 + X_{5+1} + \cdots + X_k$. It is assumed that F is also known for $X_1, X_2, \ldots, X_k$. Now, given X_{k+1}, B and F are to be updated accordingly. The complexity of the algorithm is $O(n)$ because a constant amount of work is done for every new element X_{k+1}, and there are n such elements.

Algorithm 6.3 *Bates and Constable Algorithm for Maximum Subsequence*

```
If X_{k+1} + f > b
then add X_{k+1} to f and replace B by F
else if F + X_{k+1} > 0
then add X_{k+1} to F
else reset f to be empty
```

6.3.2 Binomial Heap [5–7]

The maximum increasing subsequence can be found as the deepest branch of a binomial heap [5–7]. For example, find the largest increasing subsequence for

$$S = \{11, 17, 5, 8, 6, 4, 7, 12, 3\}$$

using binomial heap. A binomial heap H is a set of binomial trees that satisfies the following binomial heap properties:

1. Each binomial tree in H obeys the min-heap property. The key of a node is greater than or equal to the key of its parent. Each such tree is said to be *minimum-heap-ordered.*
2. For any negative integer k, there is at most one binomial tree in H whose root has degree k.

The root of a minimum-heap ordered tree contains the smallest key in the tree from the first property. The second property implies that an n-node binomial heap H consists of at most $\lg n + 1$ binomial trees.

Each binomial tree within a binomial heap is stored in the left-child, right-sibling, and parent nodes. Each node has a key field and any other satellite information required by the application. In addition, each node x contains pointers $P(x)$ to its parent, child (x) to its leftmost child and sibling (x) to the sibling of x immediately to its right. If node x is a root, then $p(x)$ = maximum likelihood (ML). If node x has no children, then child (x) = ML, and x is the field degree (X), which is the number of children of x. The roots of the binomial trees within a binomial heap are organized in a linked list that is referred to as the *root list.* The degrees of the roots strictly increase as the root list is traversed. By the second binomial heap property in an n-node binomial heap, the degrees of the roots are a subset of $\{0, 1, \ldots, \lg(n)\}$. The sibling field has a different meaning for roots than for nonroots. If x is a root, then sibling (x) points to the next root in the root list. A given binomial heap is accessed by the field head (H), which is simply a pointer to the first root in the root list of H. If binomial heap H has no elements, then head (H) = nil.

The binomial tree B_k is an ordered tree, defined recursively. The binomial tree B_k consists of two binomial trees B_{k-1} that are linked together. Some properties of binomial trees are given by the following lemma.

Lemma For the binomial tree B_k,

1. There are 2^k nodes.
2. The height of the tree is k.
3. There are exactly ${}^kC_i$ nodes at depth i, for $i = 0, 1, \ldots, k$.
4. The root has degree k, which is greater than any of the other nodes. Moreover, if the children of the root are numbered from left to right by $k - 1, k - 2, \ldots,$ 0, child i is the root of a subtree B_i.

The maximum increasing subsequence is found in the deepest branch. Once ordered, the time taken is $O(n)$. All maximum increasing subsequences are available in the binomial heap. See Fig. 6.1.

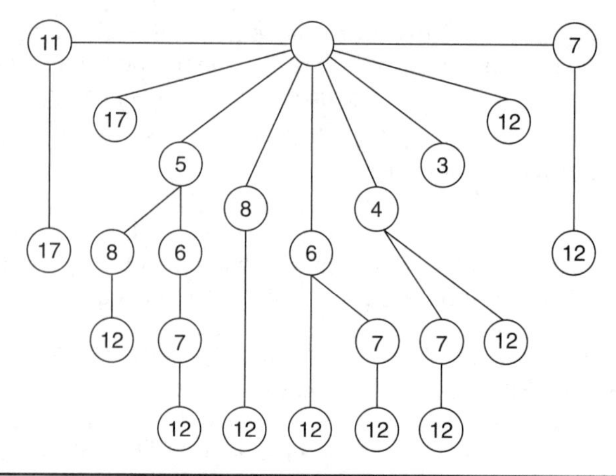

Figure 6.1 Binomial heap representation of maximum increasing subsequence.

6.4 Interpolated Markov Model (IMM)

The software package Gene Locator and Interpolated Markov Modeler (GLIMMER) is used for finding genes in bacteria and Archaea. Interpolated Markov Models (IMMs) from first to eighth order were used in this software. GLIMMER is the primary microbial gene finder at Institute of Genomic Research (TIGR) and has been used to annotate complete genomes. A special version of GLIMMER was designed for small eukaryotes (GlimmerM) and was used to find the genes in chromosome 2 of the malaria parasite *Plasmodium falciparum* [8]. It also has been trained on the plants *Arabidopsis thaliana* and *Oryza sativa* (rice), the parasite *Thieleria P arva,* the fungus *Aspergillus fumigatus,* and other organisms. The GLIMMER system consists of two main programs: (1) a training program, *build-imm,* which takes an input set of sequences and builds and outputs the IMM for them (the sequences can be complete genes or just partial open reading frames), and (2) Glimmer itself, which uses this IMM to *identify putative genes* in an entire genome. Conflicts are resolved automatically between most overlapping genes by choosing one of them. It also identifies genes that are suspected to truly overlap and flags these for closer inspection by the user.

The accuracy for 10 complete bacterial and archael genomes are shown in Table 6.1. Organisms are listed in the order in which the sequencing projects were completed. All these results were obtained by a very simple training procedure: GLIMMER was trained by first extracting all nonoverlapping open reading frames (orfs) over 500 bp (using the long-orfs program that comes with the system). The trained model then was used to find genes in the complete genome.

Organism	Genes Annotated	Percent Annotated Genes Found
Haemophilus influenzae	1738	99
Mycoplasma genitallium	483	99.4
Methanococcus jannaschii	1727	99.7
Helicobacter pylori	1590	97.5
Escherichia coli	4269	97.4
Bacillus subtilis	4100	98.3
Archaeoglobus fulgidis	2437	98.6
Borrelia burgdorferi	853	99.3
Treponema pallidum	1039	97.3
Thermatoga maritima	1877	98.8

TABLE 6.1 Accuracy for 10 Complete Genomes

6.5 Shine Dalgarno SD Sites Finding

The accurate prediction of the translation start site, i.e., the correct start codon, is important in order to analyze the putative protein product of a gene. At the initiation of protein synthesis, the ribosome binds to the mRNA at a region near the end of the mRNA called the *ribosome-binding site.* This is a region of approximately 30 nucleotides of the mRNA that is protected by the ribosome during initiation. This short mRNA sequence is called the *SD site.* The mechanism by which the ribosome recognizes the SD site is relatively simple base-pairing: The SD site is complementary to a short sequence near the end of the ribosome's 16S rRNA, one of its ribosomal RNAs. The SD site was first postulated by Shine and Dalgarno [9] for *E. coli.* Subsequent experiments demonstrated that the SD site in *E. coli* mRNA usually matches at least four or five consecutive bases in the sequence AAGGAGG (Table 6.2) and is separated from the translation start site by approximately seven nucleotides, although this distance is variable. This SD site can be used to improve start-codon prediction. The simplest way to identify whether a candidate start codon is likely to be correct is by checking for approximate base pair complementarity between the end of the 16S rRNA sequence and the DNA sequence just upstream of the candidate codon.

A greedy version of the Gibbs sampler was used in another study to find likely SD sites. Tompa [10] proposed a method to discover SD sites by looking for statistically significant patterns in the sequences upstream from the putative genes. The statistical significance is

Bacillus subtilis	CUGGAUCACCUCCUUUCUA _ _
Lactobacillus delbrueckii	CUGGAUCACCUCCUUUCUA _ _
Mycoplasma pneumoniae	GUGGAUCACCUCCUUUCUA _ _
Mycobacterium bovis	CUGGAUCACCUCCUUUCU
Aquifex aeolicus	CUGGAUCACCUCCUUUA _ _
Synechocystis spp.	CUGGAUCACCUCCUUU _ _
Escherichia coli	UUGGAUCACCUCCUUA _ _
Haemophilus influenzae	UUGGAUCACCUCCUUA _ _
Helicobacter pylori	UUGGAUCACCUCCU _ _
Archaeoglobus fulgidus	CUGGAUCACCUCCU _ _
Methanobacterium thermoautotrophicum	CUGGAUCACCUCCU _ _
Pyrococcus horikoshii	CUCGAUCACCUCCU _ _
Methanococcus jannaschii	CUGGAUCACCUCC _ _
Mycoplasma genitalium	GUGGAUCACCUC _ _

Table 6.2 End of 16S rRNA for Various Prokaryotes

measured by the t statistic. The sites with the highest t sores are unlikely to be from the background and are likely to be potential SD sites. For each possible k-mers, this approach takes into account both the absolute number N of upstream sequences containing s and the background distribution. It then calculates the unlikelihood of seeing N_s such occurrences if the sequences had been drawn at random from the background distribution. The random process used in this calculation is a first-order Markov chain based on the dinucelotide frequency of the sequences. The measure of unlikelihood used is based on the t statistics defined as follows: Let N be the number of upstream sequences that are input and P_s the probability that a single random upstream sequence contains at least one occurrence of S. Then NP_s is the expected number of input sequences containing s, and $NP_s(1 - P_s)^{1/2}$ is its standard deviation. The t score is defined as

$$t = \frac{(N_s - NP_s)}{\sqrt{NP_s(1 - P_s)}} \qquad (6.1)$$

The measure t_s is the number of standard deviations by which the observed value N_s exceeds expectations and is sometimes called the *normal deviate or deviation in standard units.* The measure τ_s is normalized to have zero and standard deviation of 1, making it suitable for comparing different motifs. The algorithm was run on 14 prokaryotic genomes. The motifs with the highest score showed a strong predominance of motifs complementary for the 3′ end of the genome's 16S rRNA.

6.6 Gene Annotation Methods

Pachter and colleagues [11] have provided a dictionary-based approach to gene annotation. The OWL and dBEST databases are used in this approach. A parse of the gene into introns and exons can be produced using the dynamic programming algorithm. Several scoring schemes for the exons are available. BLAST software [17] is often applied for the purposes of gene annotation and includes exon prediction and repeat finding. In the FLASH program, a hash table is used cleverly to keep tab of matches and positions of pairs of nucleotides in a database. The resulting information can be used to extract close matches to a given sequence. GenScan, Genie, GeneMark, fGENEH, and VEIL are statistical programs based on HMMs. GRAIL is based on nueral networks. In the PROCRUSTES program [12, 13], coding regions of a gene can be identified using protein sequences as targets. The INFO program is based on the idea of finding similarity to long stretches of a sequence in a protein database and then finding splice sites around those regions. These programs are becoming more important as the sizes of the protein and EST databases increase.

The distinct problems of sequence alignment and gene finding were treated with a unifying framework by Pachter and Lam [14]. They sought best alignment between two sequences while simultaneously annotating the regions.

The HMM developed is both a generalized HMM and a pair HMM. The former is used for gene finding, and the latter is used for sequence alignment. Such an HMM is called the *generalized pair hidden Markov model* (GPHMM, Fig. 6.2). These HMMs have been implemented successfully in GenScan and Genie. In ROSETTA, the steps of alignment and gene finding are separated. The alignment is equivalent to the Needleman-Wunsch dynamic programming algorithm discussed in Chap. 2. A program called *SLAM* was developed that implements these ideas and can be used to annotate syntenic sequences by finding coding exons and conserved noncoding sequences, or it can be used as a global alignment program that takes advantage of the biologic features of the sequences to improve the accuracy of the alignments.

There are two types of HMMs relevant to the problem: pair HMMs and generalized HMMs. Whereas one single output is generated by HMMs in each step, output in pairs were generated by PHMM, and GHMMs can generate output of different lengths (determined from a distribution) in each hidden state. The SLAM GPHMM is a combination of a PHMM and a GHMM. The main

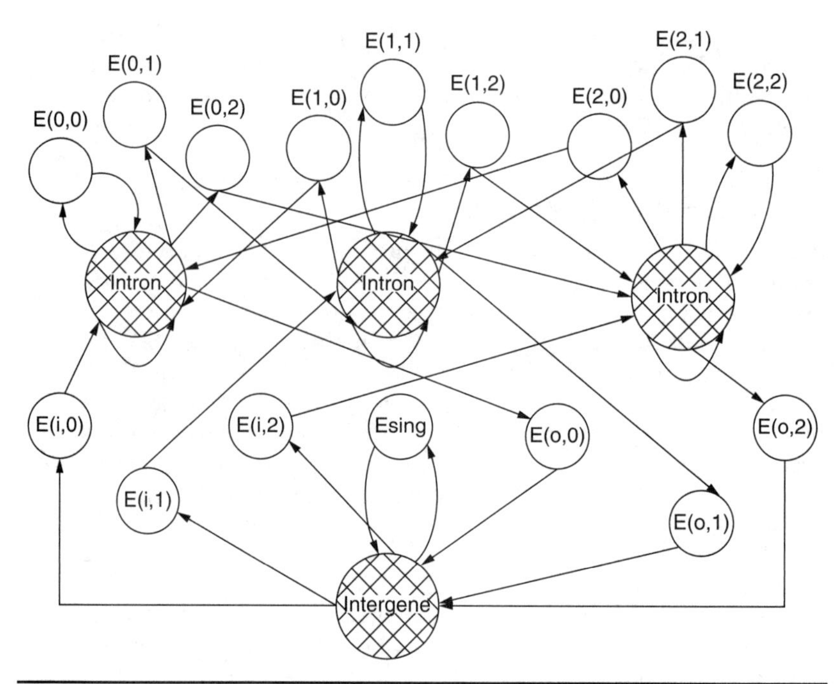

FIGURE 6.2 A GPHMM for alignment and prediction of exons using genomic DNA from two different organisms.

difference between the SLAM GPHMM model and previous HMM-based gene finders is in interpretation of the outputs of the states. The SLAM model is a PHMM, so the outputs in every state are aligned pairs of DNA bases. It is also a GHMM, meaning that a duration distribution is associated with each of the generalized states (the exon states in this case). The result of combining the two HMMs is that the generalized states now generate two sets of durations (or lengths) for the exons, one for each of the sequences.

A naive implementation of the GPHMM described has the drawback that the Viterbi algorithm has a running time on the order of $O(D^4N^2TU)$, where D is the maximum allowable length for an exon (on the order of thousands), N is the number of states, and T and U are the two sequence lengths. The memory requirements are on the order of NTU, which also scales as the product of the sequence lengths—ideally, we would like the problem to grow linearly in the length of the larger of the observation sequences. Because most alignments in the space of all possible alignments are very unlikely to be real, we adopted the approach of preprocessing to restrict the alignment search space to a set of more likely or more reasonable alignments. A set of possible alignments is called an *approximate alignment.*

Initial and transition probabilities, splice-site Variable Length Hidden Markov Models (VLMM) state duration distributions, and

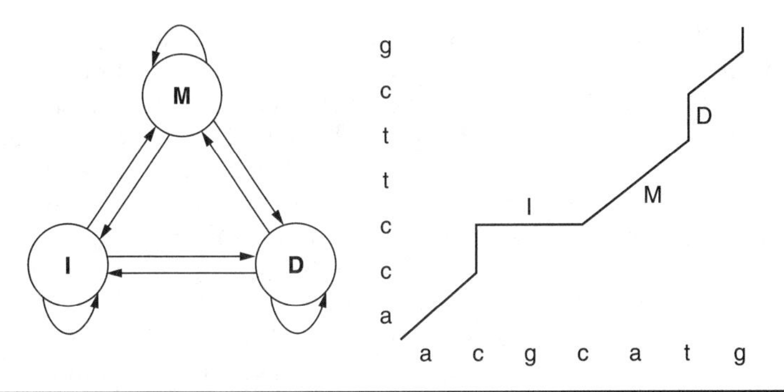

FIGURE 6.3 State space of a PHMM.

output probabilities were all obtained from appropriate training sets. Parameters were stratified by gene content. Parameter sets for different pairs of organisms can be obtained easily with the SLAM parameter toolbox, which parses GenBank files containing annotated sequences, generating all the required parameters.

Pachter and Lam [14] presented a solution to the problem of designing efficient search spaces for pair hidden Markov models (Fig. 6.3) that align biologic sequences by taking advantage of their associated features. Their approach leads to an optimization problem, for which was obtained a two-approximation algorithm, that is based on the construction of Manhattan networks, which are close relatives of *Steiner trees* (Fig. 6.4). The underlying theory was described and how their methods can be applied to alignment of DNA sequences in practice was shown, succesfully reducing the Viterbi algorithm search space of alignment PHMMs by three orders of magnitude.

The problem of designing efficient search spaces for pair hidden Markov model alignment algorithms that take advantage of the conservation patterns of biologic sequences was studied by Pachter and Lam [14]. This lead naturally to consideration of three computational problems, each of which has been studied individually in considerable detail but whose connection has not been well explored. The problems are

1. The alignment problem for biologic sequences
2. Development of efficient Viterbi algorithms for pair hidden Markov models
3. Construction of rectilinear Steiner networks

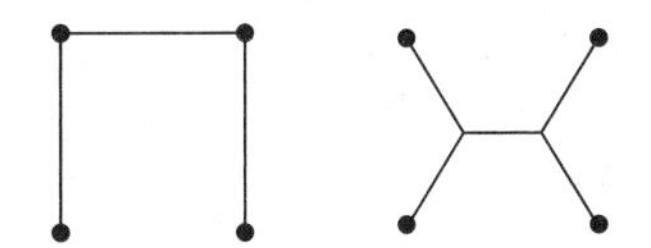

FIGURE 6.4 Minimum spanning tree and Steiner tree for a configuration of four points in a plane.

The first of these problems, the alignment of biologic sequences, is arguably the most successful application of computational biology to date. It remains a challenge to develop accurate alignment algorithms that are able to correctly align exons and other biologically interesting sequence features in large sequences. The improvements have been in the areas of speedup of alignment and more biologically meaningful alignments. For the first problem, there have been numerous investigations on how to normalize alignments taking lengths into account. "Optimal alignments" are very sensitive to the choice of parameters.

The connection between alignments, PHMMs, and Steiner trees raises a number of interesting questions that go beyond the immediate applications Pachter and Lam [14] have highlighted. Optimal networks for more complicated PHMMs, such as the GPHMMs, led to more complicated variants of the Manhattan network problem. Even the Manhattan network problem has not been "solved" in the sense that it is still unknown whether it is NP complete. The running time of the Pachter and Lam algorithm is $O(n^3)$ (worst case), where n is the number of highest-scoring pairs (HSPs), and the resulting PHMM algorithm for producing an alignment will run in time proportional to the size of the network, which in the worst case will be $O(n^2)$. It is possible to reduce the $O(n^3)$ running time for obtaining the network to $O(n \log n)$ at the expense of increasing the bound for the size of the network from twice optimal to four times optimal.

Even human genes can be predicted accurately, even in the case where only distantly related bacterial or yeast proteins are available. Gelfend and colleagues [13] achieved this by using a spliced alignment algorithm for similarity-based gene recognition. The spliced-arrangement algorithm provides 99 percent accurate recognition of human genes, i.e., average correlation coefficient of prediction 99 percent if a related mammalian protein is available. Sze and Pevzner [15] felt that although 99 percent accuarate gene predictions look like an acme of perfection, they are not sufficiently reliable for sequence annotation. They tried to develop an algorithm that either predicts an exon assembly with accuracy sufficient for sequence annotation or warns a biologist that accuracy of a prediction is insufficient and that further experimental work is required to complete the annotation. In this case, their goal is to provide biologists with accurate primer prediction. A 100 percent accurate gene prediction would greatly reduce experimental work on gene verification in large-scale sequencing projects.

Algorithms that provide a correct answer in some cases and have an option "No answer" in other cases are called *Las Vegas algorithms* in computer science. The term *Las Vegas* was introduced by Brassard and Bratley [16] to distinguish algorithms that reply correctly when they reply at all from Monte Carlo algorithms that occasionally make mistakes. Similar to many Las Vegas algorithms that benefit from the "No answer" option, Las Vegas algorithms for gene recognition use the "No answer" option to avoid unreliable predictions and benefit from reduction in experimental work in the correct answer cases.

Gelfend and colleagues [13] proposed a dynamic programming algorithm for the spliced alignment problem. The spliced alignment problem captures the major computational challenges of the similarity-search approach to exon assembly. However, in realistic situations, there exists important complications that do not seriously affect the running time of the algorithm, although they greatly increase the complexity of software implementation.

6.7 Secondary Structures of Proteins

The prediction of the secondary structure of a protein given the primary amino acid sequence distribution is one of the classic problems in bioinformatics. The secondary structure of a protein pertains to its three-dimensional stereochemical structure. As discussed in Chap. 1, this consists of α-helix, β-sheet, and γ-coil states (Fig. 6.5). With advances in sequencing technology, the number of

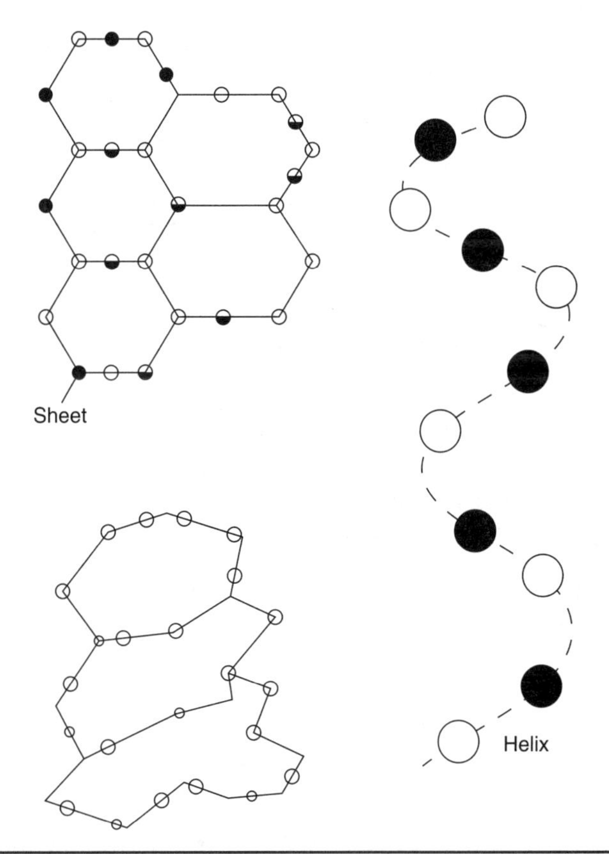

FIGURE 6.5 α-Helix, β-sheet, and γ-loop/coil states in the secondary structures of proteins.

proteomes completely sequenced increases rapidly with time. But the number of known secondary structures of proteins is fewer in number. It was realized that for five of six protein primary structures available, the secondary structures are not available. The experimental methods of obtaining the secondary structures of proteins, such as x-ray crystallography or nuclear magnetic resonance (NMR) spectroscopy, are expensive, not accurate enough, and time-consuming. A number of methods can be used to achieve this goal, such as using statistical information, physicochemical properties, sequence patterns and multilayered artificial neural networks, and/or incorporating evolutionary information from sequence families. Advanced neural network architectures have been suggested to predict the secondary structures of proteins.

The prediction problem is posed as that of predicting whether each residue in a protein forms part of an α-helix, β-sheet, or γ-loop/coil state. The secondary structure of a new protein can be found from another protein with a known secondary structure that is homologous. If no homologous proteins can be found, empirical correlations that have been developed between amino acids and local secondary structures of proteins can be used. These correlations were developed from known secondary structures at the time they were developed.

One such example is the Chou and Fasman rules [18]. Briefly stated, when four α-helix formers out of six residues or three β-sheet formers out of five residues are found clustered together in any native protein segment, the nucleation of these secondary structures begins and propagates in both directions until terminated by a sequence of tetrapeptides, designated as *breakers.* These rules were successful in locating 88 percent of α-helical and 95 percent of β-sheet regions, as well as correctly predicting 80 percent of the α-helical and 86 percent of the β-sheet residues in the 19 proteins evaluated. The accuracy of predicting the three conformational states for all residues is 77percent and shows great improvement over earlier prediction methods, which considered only the α-helix and γ-coil states.

Qian and Sejnowski [21] pioneered the use of neural networks to predict the secondary structures of proteins. They used 106 proteins from the Brookhaven National Laboratory with known secondary structures in their study. The performance measure of prediction of the secondary structures of proteins is the success rate Q_3. This is the percent of correctly predicted residues on all three types of secondary structures:

$$Q_3 = \frac{(P_\alpha + P_\beta + P_\gamma)}{N} \tag{6.2}$$

where N is the number of predicted values and P_α is the number of correctly predicted α-helix states, for example. Other performance measures such as correlation coefficients also can be used.

6.7.1 Neural Networks

Any reasonable function to any degree of required precision can be approximated by neural networks. An artificial neural network (ANN) is used in pattern recognition and knowledge acquisition and control. HMMs are closely related to or a special case of neural networks, stochastic grammars, and Bayesean networks. The structure of an ANN consists of a number of computing elements that resemble neurons and synapses of the human brain organized in a network [19–20]. Presently most of the implementations of neural networks are software-based. Interconnections higher than 2 units may lead to "higher order" or "sigma pi" networks. A number of important architectures can be recognized. These are (1) recurrent, (2) feed-forward, and (3) layered.

A *recurrent* architecture contains directed loops. An architecture devoid of directed loops is said to be *feed-forward.* Recurrent architectures are more complex. An architecture is *layered* if the units are partitioned into classes also called *layers,* and the connectivity patterns are defined between the classes. A feed-forward architecture is not necessarily layered. The number of layers is referred to as the *depth* of the network.

In the backpropagation model, the network is processed in three distinct steps. The first step is the forward sweep. In the forward sweep, the *input is given* to the input units. The output values of each unit are calculated and moved over the connections to the units in the next layer.

The units in the next layer receive the input from units in the previous layer. The output values of the units then are calculated and passed to the units in the next layer, and so on. The next step is *error calculation.* In this step, the values of the output units are compared with the desired output (*teaching*). If the difference between the actual output and the teaching is within the acceptable error range, then learning is successful. If the difference is not within an acceptable range, then an error value is calculated, and learning is unsuccessful.

The third step is *backpropagation of the error value.* In this step, if learning is unsuccessful, then the error value is propagated backward through the net. The weights of the connections between the units are adjusted to minimize the error value. The main objective of this step is to close the gap between the actual output and the desired output. These three steps are repeated until learning is successful.

The behavior of each unit in time can be described using either *differential equations* or *discrete update equations.* Typically, a unit i receives a total input X_i from the units connected to it and then produces an output $Y_i - f(x_i)$, where f is the transfer function of the unit. In general, all units in the same layer have the same transfer function, and the total input is a weighted sum of incoming outputs from the previous layer so that

$$X_i = \sum_{j \varepsilon N(i)} W_{ij} y_j + W_i \tag{6.3}$$

$$Y_i = f(x_i) = f\left(\sum_{j \varepsilon N(i)} W_{ij} Y_j + W_i\right) \tag{6.4}$$

where W_i is called the *bias* or *threshold* of the unit. W_{ij} and W_i are the parameters of the neural network (NN). Other parameters such as time constants, gains, and delays are possible. Usually, the total number of parameters is determined by the number of layers, the number of units per layer, and the connectivity between layers. The NN is said to be *fully connected* when each unit in one layer is connected to every unit in the following layer.

A normalized exponential unit is used to compute the probability of an event with n possible outcomes, such as classification into one of n possible classes. Let j run over a group of n output units, computing the n membership probabilities, and x_j denote the total input provided by the rest of the NN into each output unit. Then the final activity y_i of each output unit is given by

$$y_i = \frac{e^{-x_i}}{\sum_{k-1}^{n} e^{-x_k}} \tag{6.5}$$

$$\sum_{i=1}^{n} y_i = 1 \tag{6.6}$$

When $n = 2$, the normalized exponential is equivalent to a logistic function via a simple transformation:

$$y_i = \frac{e^{-x_1}}{\left[e^{-x_1} + e^{-x_2}\right]} \tag{6.7}$$

Any probability distribution P_i $(1 \leq j \leq m)$ can be represented in normalized exponential from a set of variables x_j $(1 \leq j \leq m)$:

$$P_i = \frac{e^{-x_i}}{\sum_{k=1}^{m} e^{-x_k}} \tag{6.8}$$

as long as $m \geq n$. This can be done in infinitely many ways by fixing a positive constant k and letting $X_i = \log(p_i) + k_j$, for $i = 1, \ldots, n$. If $m < n$, there is no exact solution, unless the p_i assumes only m distinct values at most.

The radial basis function (RBF) is another type of widely used function. Here, f is a bell-shaped function like a Gaussian function. Each RBF unit i has a reference input x_i, and f operates in the distance $d(x_i^{\infty}, x_i)$ measured with respect to some metric $y_i = f(d)$. In a spatial problem, d is usually the Euclidean distance.

Thus some of the important features of the ANN model depend on the task at hand. The process of computing approximate weights is called *learning* or *training* in the ANN paradigm. There are many ANN learning algorithms that employ the principles just described. In general, ANN learning algorithms are classified by either the tasks to be achieved or the methodologies to achieve a task: (1) autoassociation, (2) classification, (3) heteroassociation, and (4) regularity detection. ANN learning algorithms are divided into two classes: (1) supervised and (2) unsupervised.

In *supervised* learning, a network is given an input along with its desired output. On the other hand, a network in *unsupervised* learning is given only an input. After each presentation of an input, the performance is measured to tell how the network is doing. A network is expected to self-organize information by using the performance measure as guidance. Algorithms in these two categories are further divided into two groups on the basis of the input formats: binary or continuous-valued input. Taxonomy of the ANN algorithm is given in Fig. 6.6.

In Qian and Sejnowski's work [21], orthogonal encoding was used as input, with the alphabet corresponding to 20 different amino acids. A terminator symbol to encode the N and C terminals is also included, making the alphabet size 21. The input window had an optimal size, rougly of 13 amino acids. The input layer has 21 × 13 = 273 units. The typical size of the hidden layer consists of 40 sigmoidal

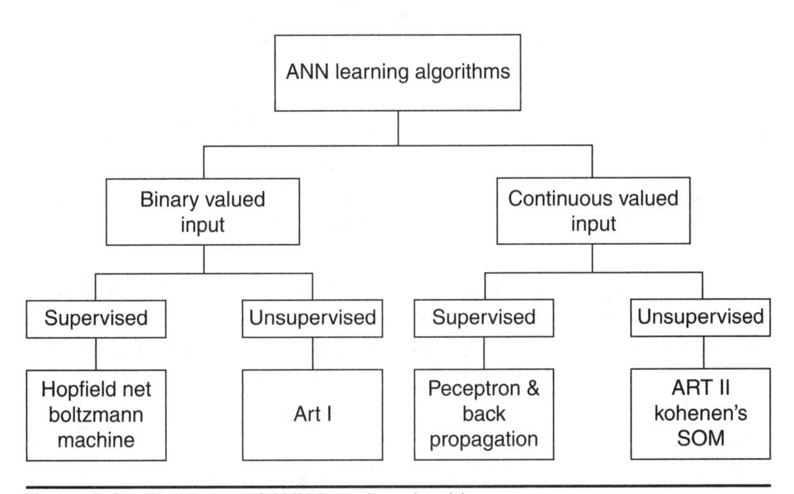

Figure 6.6 Taxonomy of ANN learning algorithms.

units. The number of parameters used for this architecture was 11,083. Three sigmoidal units were present in the output layer. The α-helix, β-sheet, and γ-coil structures were encoded. The classification was represented in the output. The networks were initialized using random uniform weights. Subsequently, the network was trained using backpropagation with the LMS error function. A more appropriate use would be the normalized exponential output layer with the relative entropy as error function. The training set is 20,000 residues in length. These are extracted from the Protein Data Bank (PDB). Many protein structures have been solved by experimentation. Peformance oscillations associated with the use of contiguous windows is avoided by using a random order of presentation when training on protein sequences. The performance increases from a 33 percent choice level to 60 percent using this architecture. Beyond this point, overfitting begins. If there is an imbalance in the amount of helix, sheet, and coil proportions from the usual number of 0.3/0.2/0.5, percentages of correctly predicted window configurations can be pair indicators of the predictive performance. The correlation coefficient can be used and is found to be a better measure of performance.

6.7.2 PHD Architecture of Rost and Sander

The most important performance improvement has been achieved by the work of Rost and Sander [22]. Their work resulted in the creation of a Profilenetwork HeiDelberg (PHD) server. The PHD method reached a performance level of 74 percent on an unknown test set. A reduction of the database of three-dimensional protein structures to a sequence of secondary structure patterns is achieved with statistical and neural network methods (Fig. 6.7). A sequence profile of a protein family, rather than just a single sequence, is used as input to a neural network for structure prediction. Each sequence position is represented by the amino acid residue frequencies derived from multiple sequence alignments as taken from the Homology-Derived Structure of Proteins (HSSP) database. The residue frequencies for the 20 residue types are represented by 3 bits each (or by one real number). To code the N- and C-terminal ends adds an additional 3 bits (or one real number). The 63 bits originating from one sequence position are mapped onto 63 (21 for real numbers) input units of the neural network

A window of 13 sequence positions thus corresponds to 819 (273) input units. The input signal is propagated through the network with one input layer, one hidden layer, and one output layer. The output layer has three units corresponding to the three secondary-structure states, *helix, strand,* and *loop,* at the central position of the input sequence window. Output values are between 0 and 1. The experimentally observed secondary structure states are encoded as 1, 0, 0 for helix, 0, 1, 0 for strand, and 0, 0, 1 for loop. The error function

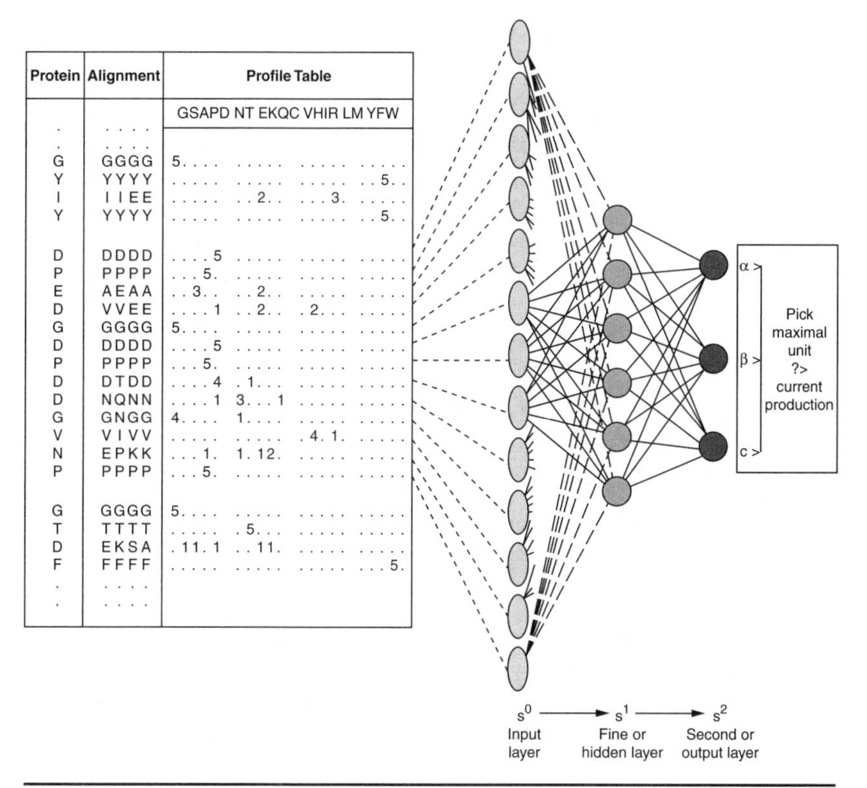

FIGURE 6.7 Neural network architecture for secondary structure prediction [20].

to be minimized in training is the sum over the squared difference between current output and target output values. The net cascade consists of the first network (sequence-to-structure), followed by a second network (structure-to-structure) to learn structural context (not shown). Input to the second network is the three output real numbers for helix, strand, and loop from the first network plus a fourth spacer unit for each position in a 17-residue window. From the $17 \times (3 + 1) = 68$ input nodes, the signal is propagated via a hidden layer to three output nodes for helix, strand, and loop, as in the first network. In prediction mode, a 13-residue sequence window is presented to the network, and the secondary-structure state of the central residue is chosen, according to the output unit with the largest signal.

More balanced predictions are achieved by the PHD server. Better accuracy is achieved from multiple sequence alignments and using evolutionary information, better prediction of the sheets achieved by balanced training, and better prediction of helix and strand lengths using structure context training. The neural network was tested on a database of 130 representative proteins with known structure. The overall improvement shown by this method is as follows:

1. The overall accuracy is 69.7 percent, three percentage points above the highest value reported so far (66.4 percent). The improvement is six percentage points relative to the best classical method tested on our database (63.4 percent, ALB).
2. Accuracy is well balanced at 70 percent helix and 64 percent strand, measured as the percentage "correct of observed." The percentages "correct of predicted," i.e., the probability of correct prediction, given a residue predicted in a particular state, are 72 percent helix and 57 percent strand.
3. The length distribution of segments is more proteinlike. Unfortunately, the length distribution is not generally given in the literature, but most methods are inferior in this regard.

There are two practical limitations to this method. Most of the gains of the PHD architecture are lost when no sequence homologues are available. These are not valid for membrane proteins and other nonglobular or water-insoluble proteins. Another limitation of the method is its limited goal. This method is useful in practice, such as for the planning of point mutations experiments, for the selection of antigenic peptides, or for identification of the structural class of a protein. Evolution may be the key to the puzzle posed of protein folding.

6.7.3 Ensemble Method of Riis and Krough [23]

The work of Riis and Krogh [23] addresses the overfitting problem by redesign of the NN architecture. Their approach has four main components:

1. The larger input (13×21) caused a large number of parameters to be cut using an adaptive encoding of amino acids. An optimal and compressed representation of the input letters is found by the NN. This technique is also referred to as *weight sharing*.
2. A different network is designed for each of the three classes. A three-residue periodicity between the first and second hidden layers was built for the case of α-helices. The second hidden layer is fully interconnected to the second hidden layer, which has a typical size of 5–10 units. A typical α-helix network contains 160 adjustable parameters. About 300–500 adjustable parameters is contained in a β-sheet or γ-coil network. Balanced training sets were used with the same number of positive and negative examples on training these architectures in isolation.
3. They use ensembles of networks and filtering to improve the prediction.

Five different networks are used for each type of structure at each position. The combining network takes a window of 15 consecutive single predictions. The input layer to the combining network has size 225. To keep the number of parameters within a reasonable range, the connectivity is restricted by having one hidden unit per position and per ensemble class. The input is locally connected to a hidden layer with 45 units. The hidden layer is fully connected to three normalized exponential output units. The error measure used is the negative log likelihood, which in this case is the relative entropy between the true assignment and the predicted probabilities.

Riis and Krogh use a weighting scheme along with the multiple alignments. The maximum entropy weighting scheme is used. Averaging operates on soft probability values produced by single-sequence prediction algorithm. A small network with a single hidden layer of 5 units is then applied to filter the consensus secondary-structure prediction derived using multiple alignment. Coil regions are less conserved and hence have higher per-column entropy in a multiple alignment. The performance is improved with an overall accuracy of 71.5 percent and with better correlation coefficients. It is comparable with the method of Rost and Sander [22]. The consensus is that there appears an upper bound on accuracy of slightly above 70 to 75 percent on any prediction method based on local information only.

6.7.4 Protein Secondary Structure Using HMMs

HMMSTR is a model for general protein sequences based on the I-sites library of sequence-structure motifs. Unlike the linear hidden Markov models used to model individual protein families, HMMSTR has a highly *branched topology* and captures recurrent local features of protein sequences and structures that transcend protein family boundaries. The model extends the I-sites library by describing the adjacencies of different sequence-structure motifs as observed in the PDB and by representing overlapping motifs in a much more compact form, achieving a great reduction in parameters. The HMM attributes a considerably higher probability to coding sequence than does an equivalent dipeptide model and predicts secondary structure with an accuracy of 74.3 percent, backbone torsion angles better than any previously reported method, and the structural context of β-strands and turns with an accuracy that should be useful for tertiary-structure prediction.

Helix-capping motifs are specific patterns of hydrogen bonding and hydrophobic interactions found at or near the ends of helices in both proteins and peptides. In an α-helix, the first four >N–H groups and last four >C=O groups necessarily lack intrahelical hydrogen bonds. Instead, such groups are often capped by alternative hydrogen-bond partners. A hydrophobic interaction that straddles the helix terminus is always associated with hydrogen-bonded capping. From a global survey among proteins of known structure, seven distinct capping motifs are identified—three at the helix N terminus and four at the C terminus. The consensus sequence patterns of these seven motifs, together with results from simple molecular modeling, are used to formulate useful rules of thumb for helix termination. Finally, we examine the role of helix capping as a bridge linking the conformation of secondary structure to supersecondary structure.

A novel method to model and predict the location and orientation of α-helices in membrane-spanning proteins was presented by Sonnhammer et. al. [24]. It is based on an HMM with an architecture that corresponds closely to the biologic system. The model is cyclic with seven types of states for *helix core, helix caps* on either side, *loop* on the cytoplasmic side, two loops for the noncytoplasmic side, and a globular domain state in the middle of each loop. The two-loop paths on the noncytoplasmic side are used to model short and long loops separately, which corresponds biologically to the two known different membrane insertion mechanisms. The close mapping between the biologic and computational states allows us to infer which parts of the model architecture are important to capture the information that encodes the membrane topology and to gain a better understanding of the mechanisms and constraints involved. Models were estimated both by maximum likelihood and a discriminative method, and a method for reassignment of the membrane helix boundaries was developed. In a cross-validated test on single sequences, our transmembrane HMM (TMHMM) correctly predicted the entire topology for 77 percent of the sequences in a standard data set of 83 proteins with known topology. The same accuracy was achieved on a larger data set of 160 proteins. These results compare favorably with existing methods.

Secondary structures such as helix, sheet, and coil can be learned by HMMs, and these HMMs are applied to new sequences whose structures are unknown. The output probabilities from the HMMs are used to predict the secondary structures of the sequences. Sonhammer et. al. [24] tested this prediction system on approximately 100 sequences from a public database (Brookhaven PDB). Although the implementation was "without grammar" (no rule for the appearance patterns of secondary structure), the result was reasonable.

6.7.5 DAG RNNs: Directed Acyclic Graphs and Recursive NN Architecture and 3D Protein Structure Prediction

Baldi and Pollastri [25] tackled protein secondary-structure prediction, which is one of the open problems in bioinformatics, by using DAG RNNs. DAG-RNNs are directed acyclic graphs and recursive neural network architectures. Protein structures are invariant after undergoing translations and rotations. This was included in the approach of Baldi and Pollastri. They proposed a machine-learning pipeline that consisted of three steps:

1. Representation of a given domain using directed acyclic graphs
2. Parameterizaion of the relationship between each variable and its parent variables by feedforward neural networks
3. Application of weight sharing within appropriate subsets of DAG
4. Connections to capture stationarity and control model complexity.

It is a three-step process. The specific class of DAG-RNN architectures is derived from lattices, trees, and other structural graphs. The overall models resulting are probabilistic. The internal deterministic dynamics allows efficient propagation of information as well as training by gradient descent to tackle large-scale problems.

All the weights of the BRNN architecture, including the weights in the recurrent wheels, can be trained in a supervised fashion using a generalized form of gradient descent derived by unfolding the wheels in space. BRNN architectural variations are obtained by changing the size of the input windows, the size of the window of hidden states that directly influences the output, the number and size of the hidden layers in each network, and so forth. Thus BRNN architectures have been used in the first state of the prediction pipeline, giving rise to the state-of-art predictors for secondary structure, solvent accessibility, and coordination number.

6.7.6 Annotate Subcellular Localization for Protein Structure

LOC3D [26], at http://cubic.bioc.columbia.edu/db/Loc3d, is both a weekly updated database and a Web server with predictions of subcellular localization for eukaryotic proteins of known 3D structure. Neural networks are used in the prediction of localization. The LOC3D database currently contains predictions for greater than 8700 eukaryotic protein chains taken from the PDB. The Web server can be used to predict subcellular localization for protein for which only a predicted structure is available from threading servers.

The native subcellular localization of a protein is important for understanding gene/protein function. Aberrant subcellular localization of proteins has been observed in the cells of patients with several disease, such as cancer and Alzheimer's disease. Attempts to predict subcellular localization either experimentally or computationally have become one of the central problems in bioinformatics.

Subcellular localization is annotated for not many of the proteins deposited in the PDB. The LOC3D database is the first comprehensive database of predicted and inferred subcellular localizations for proteins of known structure. The LOC3D database can be useful in

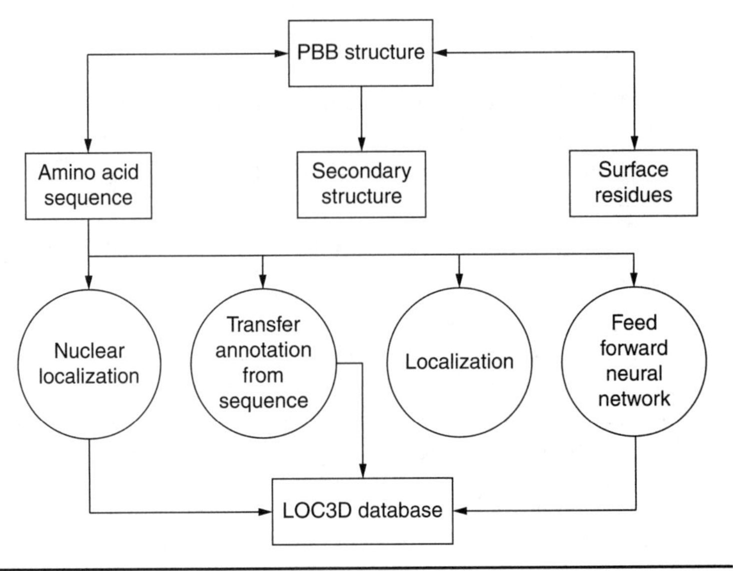

FIGURE 6.8 LOC3D database.

complementing functional information for proteins from domain databases such as SMART and PFam and functional site resources such as ELM, Protfun, and PROSITE. The LOC3D has four different paths to annotate subcellular localization (Fig. 6.8).

These are (1) Predict NLS, (2) Lochomi, (3) Lockey, and (4) Loc3DIni. From the query PDB structure, the amino acid sequence, three-state secondary structure, and solvent-accessible surface residues of the protein are extracted. In Predict NLS, the amino acid sequence is scanned for nuclear localization signals. In Lochomi, the sequence is first aligned using PSI-BLAST to a localized annotated database of proteins. If any sequence homologues are discovered, subcellular localization annotation is transferred from the homologue. Lockey infers subcellular localization based on keyword entries. These three programs are based solely on the amino acid sequence of the protein and do not use any structural information. Subcellular localization is predicted by a system of neural networks in LOC3 Dini. The NNs are trained in a number of global features such as amino acid composition, secondary structure composition, and surface residue composition. The final localization annotation in the LOC3D database is taken from individual methods.

LOC3 Dini is a prediction system that predicts subcellular localization from sequence and structure using NNs. Subcellular localization is predicted using a number of global features of protein sequence and structure. The LOC3 Dini system consists of three layers and sorts proteins into one of four localization classes: extracellular, cytoplasmic, nuclear, and mitochondrial.

1.0 The first layer consists of four dedicated of neural networks that use particular features from protein sequences, alignments, and secondary structures to presort proteins into L/not L, where L= cytoplasmic, nuclear, extracellular, or mitochondrial. The features used include amino acid composition, composition of surface-accessible residues, and composition of amino acid residues in one of the three secondary structure states (helix, sheet, or coil). Evolutionary information was incorporated by replacing the amino acid with profile-based amino acid composition.

2.0 The second layer consists of neural networks combining output from networks trained on different input features. The third layer uses a simple jury decision to assign one of four localization states to each protein. Major sources of improvement over publically available methods originating from using (1) secondary structure information, (2) solvent accessibility, and (3) evolutionary information from sequence profiles as input to the neural networks. The final four-state classification accuracy of the system was 76.5 percent. This is greater than 10 percentage points higher than systems using only amino acid composition.

Summary

The relative entropy site-selection problem is NP complete. Hertz and Stormo presented a greedy approach to develop an efficient algorithm for the relative entropy site-selection problem. Profiles with lower relative entropy scores than d will be discarded. An iterative approach is used in the Gibbs sampling method for the solution to the relative entropy site-selection problem. The maximum-subsequence problem is a corollary of the problem of finding the coding regions in DNA. Bates and Constable suggested an algorithm that solves the maximum-subsequence problem using the principle of recursion. Sharma has shown that maximum subsequence is found in the deepest branch of the binomial heap. The time taken is $O(n)$. All subsequences are available in the binomial heap. The interpolated Markov model (IMM) is implanted in GLIMMER. Markov models from first to eighth order are used in this procedure. For prediction of the translation start-site codon, the SD sites problem can be solved by using the t statistic and measurement of statistical significance. OWL and dBEST databases are used in the dictionary-based approach to gene annotation. The problems of sequence alignment and gene finding were treated with a unifying framework by Pachter and Lam. The GPHMM is both a generalized HMM and a pair HMM. Manhattan networks and Steiner trees are used in the optimization problem of designing efficient search spaces for HMMs. The Viterbi algorithm search space was reduced from $O(D^4N^2TU)$ by three orders of magnitude. The Pachter algorithm consumes $O(n^3)$, where n is the number of highest-scoring pairs. It can be reduced to $O[n \lg(n)]$ at the expense of increasing the bound for the size of the network from twice optimal to four times optimal. The spliced alignment algorithm for similarity-based gene recognition is reviewed. The Las Vegas algorithm provides the option of "No answer."

The protein secondary structure—α-helix, β-sheet, or γ-coil—given the primary sequence of the protein can be determined using neural networks. Although the number of proteomes sequenced increases rapidly with time, the number of known secondary structures is not commensurate with growth of the proteomes. Empirical correlations have been developed between protein primary structure and protein secondary structure, such as in the Chou and Fasman rules. Pioneering work on predicting protein secondary structure using neural networks was that of Qian and Sejnowski. They used 106 proteins in their study. The fundamentals of neural networks were reviewed. Rost and Sander came up with the most important performance improvement by using evolutionary conformation in their PHD server to predict protein secondary structure. The NN had 819 input units, one hidden layer, and one output layer. The output layer had 3 units. The work of Riis and Krogh is a redesign of NN architecture to solve the overfitting problem. The NN was designed with a larger input layer, balanced training sets, 160 adjustable parameters for the α-helix network, 300 to 500 adjustable parameters contained in the β-sheet or γ-coil network, ensembles of networks, and filtering for improved prediction. HMMs can be used to predict protein secondary structure. Baldi and Pollastri used DAG-RNNs for protein secondary-structure prediction. It is a three- step process. Weights of the BRNN architecture, including the weights in the recurrent wheels, can be trained in a supervised fashion. Native subcellular localization of a protein is important for an understanding of gene/protein function.

References

[1] T. Akutsu, H. Asimira, and S. Shimozono, "An approximation algorithm for local multiple alignment." In *RECOMB 2000, Proceedings of the 4th Annual International Conference on Computational Molecular Biology.* Tokyo, Japan: ACM, 2000.

[2] G. T. Hertz and G. D. Stormo, "Identifying DNA and protein patterns with statistically significant alignments of multiple sequences." *Bioinformatics.* 15 (1999), 563–577.

[3] C. E. Lawrence, S. F. Altschul, M. S. Boguski, J. S. Liu, A. F. Nuewald and J. c. Wooton, "Detecting Subtle Sequence Signals- A Gibbs Sampling Strategy for Multiple Alignment", *Science.* 262, 8, (1993), 208–214.

[4] J. C. Bates and R. L. Constable, "Proofs as programs," *ACM Trans. Programming Languages and Systema* 7 (1985), 113–136.

[5] K. R. Sharma, "Binomial tree representation for the maximum subsequence problem," Central Regional Meeting of the ACS, Pittsburgh, PA, 2003.

[6] K. R. Sharma, "Binomial tree representation of maximum increasing subsequence problem," 41st Annual Convention of Chemists Meeting, Delhi University, New Delhi, India, 2004.

[7] K. R. Sharma, "New data structures in bioinformatics to improve search cost," AIChE Spring Meeting, New Orleans, LA, 2004.
[8] A. L. Delcher, D. Harmon, S. Kasif, et al., "Improved microbial gene identification with GLIMMER." *Nucleic Acids Res.* 27 (1999), 4636–4641.
[9] J. Shine and L. Dalgorno, "The 3′ terminal sequence of *Escherichia coli* 16S ribosomal RNA: Complementary to nonsense triplets and ribosome binding sites," *Proc. Natl. Acad. Sci. U.S.A.* 71 (1974), 1342–1346.
[10] M. Tompa, "An exact method for finding short motifs in sequences with application to the ribosome binding site problem." In *Proceedings of 7th International Conference an Intelligent Systems for Molecular Biology.* Heidelberg: AAAI Press, 1999, pp, 262–271.
[11] L. Pachter, S. Batzoglou, V. I. Spitkovshy, et al., "A dictionary based approach for gene annotation." In *Proceedings of the Third Annual International Conference on Compuational Molecular Biology.* Cologne: ACM, 1999, pp. 285–294.
[12] L. Pachter, M. Alexanderssan, and S. Cawley, "Applications of generalized pair HMM to alignment and gene finding problem." In *Proceedings 5th Annual International Conference on Computational Biology.* Montreal, Canada: ACM, 2001, pp. 241–248.
[13] M. S. Gelfend, A. Mironov, and P. A. Pevzner, "Gene recognition via spliced alignment," *Proc. Natl. Acad. Sci. U.S.A.* 93 (1996), 9061–9066.
[14] L. Pachter and F. Lam, "Picking alignments from Steiner trees." In *Proceedings of the Sixth Annual International Conference on Computationsl Biology.* New York, NY: ACM, 2002, pp. 246–253.
[15] S. H. Sze and P. A. Pevzner, "Las Vegas algorithms for gene recognition: Suboptimal and error-tolerant spliced alignment." In *Proceedings of 1st Annual International Conference on Computational Molecular Biology.* Sante Fe, NM: ACM, 1997, pp. 300–309.
[16] G. Brassard and P. Bratley, *Fundamentals of Algorithmica.* Englewood Cliffs, NJ: Prentice-Hall, 1996.
[17] S. F. Altschul, W. Gish, W. Miller, et al., "Basic local alignment search tool," *J. Mol. Biol.* 215 (1990), 400–403.
[18] P. Y. Chou and G. D. Fasman, "Prediction of protein conformation," *Biochemistry.* 13 (1974), 222–245.
[19] J. L. McClellland, D. E. Rumelhart, and G. E. Hilton, "The appeal of parallel distributed processing." In D. E. Rumelhart and J. C. McCelelland (eds.), *Parallel Distributed Processing: Explorations in the Microstructure of Cognition.* Boston: MIT Press, 1986.
[20] P. Baldi and S. Brunak, *Bioinformatics: The Machine Learning Approach.* Boston: MIT Press, 2001.
[21] N. Qian and T. J. Sejnowski, "Predicting the secondary structure of globular proteins using neural network models," *J. Mol. Biol.* 202 (1988), 865–884.
[22] B. Rost and C. Sander, "Improved prediction of protein secondary structure by use of sequence profiles and neural networks," *Proc. Natl. Acad. Sci. U.S.A.* 90 (1993), 7558–7562.
[23] S. K. Riis and A. Krogh, "Improving prediction of protein secondary structure using structured neural networks and multiple sequence alignments," *J. Comput. Biol.* 3 (1996), 163–183.
[24] E. L. Sonnhammer, G. von Heijne, and A. Krogh, "A hidden Markov model to predict trans-membrane helices." In *Proceedings of the 6th International Conference on Intelligent Systems for Molecular Biology*, Montreal, Canada: Quebec,Vol. 6., 1998, pp. 175–182.
[25] P. Baldi and G. Pollastri, "The principled design of large-scale recursive neural network architectures—DAG-RNNs—and the protein structure prediction problem," *J. Machine Learning Res.* 4 (2003), 575–602.
[26] R. Nair and B. Rost, "Loc3D: Annotate subcellular localization for protein structures," *Nucleic Acids Res.* 31 (2003), 3337–3340.

Exercises

1.0 Show that the longest increasing subsequence for $S = \{11, 17, 5, 8, 6, 4, 7, 12, 3\}$ can be found by keeping track of its indices in $O[n \lg(n)]$ time.

S_i	11	17	5	8	6	4	7	12	3
i	1	2	3	4	5	6	7	8	9
Length of h	1	2	1	2	2	1	3	4	1

Predecessor 1 3 3

So the maximum increasing subsequence $\{5, 6, 7, 12\}$.

2.0 Develop the relative entropy of the site-selection procedure introduced by Hertz and Stormo for the following four sequences:

TGCAATA

TT ATCGG

CAATA AA

TGTGCGC

3.0 What is the implication of negative number occurring in the binomial heap of all maximum increasing subsequences problem?

4.0 Define the mutual information of three pairs (x, y, z) and random variables.

5.0 Define score as $\log_4$ CK/CB. Would the choice of base 4 for DNA be a better representation than the base 2?

6.0 Discuss the limitation of the recursion principle for the maximum subsequence problem. Start with one element at different locations in the sequence. Do you get the same answer?

7.0 Show how the interpolated context model tree will look like for three pairs.

8.0 What is the reason for the value 0.5 used as a cutoff for the different definitions of λ?

9.0 What if the overlap of A and B starts at the same location? What remains the same?

10.0 Under what circumstances will the t statistics be less desirable. How about a periodic probability distribution? Do you need a generalized normal distribution? If so, what is the equivalent of the z score?

11.0 What if in the Steiner problem the points fall on a square grid or a rectangular grid?

12.0 Construct an alignment and state space for the following sequences:

G CGATAT

C GGTTAG

13.0 Can you have a curvilinear Steiner tree? When?

14.0 Discus the merits and demerits of glueing in space requirements reduction.

15.0 Discuss the spliced alignment with nucleotide comparison with Δ mismatch = –2, Δ match = 2, Δ indel = 0.

16.0 When the databases reach the level of petabytes, which system will be preferred? What are ORACLE Company's latest forays in this area? When is a supercomputer apt for the occasion?

17.0 Some investigators use neural networks to predict secondary structures, whereas others use HMMs. What is the difference in approach? What is predicted, and what is learned?

18.0 The Ramachandran plot results in α-helices and β-sheets given the dihedral angles. The primary sequence of the protein is entered as input into a neural network, and the output is the geometric secondary structure of the protein. What are the similarities of the two methods? What are the differences?

19.0 Can you design a neural network to predict the tertiary structure of a protein?

20.0 Show using a neat schematic what would be the strategy to obtain the protein secondary structure from the nucleic acid sequence distribution.

21.0 Can two protein structures be aligned if their three-dimensional structures are known? Will such an alignment be easier to achive given that there are only three possibilities, α-helix, β-sheet, and γ-coil, compared with the primary sequence distribution with 20 different amino acids as the alphabet.

22.0 What is the relation between disease and protein secondary structure?

23.0 Can drugs be designed using protein secondary structure?

24.0 There are experimental methods to obtain the protein secondary structure. Which is preferred, the experimental methods or the neural networks? Why?

25.0 Are computer calculations needed to calculate the protein secondary structure using the Chou Fasman rules?

26.0 Is there a secondary structure to DNA as in protein? Why?

27.0 Design the architecture of an NN to learn the standard genetic code. How many output units, input units, and intermediate units are needed?

28.0 Why is backpropagation preferred in Exercise 27.0?

29.0 How would your design change if reverse transcription, i.e., from RNA to DNA, is sought? Can you start from a protein?

30.0 Devise an NN to learn a DNA sequence and output a suffix tree.

31.0 Design an HMM to learn the standard genetic code.

32.0 Devise an NN to learn a polypeptide sequence. Compare this with an HMM that can perform the same task.

33.0 Can a phylogenetic tree be constructed using a neural network?

34.0 Can neural networks be used to obtain a multiple-sequence alignment?

35.0 Can a neural network be designed to replace the effect of the affine gap penalty model introduced in Chap. 2?

36.0 Can you retrieve a sequence alignment using a neural network?

37.0 Can protein family classification be achieved using neural networks? Provide a sketch of your strategy.

38.0 Can you achieve the same results as the wheel HMM shown in Fig. 5.13 using a neural network? How many hidden layers are needed to find the periodicity in DNA?

39.0 Why are HMMs preferred to neural networks in database mining?

40.0 Why are HMMs preferred to predict the Chargaff parity rules compared with neural networks?

41.0 Can local alignment of two sequences be performed by neural networks?

42.0 What is a structural alignment? Design a neural network to achieve the structural alignment.

43.0 When are "sigma-pi" networks used?

44.0 *PHDsec.* The secondary structures of proteins are predicted by PHDsec using neural networks. The Internet link to the Predict protein site is www.predictprotein.org. Show that the neural network used by this software is of the feed-forward type. Discuss the accuracy level reached, the architecture of the ANN (e.g., number of hidden layers), and the weighting functions used in this approach.

45.0 Why is a feed-forward architecture used in PHDsec discussed in Exercise 44.0? What tasks are performed by the NN?

46.0 *DISULFIND.* The disulfide bridges in the microstructure of a polypeptide are predicted in DISULFIND. The Internet hotlink to this site is http://cassandra.dsi.unifi.it/cysteines/index.html. Show that the architecture of the neural network used is bidirectional and recurrent. What tasks are performed by the NN?

47.0 What is the accuracy reached in DISULFIND discussed in Exercise 46.0? Discuss the architecture of the NN used.

48.0 *SAM-T99.* The secondary structures of proteins can be predicted using HMMs in SAM-T99. This was developed at the University of California, Santa Cruz, and the Internet hotlink is www.soe.ucsc.edu/compbio/HMM-apps. What order of HMM is used? What is achieved in the hidden layers?

49.0 What is the accuracy level reached in SAM-T99 discussed in Exercise 48.0? What can you suggest to improve the time taken and space needed using HMMs?

50.0 *JPRED.* Protein secondary structure and solvent accessibility are predicted by JPRED. Show that the NN used has three layers and is fully connected. Profiles generated by HMMs and PSI-BLAST are used by the software. Discuss the accuracy reached in this approach.

51.0 Discuss the architecture of the NN used in JPRED in Exercise 50.0.

52.0 Can neural networks be designed to measure the periodicity in protein primary structure? If so, how does this help in elucidation of protein secondary structure?

53.0 The forward algorithm was developed to solve the evaluation problem with increased time efficiency and storage needs. In a similar fashion, what ought to be the strategy for neural networks and the primary structures of proteins?

54.0 What is the equivalent of the Viterbi algorithm to the decoding problem in the construction of an HMM to represent sequences to the neural network representation of protein primary and secondary structures?

PART 3

Measurement Techniques

CHAPTER 7
Biochips

Objectives

The objectives of this chapter are to

- Learn what a biochip is and prepare a microarray slide.
- Draw parallels between biochips and the microprocessor industry.
- Learn the five steps in the microarray cycle.
- Employ microarray detection using a confocal scanning microscope.
- Know the criteria for microarray surfaces.
- Understand optimal probe, optimal target concentrations.
- Learn phosphoramadite synthesis
- Be familiar with the three manufacturing methods for ink-jet printing, mechanical microspotting, and photolithography.
- Be familiar with *t* test statistics and normalization of gene expression data.
- Read the case study in the detection of cancer.

7.1 Introduction

Microarrays can be used to understand disease states by enabling the analysis of gene expression patterns, sequence variation, and other biochemical reactions. According to Schena [1–4], in another 50 years, human disease will be eradicated. Although Pauling was the first to correlate gene mutations, altered proteins, and disease, the biochip technique that is rapidly gaining worldwide acceptance can lead to a better understanding of disease mechanisms and suitable drug designs to effect cures.

Professor Ron Davis wanted his Ph.D. student Mark Schena to study the function of transcription factors in the flowering plant *Arabidopsis thaliana* based on solid-state assays over a cup of coffee. They decided to use glass as substrate because it offered less background fluorescence and better signal detection at the photomultiplier tube diode arrays. Other materials used by prior investigators were nylon and nitrocellulose. The company Affymetrix joined hands with the academics, and the rest is history. With stark similarity to the growth of the computer chip and microprocessor industry, microarray technology is gaining momentum. The concepts of miniaturization, automation, and parallelism are used. More genes per minute can be scanned in the slides with tiny dots arranged in uniform rows and columns.

When the work on microarrays was first presented at a conference in the Netherlands in 1994, the audience howled with laughter. Schena presented the first microarray enzymatic labeling procedure, demonstrating the feasibility of preparing fluorescent probes from yeast and plant messenger RNA. Some luminaries in the field noted that repetitive sequences in the human genome would prevent the use of microarray assays for human studies. The first human microarray data were presented at the Stanford Sierra retreat in October 1995. The field has exploded in size since the appearance of the paper in *Science* magazine in 1995 [2], and several thousands of papers in microarray technique and statistical analysis of the same have been published. Microarray analysis will lead to a better understanding of the genetic, molecular, and cellular processes common to aging, as well as how these processes may differ in individuals.

7.1.1 Microarrays, Biochips, and Disease

A *microarray* is a small analytical device that allows genomic exploration with speed and precision unprecedented in the history of biology. Glass chips containing thousands of genes are used to examine fluorescent samples prepared by labeling mRNA from cells, tissues, and other biologic sources. Molecules in the florescent sample react with cognate sequences on the chip, causing each spot to glow. The intensity of the glow is proportional to the activity of the expressed gene (Fig. 7.1). The entire genome can be analyzed in a single experiment. Since patterns of gene expression correlate strongly with function, microarrays can be used to generate unprecedented information on human disease, aging, drug and hormone action, mental illness, diet, and many other clinical matters. Microarrays can be used to find alterations in gene sequences. This can usher in a new era of genetic screening, testing, and diagnostics. Tissue and protein microarrays are miniaturizations of traditional histologic and biochemical assays. This speeds up the analysis of tumor specimens, protein-protein interactions, and enzymes.

A microarray is an ordered array of microscopic elements on a planar substrate that allows the specific binding of genes or gene products. The word *microarray* is derived from the greek word *mikro,* meaning "small" and the French word *arayer,* meaning "arranged." Microarrays are also called *biochips, DNA chips,* and *gene chips.* They contain collections of small elements or spots arranged in rows and

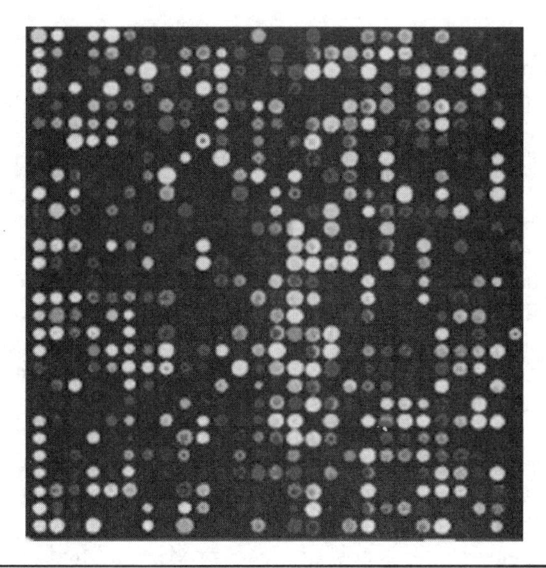

Figure 7.1 Example of an approximately 40,000-probe spotted oligo microarray.

columns. To qualify as a microarray, the analytical device must be (1) ordered, (2) microscopic, (3) planar, and (4) specific.

An *ordered array* is any collection of analytical elements configured in rows and columns. Each row of elements must form a straight line horizontally across the substrate, and each column of elements must form a straight line vertically down the substrate in a manner perpendicular to the rows. Ordered elements must have a uniform size and spacing and a unique location on the microarray substrate. *Microscopic* is defined as any object smaller than 1 mm. Microarrays manufactured using photolithography produce 15- to 30-μm features. Most tissue microarrays contain spots of 200 to 600 μm. Microarray elements are collections of target molecules that allow specific binding of probe molecules, including genes and gene products, and a typical printed DNA spot contains approximately 10^9 molecules attached to the glass substrate. The microarray target material can be derived from whole genes or parts of genes and may include genomic DNA, cDNA, mRNA, protein, small molecules, tissues, or any other type of molecule that allows quantitative gene analysis. Target molecules include natural and synthetic derivatives obtained from a variety of sources, such as cells, enzymatic reactions, and machines that carry out chemical synthesis. Synthetic oligonucelotides, short single-stranded molecules that contain chemical syntheses, provide an excellent source of target material.

Microscopic elements enable a density greater than 5000 elements/cm^2, rapid kinetics, and the analysis of entire genomes on a single chip, i.e., miniaturization and automation. Filter arrays and other nonmicroarray formats made with larger elements prevent miniaturization and automation and do not allow whole-genome analysis in a miniature format.

A *planar* substrate is parallel and unbending support such as glass, plastic, or silicon onto which a microarray is configured. Glass is the most widely used substrate material owing to the many advantages offered by SiO_2. Planar materials are flat over the entire surface. Flat supports are amenable to automated manufacture and high-quality manufactured microarrays. They allow for accurate scanning and imaging and rely on a uniform detection distance between the substrate and the detector. Impermeable to liquids, they allow for small feature size and low reaction volumes.

Specific binding refers to unique biochemical interactions between probe molecules in solution and their cognate target molecules on the microarray. Each microarray spot/target should bind essentially to a single species in the labeled probe mixture to provide the most accurate measure of genes or gene products. Microarray assays exploit a one-target, one-probe-molecule paradigm, and assay precision can be enhanced using multiple microarray elements per gene. Between 15 and 25 nucleotide target sequences define the minimal target length required to achieve single-gene specificity.

Microarray technology development used the combined expertise of different disciplines such as biology, chemistry, physics, engineering, mathematics, and computer science. The correlation between gene mutations, altered proteins, and disease was made first by Pauling in 1949. Pauling showed that hemoglobin from sickle-cell patients differs from hemoglobin from healthy individuals in that it migrates aberrantly in gel electrophoresis assays. This finding was correctly attributed to a change in the surface charge of the molecule. By examining normal individuals, carriers, and patients with sickle-cell disease, Pauling concluded that changes in the hemoglobin gene were responsible for the altered protein, and this was verified later in gene sequencing studies. This landmark journal article paved the way for the molecular genetic analysis of human disease and provided a conceptual foundation for the use of microarrays in genetic screening, testing, and diagnostics.

The discovery of the double-helical structure of DNA by Watson, Crick, and Williams is the chemical basis of microarray hybridization reactions. The discovery of polymerase chain reactions (PCRs) with the catalytic activity of the enzymes DNA polymerase, RNA polymerase, and reverse transcriptase has contributed to microarray analysis. Kornberg discovered polymerase. Reverse transcriptase catalyzes the synthesis of DNA [5]. Last year, the Nobel Prize went to Roger Kornberg, the son of Arthur Kornberg, for elucidating the molecular basis of transcription. Maxam, Gilbert, and Sanger developed DNA sequencing technology independently. Berg received the Nobel Prize in 1980 for his fundamental studies of the biochemistry of nucleic acids. Bergy developed recombinant DNA technology.

The growth of the microarray industry has a striking resemblance to the history of the microprocessor industry. Computer chip discovery was credited to Shockley, a cocreator of the transistor and recipient of the Nobel Prize in physics in 1956. He is also called the "father of silicon valley." He founded the Shockley Semiconductor Laboratories at Palo Alto, California. It transformed northern California from a sleepy pastoral community into a world center for technological innovation. His aversion to silicon lead to the "traitorous eight," a group of disgruntled employees who left Shockley's company to start Fairchild Semiconductor. Noyce and Moore were two of its founders.

The new company quickly exploited silicon-based fabrication methods and manufactured the *integrated circuit*. The first commercial integrated circuits were manufactured in 1961, Moore's law was formulated in 1965, and the computer mouse invented in 1962. Moore, the head of research and development at Fairchild, noticed that transistor density and computing power were doubling every 12 to 18 months. Noyce and Moore left Fairchild Semiconductors to found Integrated Electronics (Intel) in Santa Clara.

Intel released the first commercial microprocessor in 1971, and the 4004 chip contained 2300 transistors capable of performing approximately 100,000 calculations per second (108 kHz). As Moore predicted, modern chips have greater computing power. The Pentium IV chip, released in 2000, contains 42 million transistors capable of carrying out 1.5 billion calculations per second (1.5 GHz) with 180-nm circuit lines.

Microarrays similarly have grown in analytical power and have decreased in feature size. The first plant microarrays printed in 1995 contained 96 genes with 200-μm features, compared with the highest-density microarrays manufactured in 2001, which contain 30,000 genes with 16-μm features. Microarray gene content has increased more than 300-fold in 6 years, doubling once every 8 months during the 6-year period. In due course, the feature size of the microarray will be in the nanometer range. New scanning devices such as x-rays are needed because the wavelength of light is 400 nm, and the optical scanning microscopes currently used to image the microarrays may not be sufficient. In addition to analytical power, other similarities between microprocessors and microarrays are the parallelism, miniaturization, and automation.

The NanoPrint Microarrayer is a robust and customizable platform for all microarray manufacturing applications regardless of the type of biomolecule. The NanoPrint systems manufacture high-quality, precision microarrays using TeleChem's ArrayIt brand patented and widely used Professional-946, and Stealth Style Micro Spotting Pins. The NanoPrint uses superior linear drive motion control technology and proprietary Warp1 controllers from Dynamic Devices. The NanoPrint is compatible with all standard microarray surfaces made by ArrayIt and other vendors. The system is easily configured to print microarrays into the flat bottoms of 96-well plates by taking advantage of its flexible deck configuration and easy-to-use software interface. The Microarray Manager Software combines unparalleled power and simplicity into a graphic Windows-based package.

Features include a method-creation wizard, user and version control management, custom calibration of the slide and microplate positions, complete sample tracking, support of input-output data files, custom array designs, speed profiles and wash protocols, automatic method validation, runtime sample and spotting views, and a simulation mode and easy-to-use graphic reprint wizard. The high-speed, high-precision linear servo control system of the NanoPrint results into superior instrument performance in both speed and precision. Combined with the efficient benchtop design, user-configurable worktable, humidity and dust control, a host of available options, and the flexible and sophisticated software, the NanoPrint system is the complete solution for high-performance microarray printing.

7.1.2 Five Steps and Ten Tips

The microarray analysis life cycle consists of five steps, as shown in Fig. 7.2. These steps are formulation of a biologic question, sample preparation, biochemical reaction, detection, and data analysis and

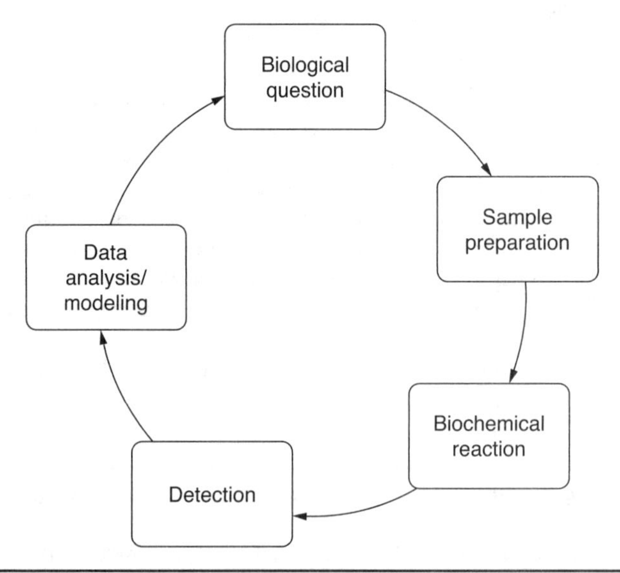

FIGURE 7.2 Five steps in microarray analysis life cycle.

modeling. A biologic question has to be formulated prior to embarking on a microarray study. For example, how do gene expression patterns in a normal human and a patient with cancer differ from each other? The goal of the project is to better understand the mechanism of cancer disease affliction.

The second step is sample preparation. This includes DNA and RNA isolation [7] and purification, target synthesis, probe amplification and preparation, and microarray manufacture. The biochemical reaction involves the incubation of the fluorescent sample with the microarray to allow productive biochemical interactions to occur between target and probe molecules. DNA microarrays use hybridization for this step. Protein microarrays use protein-protein interactions for this step.

The fourth step is the detection step. This involves use of a confocal scanning microscope to obtain the image of the microarray during gene expression on a photomultiplier tube using diodes. Lenses and mirrors are used to effectively illuminate the sample and detect the key reactions of interest. Captured images are analyzed and modeled to complete the fifth step. Microarray manufacture can be achieved using different methods, such as photolithography, ink-jet printing, and mechanical microspotting.

The following 10 tips will ensure success in microarray analysis [1]:

1. *Follow the protocol.* The experimental recipes have been optimized within a specific set of reagents, surfaces, fluorescent labels, tools, methods, and techniques. The recipe has to be followed to the letter.
2. *Read the mannual.* Microarray manufacturers estimate that greater than 50 percent of the damage that occurs to expensive microarray instruments is incurred in the first 24 hours of use. It is prudent to read the manual prior to use of the instrument.
3. *Think small.* The nucleic acid concentration in a microarray hybridization reaction containing 1000 ng of fluorescent probe in 5 μL (200 ng/μL) is 40,000 times greater than a filter hybridization containing 100 ng of probe in 20 mL. Quantitative gene expression data from 10,000 genes can be obtained in a 5-minute scan at a rate of 2000 genes/min using a microarray slide compared with 2 weeks required to measure a single gene using a filter blot. There is an increase by a factor of 80 million over traditional methods.
4. *Keep it clean.* Even a small amount of contamination will alter the microarray reaction and skew the data. Protective gloves should be worn at all times, and clean rooms are recommended for exacting procedures.

5. *Keep it warm, and keep it hydrated.* Elevated background fluorescence can harm the data. Background fluorescence can be minimized by using elevated reaction temperatures and proper hydration. Water evaporates at 0.1 μL/min at ambient conditions, and a low-volume microarray reaction can lose a significant percentage of its volume quickly if steps are not taken to minimize evaporation. Thus keeping it at an elevated temperature and hydration is a challenge.
6. *Think globally.* A holistic view of biologic systems is required. Traditional studies focus on one gene. Global interactions of genes and proteins are more important. The global view of the cell afforded by gene expression studies using microarrays has to be taken into account.
7. *Do the small experiments first.* It is prudent to perform a pilot study before scaling it up to the entire genome.
8. *Confirm as you go.* It is recommended to confirm the identity of a small number of genes by microarrays before a large number of precious samples are achieved. Confirming microarray analysis pathway early saves a lot of heartache later on.
9. *Look early.* Given the complexity of cell signaling pathways, looking early after stimulation, i.e., within 1 to 4 hours will maximize the chances of identifying the primary response genes and will yield a gene fingerprint specific to a particular response.
10. *Don't panic*: One way to combat the microarray data flood that results in panic is to use data quantitation, mining, modeling tools, and focused experimentation to narrow the list of candidate genes before embarking on detailed study of each gene.

7.1.3 Applications of Microarrays

One of the important applications of microarrays is the study of *gene expression.* Eighty-one percent of the scientific publications on microarrays contain such studies. Researchers in the United States have contributed 71 percent of the microarrays, but scientists from nine other nations, including Japan, the United Kingdom, Germany, Canada, France, Australia, Sweden, China, and Finland, have provided nearly 33 percent of the publications.

By measurement of gene expression levels as a function of cell and tissue type and storing the results in databases, a deeper insight into multicellular development and a better understanding of pathologic cellular events can be achieved. Human brain tissue has been most actively studied to date, but other tissues, including liver, breast, prostate, lung, colon, kidney, heart, bladder, and skin, have also been studied. The key to longevity-causing genes can be obtained from microarray studies. The onset and progression of *human disease* are determined by a complex set of factors that include genetics, diet, the environment, and the presence of infectious agents. Microarray analysis is unique in its ability to detect each of the contributing factors.

Oncologic studies have accounted for 83 percent of microarray publications. Diabetes, cardiovascular disease, Alzheimer's disease, stroke, AIDS, cystic fibrosis, Parkinson's disease, autism, and anemia are under intense investigation using microarray analysis by scientists around the world. Through the study of differential gene expression using microarrays, the mechanism of cancer formation can be better understood—and the cure will soon follow. The ultimate goal of microarray analysts is to eradicate every human disease by the year 2050.

Many drugs impart their therapeutic action to specific cellular targets, inhibiting protein function and altering gene expression. In principle, microarrays can be used for *drug discovery* and clinical trials by generating gene-expression profiles in patients undergoing drug treatment. Many illness result in specific changes in gene expression, and drugs that reverse these changes are expected to ameliorate the disease. The cost of drug development may be cut down and safer medicines may be produced on account of microarray studies. Microarrays can be used for patient genotyping and dividing the population into drug responders and nonresponders.

Microarrays can be used in *genetic screening and diagnostics*. Thousands of disease-causing sequence variants are known, and affordable microarray screens for these diseases are of tremendous scientific and commercial interest. Microarray screening can be used to distinguish the population as normal, carrier, and disease genotypes. Treatable and curable genetic diseases can be identified at an early stage. Genetic testing kits can reduce health care costs. The commonly inherited diseases, such as cystic fibrosis, sickle-cell anemia, Tay-Sachs disease, and breast cancer, can be studied using microarrays, and the genomic information can be provided to the public by confidential access.

Protein chips can be used to obtain the polypeptide sequence distribution. Metabolomics, like genomics and proteomics, is the complete functional annotation of the genome. The functions of the organism are triggered by the signals from the proteins that are generated by the DNA. Microarray analysis can be used in metabolomics.

Gene chip technology is a practical method for determining the sequence of genetic building block. It can speed up searches for disease-related genetic changes. Using gene chips and analysis, a team of scientists at Johns Hopkins University was able to accurately determine the order of 2 million blocks of each of 40 individuals' genomes in just a year. This is in a fraction of the time required by traditional technology. Only 10 errors of every 10,000 points were detected. Researchers at Washington University School of Medicine in St. Louis helped to explain how genes dictate our biologic clock. The circadian rhythm was studied using microarray analysis. How do you feel when you get up at 4:00 a.m. compared with 4:00 p.m.? Events such as this are driven by the internal clock, connected to external cues such as the sun.

So far products of eight different genes have been discovered to be essential to operations of this clock. Three laboratories in collaboration with Affymetrix have identified 22 genes that appear to be rhythmically regulated by the internal clock of the *Drosophila* fly. *Drosophila melanogaster* has 14,000 genes. Microarrays can be used to prepare a comprehensive list of all the active genes in a tissue sample. The fly was exposed to light for 12 hours, followed by darkness for 12 hours. The cycle continued for a total of 96 hours. Genetic analyses were performed on half the flies at six different time-points on the fifth day. Seventy readings of 14,000 genes were taken, and a million individual measurements were completed. Sophisticated computer statistical analyses were performed, and the team determined that between 72 and 200 of the flies' 14,000 genes showed significant rhythm of gene expression in normal flies living in a daily light-dark cycle. Oscillating genes also were detected. So were mutant flies.

The chip is embedded with DNA molecules instead of electronic circuitry. It is designed to probe a biologic sample for genetic information that indicates whether the person has a genetic predisposition for certain diseases. A University of Houston scientist has developed a chemical process for building a device that could help doctors predict a patient's response to drugs or screen patients for thousands of genetic mutations and diseases, all with one simple lab test. This is a highly parallel technology—10,000 experiments can be performed at once.

Aging of the human retina has been found by researchers to be accompanied by distinct changes in gene expression. Using commercially available DNA slides, a team of researchers directed by Swaroop has established the first-ever gene profile of the aging human retina, an important step in understanding the mechanisms of aging and its impact on vision disorders. In the *Journal of Investigative Ophthalmology and Visual Science*, Swaroop and colleagues show that retinal aging is associated, in particular, with expression changes of genes involved in stress response and energy metabolism. The term *gene expression* means that in any given cell, only a portion of the genes is expressed or switched on. For example, a person's pancreas and retina have the same genes, but only the pancreas can turn on the genes that allow it to make insulin.

Swaroop believes that these findings will help scientists to understand whether age predisposes one to changes in the retina that, in turn, lead to age-related diseases. For vision researchers, one of the most pressing disorders is age-related mascular degeneration (AMD), a progressive eye disease that affects the retina and results in the loss of one's fine central vision. Microarray technology is an important tool for *gene profiling* because it allows rapid comparison of thousands of genes, something that was unheard of even few years ago.

7.2 Microarray Detection

7.2.1 Fluorescence Detection and Optical Requirements

All microarrays require fluorescence scanning to facilitate reliable imaging of the gene expression pattern or the problem at hand. The confocal laser scanner delivers the highest image and data quality. Commercial devices such as ScanArray are used currently. In future, as the minimum feature size of the microarray dot size reaches the nanometer range, x-ray scanners may have to be developed because the wavelength of light is 400 nm. The substrate is chemically treated glass in the form of a 25 × 75 mm slide. DNA arrays incorporate samples tagged with multiple fluorescent probes. Differential gene expression leads to a ratiometric approach and renders absolute calibration unnecessary. The glass substrate gives minimal background fluorescence and hence is a good choice.

Fluorescence in biologic detection is a vast topic and has been discussed comprehensively elsewhere [6]. Fluorescent light is emitted from a dye or fluorophore that is illuminated by excitation light. The fluorescence emission wavelength is always longer than the wavelength of excitation light. For example, fluorescein isothiocynate (FITC) exhibits a excitation curve peak at 494 nm and an emission peak at 518 nm. The wavelength difference between the emission and excitation peaks is 24 nm. Typical for most dyes used in microarrays, this wavelength difference is called the *Stokes shift*.

The optical requirements of a detection instrument are as follows:

Excitation. A number of sources are possible for providing the excitation. These are lasers, arc or filament lamps, and light-emitting diodes (LEDs). Excitation wavelength range cannot overlap with that of the emitted wavelength range. Flood illumination may not be desirable on account of the nonuniformity introduced. Excitation wavelengths may be chosen based on the dyes used. The wavelength should be smaller than that of the dye. Excessive light may cause harm to the sample. This is so because of the onset of photobleaching.

Emission light collection. The fluorescent light is collected using an objective lens. The angle of collection is critical. Fluorescent emissions are spherical in nature. The light-collection angle of the lens is often characterized by the numerical aperture (NA). An NA of 1.0 describes a lens that collects light over an entire hemisphere, corresponding to a light-collection efficiency of 50 percent. Most confocal laser microarray scanners have NAs between 0.5 and 0.9. CCD-based array scanners have NAs between 0.2 and 0.5.

Spatial addressing. The sample is divided into pixels. Pixel size needs to be smaller than the dot size. Scanners for 100-μm-diameter microarray dots commonly used pixel sizes between 5 and 20 μm. As the microarray technology develops into the nanometer feature size range, pixel size in the nanorange will be a challenge.

Excitation/emission discrimination. Microarray fluorescence emission power is orders of magnitude smaller than the excitation power. An optical device that can delineate the two sources of light is needed. Most objective-lens–based microarray scanners are *epi-illuminated*. In epi-illuminated systems, the excitation and emission beams follow the same path through the objective lens to and from the sample but in opposite directions. A *beamsplitter* is used to separate the mixture of light. One type of beamsplitter is a color-separating *dichroic* or *multichroic* interference filter that reflects the excitation beam and transmits the emission beam. This device can handle two or three different excitation/emission wavelength pairs. More than four wavelengths will make it a difficult separation task even for a multichroic lens. Since all devices are real and far from ideal, emission filters are placed in the path of the light beam. A geometric beamsplitter can be used where the excitation beam and emitted beam do not mix or cross each other.

Detection: Detectors found in array scanners include photomultiplier tubes (PMTs), charged coupled devices (CCDs) arrays, and avalanche photodiodes (APDs). In the visible wavelength range, PMTs are the most sensitive detectors. PMT sensitivity falls rapidly between the red and near-infrared ranges. A CCD does not posses the inherent low noise amplification of a PMT and therefore needs external amplification. It has a high NA (0.6–0.9), limiting the optical signal available for collection. CCD inclusion is impractical in the confocal scanning arrangement.

7.2.2 Confocal Scanning Microscope

Confocal scanners [8] have two focal points (Fig. 7.3) configured to limit the field of view in three dimensions. They image a small area with an aim of point resolution using pixels. The collimated laser beam is reflected from the beamsplitter into the objective lens. The laser beam fills only a fraction of the lens. The degree of fill depends on the choice of the lens NA and pixel size. The laser beam in focused on the sample, where it induces spherical fluorescence in all directions. The excitation beam also reflects back up toward the detector. The objective lens

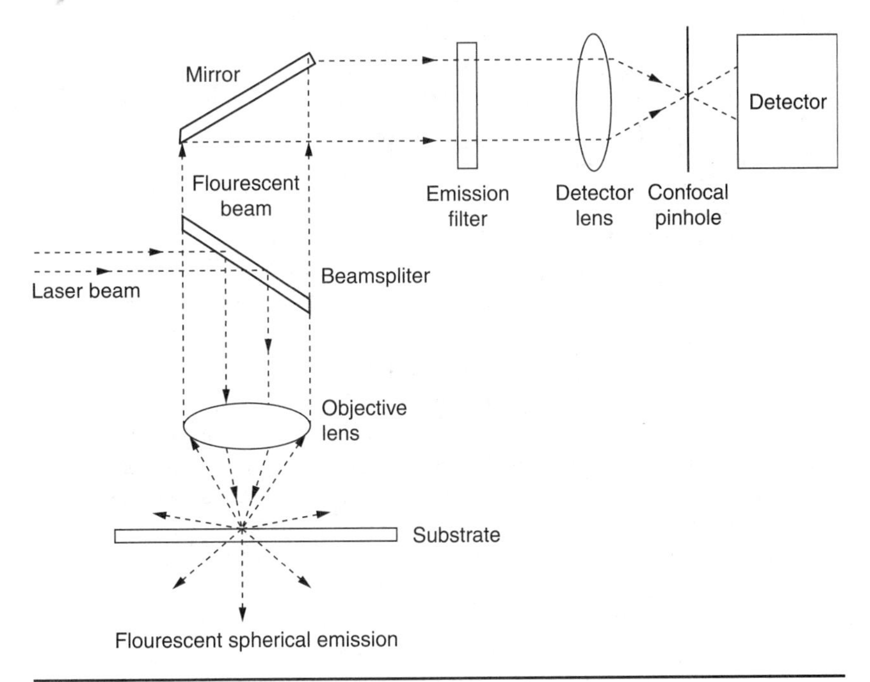

FIGURE 7.3 Confocal scanning arrangement in a microarray scanner.

collects a fraction of the spherical fluorescence emission and collimates it into a parallel beam. It also collects the reflected laser light, which is three to seven orders of magnitude higher in intensity than the fluorescent light. The return beam is again directed to the beamsplitter, which reflects most of the laser light back toward the laser source and transmits most of the fluorescent beam toward the detector. A mirror then reflects the system without any optical functionality, followed by the emission filter, which selects a narrow band of fluorescence and rejects all remaining laser excitation light. The pinhole arrangement facilitates the depth of focus of the objective lens, coinciding with the imaging in the detector.

Restricted depth of focus is a disadvantage of the confocal scanning arrangement. It has a moving substrate scanner. Using a moving lens and a moving substrate, higher light collection efficiencies can be obtained. Useful microarray scanners must detect low levels of fluorescence in the picowatt range. At these low levels, almost all materials fluoresce—the glass substrate, the chemicals comprising the substrate's surface coating, sample washing chemicals, lenses, filters, and even DNA molecules. The scanning instrument needs to maximize detection of the target dye's emission while minimizing detection of all the other fluorescence sources. The reflected and scattered light must be rejected even though it is

1 million times brighter than the dim fluorescent light. A PMT can detect a single photon or a beam of light that is low in power. PMT amplifies the photon event into an electron event. By varying the tube high voltage, the PMT sensitivity or gain increases by a range of several hundred to one.

Some of the instrument performance measures are as follows:

1. *Number of lasers and fluoresence channels.* A single excitation laser may excite several dyes and can be used with emission filters. Crosstalk between multiple dyes has to be minimized.
2. *Detectivity.* Detectivity is the minimum dot fluorescent brightness that can be distinguished from the background when the sensitivity is set so that the brightest element of the sample produces an intensity level at full scale. Dye molecules per unit area (fluors/μm^2) may be the unit of measure. Directivity for the array preparation process is often defined by the dimmest dot in the dilution series that can be detected.
3. *Sensitivity.* Instrument conversion efficiency of light power to a digital value at a particular wavelength is called the *sensitivity.* It is a measure of a "gain" of the instrument. Sensitivity is independent of properties of the sample.
4. *Crosstalk.* When scanning samples with multiple dyes, crosstalk can occur. Crosstalk is the excitation and detection of dye with thc "wrong," or unintended, excitation wavelength and emission filter. In differential gene expression, crosstalk negatively distorts the expression ratio between two channels. It is minimized by the use of narrow-band emission filers centered on the dye peaks with good attenuation of out-of-band wavelengths.
5. *Resolution.* Spatial resolution of a microarray scanner is usually expressed as a pixel size, with 5, 10, and 20 μm being common in commercial devices. Each microarray dot has to be imaged into many pixels. Edge effects and other defects can be rejected at the quantitation stage. Pixel dimension should be no larger than one-eighth to one-tenth the diameter of the smallest microarray dot to be imaged.
6. *Field size.* Field size, the area on the substrate that can be scanned, must match the array-making process. The larger the scan area, the more dots there are that can be placed on each sample. Usually a 1- to 1.5-mm border around the periphery of the slide is not used because it may be clipped or not flat. Maximum usable area is about 22×73 mm.
7. *Uniformity.* Uniformity is a measure of the consistency of fluorescence emission and detection across the field. Uniformity of light collection throughout the image field is of

particular concern in confocal scanners. Scanner uniformity within ±10 percent is sought by the users.

8. *Image geometry.* Image quantitation software is used for postprocessing of the image data. The image size, *x-y* orthogonality, and pixel placement linearity are important considerations. Tolerance of ±2 percent in image size and linearity is allowed. There may be some errors owing to random geometry. Jitter manifests as vertical lines in the image.
9. *Throughput.* It is a measure of the number of samples scanned in a day. It depends on the resolution, image field size, and number of channels. Some CCD camera–based scanners exhibit high throughput. Dim samples are an important consideration. Throughput for multichannel scanning can be increased dramatically by incorporating color-separating beamsplitters in the emission path. Multiple detectors can scan multiple colors simultaneously, and multiple signal-processing modules are used. The specification of first-generation scanners in a single-color, 20 × 60 mm field is 5 to 15 minutes at 10-μm resolution.
10. *Superposition of signal sources.* The image viewed on the scanner's monitor is not a simple image of dye fluorescence in the microarray dots. It is a superposition of several images, of which only one is desired. The image acquired has to be postprocessed, and some salient considerations are (1) fluorescence of the target dye being scanned, (2) photon statistical noise, (3) fluorescence of the background owing to other chemicals and the glass, (4) laser light reflection, and (5) electronic noise.

7.3 Microarray Surfaces

High-quality surfaces are needed for the preparation of microarray samples. How well the molecules attach to the surface determines the efficiency of the biochemical reactions, the precision of detection, and the quality of the resulting data. A microarray experiment is only as good as the surface used to create it. An ideal microarray surface has to be (1) dimensional, (2) flat, (3) planar, (4) uniform, (5) durable, (6) inert, (7) efficient, and (8) accessible.

There exists an *optimal target concentration*. This is the number of target molecules per unit volume of printed sample that provides the strongest signal in a microarray assay. *Optimal target density* is the number of target molecules per unit area on a microarray substrate that provides the strongest signal in a microarray assay. Experiments were conducted, and microarray signals are plotted as a function of the target molecule concentration. A 15-base oligonucleotide was printed on a microarray substrate at a concentration range of 1 to 100 μM. Hybridization with probe solution containing a fluorescent 15-mer complementary to the target sequence was performed. The scanning was measured at different target concentrations of 1, 3, 10, 30, 50, and 100 μM. Examination of the results revealed that the fluorescent intensity increased steadily in the range of 1 to 10 μM target and reached a peak intensity at 30 μM oligonucleotide, at which point the signal leveled off and decreased significantly as the target concentration reached 100 μM. At the optimal target concentration, the number of target molecules bound to the microarray surface area can be calculated. Assuming that 30 percent of the printed oligonucleotide couples to the substrate and that a typical printed droplet is 300 pL, a 30-μM solution of oligonucleoides gives 2.6 lakh oligonucleotide molecules per square micron of the substrate. This is the optimal target density. Additional calculations reveal that 2.6 lakh molecules/μm^2 correspond to 1 oligonucleotide per 400 $Å^2$ or 1 target molecule per 20 Å in a single dimension. It is interesting that a single-stranded DNA is 12 Å in diameter. The probe-target duplexes (Fig. 7.4) would be approximately 24 Å in diameter. Owing to major and minor grooves, the effective diameter is 20 Å. A spacing of 1 target per 20 Å defines the optimal target concentration. More material would cause steric hindrance in the packing. Insufficient target density means too few molecules available for hybridization. Physical interference at higher concentrations cause damage and a fall in signal intensity. In a similar fashion, *optimal probe concentration* is the number of probe molecules per unit volume of sample that provides the strongest signal in a microarray assay.

FIGURE 7.4 Target DNA molecules hybridized with probe molecules with fluorescent tags and attached to the substrate via linker molecules.

Probe concentrations greater than the optimal concentration are useful under certain circumstances. Target (T) molecules on the microarray surface form productive interactions with probe (P) molecules in the solution to form probe-target (T-P) pairs. The generalized biochemical reaction for target-probe binding can be given as

$$T + P \rightarrow T\text{-}P \tag{7.1}$$

The rate of formation of target-probe products depends on the concentration of the two reactants and can be expressed as the product of the concentration of T and P times a proportionality constant k:

$$\text{Rate} = -k[T][P] = d[T]/dt = d[P]/dt \tag{7.2}$$

As indicated by Eq. (7.2), the reaction between target and probe is a second-order biochemical reaction. The constant k is the rate constant. Under optimal experimental conditions, the printed microarray will contain a much larger number of target molecules than are required to form T-P pairs during the course of the reaction. *Target excess* is a kinetic condition in a microarray assay in which the concentration of target molecules on the surface exceeds the concentration of probe molecules in solution. Under target-excess conditions, the concentration of target molecules is relatively constant and can be lumped with the reaction rate constant term k. Thus

$$\text{Rate} = -k'[P] \tag{7.3}$$

where k' denotes the fact that the constant target concentration has become part of this term. As can be seen by Eq. (7.3), the reaction rate becomes a pseudo-first-order expression. Integrating with respect to time,

$$[P]/[P_0] = \exp(-k't) \tag{7.4}$$

The probe molecules get consumed during the course of the reaction in an exponential fashion. Doubling the concentration of a microarray probe solution will double the rate of the reaction. Because faster rates result in more target-probe pairs per unit time and greater [T-P] means greater signal, it is desirable to use as much probe material as possible in any given microarray experiment as long as the performance of the assay is not compromised. The probe concentration that gives the strongest microarray signals is known as the *optimal probe concentration*. The linear portion of the graph is called the *linear range* of the assay. A *saturated* condition occurs when the microarray target element in which most or all of the target molecules are located contains bound probe molecules. *Selective target saturation* refers to a microarray assay condition in which a subset of the target elements becomes largely or fully bound, leading to a loss of quantitation. *Signal compression* is a microarray assay condition in which the fluorescent readings underestimate the number of molecules present on the target element or in the probe mixture, leading to a loss of assay quantitation.

A glass surface is preferred as the substrate because of the low background fluorescence generated from it. The smoothness of the glass can be measured using a scratch and dig specification. There are different types of glass. The structure of the glass is SiO_2 tetrahedra. The smoothness of the glass surface can be accessed at high resolution using atomic force microscopy (AFM). The AFM technique employs a fine silicon tip that traces back and forth across the surface, detecting and recording surface irregularities as it moves. Three-dimensional images are produced in AFM scans. A typical microarray glass substrate subject to AFM analysis reveals a maximal roughness of 5.3 nm over a 4-μm^2 area, corresponding to a distance of approximately 40 Si—O bonds or about twice the diameter of duplex DNA.

Etching refers to a chemical process used to score glass surfaces for the purpose of labeling and indentification. The glass surface may be treated by using either amine or aldehyde. Silane reagents are used for this purpose. The reaction of glass with three-aminopropyl trimethoxysilane is a typical treatment reaction. The overall positive charge of amine microarray surfaces allows attachment of printed biomolecules that carry negative charges. Attachment occurs primarily via electrostatic interactions or attractive forces between positive charges on the amine groups and negative charges on biomolecules such as nucleic acids. Attachment of nucleic acids to an amine surface occurs via interactions between negatively charged amine groups. The DNA phosphate backbone can be attached along the side of the chain with the microarray glass substrate.

Denaturation is the process of converting DNA into single strands. Aldehyde surface treatment uses a spacer arm and an amino linker. The substituted amine attaches by covalent coupling. Covalent coupling is an attachment scheme that involves electron sharing between target molecules and the microarray substrate. Molecules couple to an aldehyde surface in a directional manner such that the end of the molecule containing the amino linker bonds to the microarray surface. Proper reaction conditions and blocking agents all but eliminate background fluorescence with aldehyde surfaces.

Steric availability is a desirable spatial configuration such as end attachment that maximizes the physical accessibility of target molecules to incoming probe molecules. *Blocking agents* are chemical or biochemical agent such as borohydrate or bovine serum albumin used to inactivate reactive groups on a microarray substrate to prevent nonspecific reactivity.

7.4 Phosphoramadite Synthesis

Oilgonucleotides are short chains of single-stranded DNA or RNA. Single-stranded oligonucleotides provide another common source of target sequences for nucleic acid microarrays. Microarrays of oligonucleotides can be prepared using *delivery* or *synthesis* methods. In the delivery strategies, oligonucelotides made offline are prepared using standard phosphoramadite synthesis, suspended in a suitable printing buffer, and formed into a microarray using a contact or noncontact printing technology. In the synthesis approaches, oligonucleotides are made in situ one base at a time, and many synthesis cycles are used until the microarrays are complete. Owing to reduced coupling efficiency and large synthesis time, the length of the oligonucleotides is only 5 to 25 nucleotides. The main advantages of oligonucleotide targets are increased specificity and the capacity to work directly from sequence database information. Two disadvantages of oligonucleotide targets are the requirement for sequence information prior to manufacture and the loss of signal when using certain types of fluorescent probes.

The chemistry used in the phosphoramadite synthesis in the industry was developed by Caruthers in the early 1980s. Phosphoramadite-based oligonucleotide synthesis underlies most of the synthetic DNA market. The DNA market includes 75 commercial vendors worldwide and annual revenues totaling hundreds of millions of dollars. The oligoncucleotides of any sequence can be built from the four DNA building blocks. The four DNA bases used most often are known as *cyanoethyl phosphoramidites.* Each base is identical to its natural counterpart except for the presence of several chemical substituents that protect the phosphoramidites during synthesis and activate the 3′ phosphate for chemical coupling.

Three of the phosphoramadite bases, A, C, and G, contain a reactive primary amine on the purine or pyrimidine ring and therefore require a protecting group on the amine to avoid damaging this position during synthesis. A benzoyl protecting group is typically used for bases A and C, whereas an isobutyryl group is usually employed on G. The fourth base, T, does not contain a primary amine on the pyrimidine ring and thus does not require a protecting group. All four phosphoramidite bases also contain a dimethoxytrityl (DMT) group on the 5′ hydroxyl that blocks the 5′ hydroxyl from chemical coupling until it is intentionally deprotected during synthesis. Selective deprotection allows synthesis to proceed in a stepwise manner. The 3′ phosphate is protected against side reaction and activated for nucleophilic attack by the presence of β-cyanoethl and diisopropyl groups, respectively. The protecting groups are removed at the end of synthesis, yielding an oligonucleotide that is identical to native DNA.

The synthesis process proceeds in a 3′ and 5′ direction as follows: The initial step in oligonucelotide synthesis involves coupling the first base to the solid support. Oligonucleotides can be synthesized on a variety of different supports, but the most common matrix is controlled-pore glass (CPG). CPG contain pores of identified diameters inside of which synthesis occurs. A deprotection step in oligonucelotide synthesis allows the 5′ hydroxyl to act as a nucleophile, attacking the 3′ activated phosphate group of the second base that is added to the activated CPG matrix by coupling to the first base. The result is dinucelotide bond formation in the 3′ to 5′ direction. After the coupling step, unreacted 5′ hydroxyl groups are inactivated or capped by acetylation to prevent these bases from reacting with phosphoramidites in subsequent coupling steps. Capping prevents the formation of frame-shift oligonucelotides that are missing one or more bases compared with the full-length product, a process that occurs if unreacted 5′ hydroxyls are not capped before the next coupling cycle. After capping, the phosphate trimester of the newly formed dinucleotide is oxidized to the phosphate form to stabilize the phosphate linkage.

The four-step process of deprotection, coupling, capping, and oxidation is the basis of phosphoramidite synthesis and is shown in Fig. 7.5. An oligonucleotide of a known sequence is synthesized by repeating the cycles a few times and using the right bases and reagents efficiently. Each four-step cycle takes 5 to 7 minutes, enabling synthesis of a synthetic 70-mer in less than 8 hours. Following synthesis, the nascent oligonucleotides are treated overnight with ammonium hydroxide to remove the protecting groups from the base and phosphate groups and to cleave the oligonucelotides from the CPG support. With coupling efficiencies exceeding 99 percent per cycle,

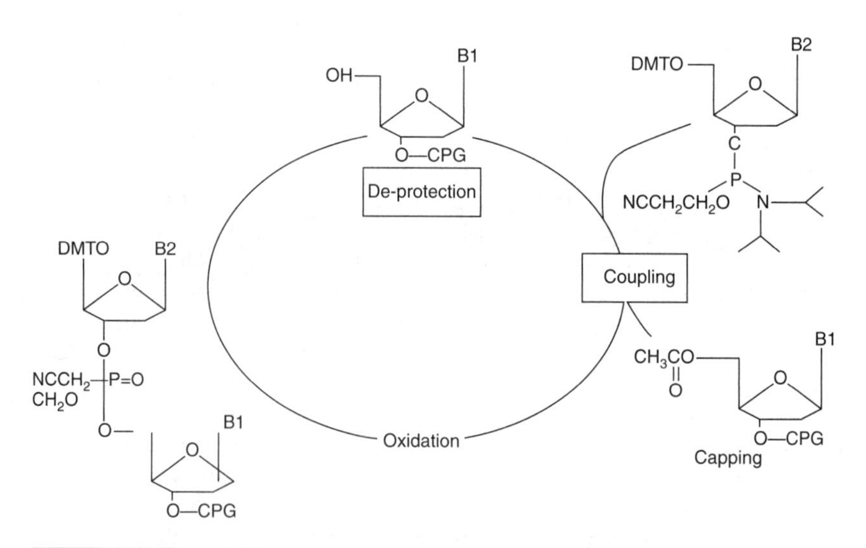

FIGURE 7.5 The four-step process of oligonucleotide synthesis on CPG.

a synthetic 70-mer preparation would contain more than 60 percent full-length product. Full-length oligonucleotides can be purified away from shorter products using polyacrylamide gel electrophoresis (PAGE) or high-pressure liquid chromatography (HPLC).

7.5 Microarray Manufacture

New superconductors are prepared using combinatorial mixtures of components. Combinatorial synthesis programs are the state-of-the-art mode for discovery of novel drug leads. And in biology, arrays of unique sequences are used commonly to assay the genetic state of cells. In all cases, small volumes of liquids must be metered precisely at high rates of speed. Technology derived from ink-jet printing has been applied to meet such liquid-handling needs. *Ink-jet printing, mechanical microspotting,* and *photolithography* are the three primary methods of manufacture of microarray slides.

There are two modes of DNA microarray fabrication using ink-jet technology. First is the step-step synthesis of DNA by applying reactive nucleotide monomers to individual surface sites. Second is the spotting and immobilization of presynthesized DNA. Ink-jet technology has been used for over 20 years to control delivery of small volumes of liquid to defined locations on two-dimensional surfaces. Different droplet-generating devices are available, such as piezoelectric capillary, piezoelectric cavity, thermal, acoustic, continuous-flow, etc. Drop diameters of 25 μm at up to 10 kHz can be readily achieved using piezoelectric devices. Smaller-diameter droplets and higher frequencies can be generated using piezoelectric cavity devices, and even higher with thermal devices. No nozzle is used in the acoustic device, which possess high rates of drop formation (5 mHz) and small drop diameters (<1 μm). In continuous-flow droplet-generating devices, stream or liquid is broken into distinct droplets by oscillatory pressure. A typical device consists of static-pressure ink reservoir, a small-diameter orifice, and a pressure-generating element. The orifice plays a significant role in determining the diameter of the droplets ejected from the device. Drops can be generated on demand, and ink can be consumed efficiently. Another method of printing uses a continuous stream of droplets directed via an electric or magnetic field onto a print area or, alternatively, a gutter where the ink is recycled. The printing mode is quite robust because the jet is primed by pressurizing the liquid reservoir. Nozzle-less acoustic jet is an interesting development in technology. The drop size can be derived by equating the forces acting on the surface of the drop from the internal and external pressures:

$$\Delta P(4\pi)R^2 = \sigma 2\pi R \tag{7.5}$$

or

$$R = \sigma/2\Delta P \tag{7.6}$$

A piezoelectric capillary jet consists of a glass capillary fixed with an orifice and surrounded by a cylindrical piezoelectric. Droplet formation with piezoelectric capillary jets is accomplished by alternately expanding and contracting the piezoelectric element to generate shock pulses in the fluid chamber. When appropriately tuned to the characteristics of the liquid, pressure pulses sufficient to eject droplets from the nozzle can be generated. Drop formation rates can be up to 10 kHz. Droplet formation is easier at certain frequencies. The size of droplets from these devices depends on the diameter of nozzle, the magnitude of the driving force, and the physical properties of the liquid in use. Care must be taken when manufacturing high-quality nozzles and in supplying the appropriate waveform to obtain droplets that are satellite-free and propagating perpendicular to the nozzle plate. Two commercial instruments that are built using this concept are the CombiJet and the GeneJet.

The CombiJet can be used to *synthesize DNA* microarrays by delivering reagents for phosphoramidite oligonucleotide synthesis to defined locations on glass substrates. GeneJet is used to manufacture DNA microarrays by spotting *presynthesized DNA fragments*. Localized DNA synthesis is achieved by using jets to deliver reagents for one of two reactions in the phosphoramidite oligonucleotide synthesis cycle. The first uses this single-jet device to deliver reagent to deprotect the 5′ hydroxyl position at specific regions on the two-dimensional surface. Oxidation of the phosphor and coupling of one of the four bases are done in bulk chemical treatment of the entire surface. The second method uses five jets, one for each of the four phosphoramidites and one for the activating reagent. The CombiJet III was designed to fully automate all steps of DNA microarray synthesis. During in situ synthesis of DNA, no purification is possible. All the reactant products have to remain on the surface. The quality of the material in each locus thus is determined by the *stepwise coupling efficiency*.

In order to evaluate the coupling yield, a set of 64 spots of identical sequence was synthesized with a cleavable attachment to the surface. At the end of 15 cycles, the slide was subjected to a gas-phase base reaction to disrupt the surface treatment. The oligos were collected by washing the surface. After complete removal of the remaining protecting groups, the oligo product was end labeled with [^{32}P]phosphate and subjected to PAGE. The banding patterns of the oligo products were analyzed quantitatively to derive an average stepwise yield of 91 percent.

To increase the efficiency of hybridization and the DNA attachment of the surface, *linker molecules* can be used. Linkers attach themselves to the substrate on one of its end and to the target molecule on the other. One example of a linker molecule is polyethylene glycol polymers. Solvents that are compatible both with the ink-jet hardware and the particular chemical reactions desired are difficult to find. Acetonitrile was used as a solvent for phosphoramidites. For the deprotection reagent, di- or trichloroacetic acid is common. The volatility of these solvents makes them less suitable. Less volatile dibromomethane was used in place of dichloromethane to reduce the loss of solvent during preparation. This approach offers flexibility and is low in cost.

To scale up the microarray synthesis into commercial practice, some technical hurdles have to be overcome. With a cycle time for the instrument of 10 minutes, an array of 18-mer is printed in 3 hours. The next-generation instrument will be expected to print more than one array at a time. Robustness of jetting has to be improved. Jet-to-jet variability has to be reduced. Sensitivity of drop size to nozzle characteristics needs to be reduced. Change from a uniform glass substrate to a patterned region is desirable.

Deposition of presynthesized biologic material is another method of fabrication. The GeneJet III device can use up to eight jets to aspirate samples from 384- or 1536-well microtiter plates and apply them to microarrays. The instrument has five independent axes. The jets are connected by solenoid valves. Monitoring by video camera can be used to deliver droplets free of satellites. The equipment is operated in two modes of printing—start-stop mode and print-on-the fly mode. Appropriate software is used in the control of the instrument during its operation. Based on the concentration of material and the expected amount of cross-linking to the surface, 5 to 50 attomoles of material are available in each spot. No shearing of DNA strands has been observed with material up to 2000 base pairs in length. Viscosity limits the length of the DNA that can be studied. At the desired concentration, such as 1 μg/μL, DNA of 5 kb and larger likely will be too viscous for a small-orifice (30-μm) jet. Viscosity reduction by adding cosolvents may alleviate the problem.

The total time to print a batch of arrays includes the setup time, the time spent cleaning and loading the liquid deposition devices, and the print time itself:

$$\text{Total time} = \text{print time} + \text{fill time} + \text{setup time} \tag{7.7}$$

All time associated with movement and deposition of spots is included in the print time. Fill time includes all wash steps for deposition device, time for loading the device, and time for testing the load and getting into position for printing. Setup time includes the time to load the array substrates and microtitre plates, etc. containing array element material, as well as time to offload the instrument when the batch run is complete.

For either the pin or jet instruments, the sum of the print time and fill time can be expressed in terms of the number of instrument cycles and the time per cycle for each component:

$$\text{Print time} + \text{fill time} = \text{number of cycles} \times (\text{print time/cycle} + \text{fill time/cycle}) \tag{7.8}$$

$$T = C(P \times T_{\text{cycle}} + T_f) + T_s \tag{7.9}$$

where T is total time, C is cycles, $P \times T_{cyc}$ is print time per cycle, T_f is fill time per cycle, and T_s is setup time. For the pin tool,

$$P \times T_{cycle} = NT_c \tag{7.10}$$

where N is the number of arrays printed and T_c is contact time per array, including motion

$$C = \frac{G}{P} \tag{7.11}$$

where G is the number of genes and P is the number of pins. Combining terms, the total print time for a batch-mode pin device thus is

$$T_p = G/P(NT_c + T_{f,p}) + T_{sp} \tag{7.12}$$

where R is the number of rows of arrays on a platter, T_1 is the print time per line, and J is the number of jets. By arranging the arrays as a square, the number of rows can be calculated:

$$R = (N)^{1/2} \tag{7.13}$$

Combining the preceding terms, the total time for batch printing with the jet approach is

$$T_j = G/J(RT_1 + T_{f,j}) + T_{s,j} \tag{7.14}$$

The time for printing for the jet approach depends on the number of rows in the arrays, and time for pin printing depends on the number of arrays. The time grows linearly with the number of arrays in the pin tool and changes with the square root of the number of arrays in the jet instrument. For an equal number of jets and pins, the jet instrument always will have the time advantage. The crossover point is independent of the number of genes printed.

7.6 Normalization for cDNA Microarray Data

There are many sources of systematic variation in microarray experiments that affect the measured gene expression levels. *Normalization* is the term used to describe the process of removing such variation, e.g., for differences in labeling efficiency between the two fluorescent dyes. In this case, a constant adjustment is commonly used to force the distribution of the log ratios to have a median of zero for each slide. For cDNA microarrays, the purpose of dye normalization is to balance the fluorescence intensities of the two dyes green Cy3 and red Cy5 dye as well as to allow the comparison of expression levels across experiments. Dye bias can be seen most obviously in an experiment where two identical mRNA samples are labeled with different dyes and subsequently hybridized to the same slide. The bias can stem from a number of factors, including physical properties of the dyes (e.g., heat and light sensitivity and relative half-life), efficiency of dye incorporation, experimental variability in probe coupling and processing procedures, and scanner settings at the data-collection step. The relative gene expression levels measured as log ratios from replicate experiments may have different spreads owing to differences in experimental conditions. Some scale adjustment then may be required so that the relative expression levels from one particular experiment do not dominate the average relative expression levels across replicate experiments.

Speed [9] developed a normalization procedure using gene expression data from lipid metabolism in mice. He attempted to identify genes with altered expression in apoliprotein AI knockout mice with low high-density lipoprotein (HDL) cholesterol levels compared with inbred C57B1/6 control mice. The normalization procedure depends on the experimental setup. Three situations are identified:

1. Within-slide normalization
2. Paired-slide normalization
3. Multiple-slide normalization

A number of considerations influence this decision, such as the proportion of genes that are expected to be expressed differentially in the red and green samples and the availability of control DNA sequences. Three types of approaches were described:

1. *All genes in the array.* Frequently, biologic comparisons made on microarrays are very specific in nature, i.e., only a small proportion of genes are expected to be differentially expressed. Therefore, the remaining genes are expected to have constant expression, and so can be used as indicators of the relative intensities of the two dyes. Almost all genes on the array may be used for normalization.
2. *Constantly expressed genes.* Instead of using all genes on the array for normalization, a smaller set of genes called *housekeeping genes* has constant expression across a variety of conditions, e.g., β-actin. Although it is very hard to identify a set of housekeeping genes that does not change significantly under any conditions, it may be possible to find sets of "temporary" housekeeping genes for particular experimental conditions.
3. *Controls.* An alternative to normalization by housekeeping genes is used to spike controls or a titration series of control

sequences. In the spiked-controls method, synthetic DNA sequences or DNA sequences from an organism different from the one being studied are spotted on the array (with possible replication) and included in the two mRNA samples in equal amounts. These spotted control sequences thus should have equal red and green intensities and could be used for normalization. In the titration-series approach, spots consisting of different concentrations of the same gene or expressed sequence tag (EST) are printed on the array. These spots are expected to have equal red and green intensities across the range of intensities. Genomic DNA that is supposed to have constant expression levels across various conditions may be used in the titration series. In practice, however, genomic DNA is often too complex to exhibit much signal, and setting a titration series that spans the range of intensities for different experiments is technically very challenging.

The apo AI experiment was carried out as part of a study of lipid metabolism and artheroscelerosis susceptibility in mice. Apoliprotein AI is a gene known to play a pivotal role in HDL metabolism. The treatment group consisted of 8 mice with the apo AI gene knocked out, and the control group consisted of 8 normal C57B1/6 mice. For each of these 16 mice, target cDNA was obtained from mRNA by reverse transcription and labeled using a red fluorescent dye, Cy5. The reference sample used in all hybridizations was prepared by pooling cDNA from the 8 control mice and was labeled with a green fluorescent dye, Cy3. In this experiment, target cDNA was hybridized to microarrays containing 6384 cDNA probes, including 200 related to lipid metabolism. Each of the 16 hybridizations produced a pair of 16-bit images that were processed using the software package Spot. The main quantities of interest produced by the image-analysis methods are the (R, G) fluorescence intensity pairs for each gene on the array.

After image processing and normalization, the gene expression data can be summarized by a matrix X of log-intensity ratios $\lg_2(R/G)$ with p rows corresponding to the genes being studied and $n = n_1 + n_2$ columns corresponding to the n_1 control hybridizations (C57BI/6) and n_2 treatment hybridizations (apo AI knockout). In the experiment considered, $n_1 = n_2 = 8$ and $p = 5548$. Differentially expressed genes were identified by computing t statistics. For genes, j, the t statistic comparing gene expression in the control and treatment groups, is

$$t_j = \frac{(x_{2j} - x_{1j})}{\sqrt{\dfrac{s_{1j}^2}{n_1} + \dfrac{s_{2j}^2}{n_2}}} \tag{7.15}$$

where x_{1j} and x_{2j} denote the average background corrected and normalized expression levels of gene j in the n_1 control and n_2 treatment hybridizations, respectively. Similarly, s_{1j}^2 and s_{2j}^2 denote the variances of gene j's expression levels in the control and treatment hybridizations, respectively. Large absolute t statistics suggest that the corresponding genes have different expression levels in the control and treatment groups. The statistical significance of the results was assessed based on p values adjusted for multiple comparisons.

Global normalization methods assume that the red and green intensities are related by a constant factor. That is R = kG, and in practice, the center of the distribution of log ratios is shifted to zero:

$$\log_2(R/G) \rightarrow \log_2(R/G) - c = \log_2[R/(kG)] \qquad (7.16)$$

A common choice for the location parameter $c = \log_2(k)$ is the median or mean of the log intensity ratios for a particular gene set. Global normalization methods are mentioned in the preprocessing steps in a number of papers on the identification of differentially expressed genes in single-slide cDNA microarray experiments. In many cases, the dye bias appears to depend on spot intensity, as revealed by plots of the log-ratio M versus overall sport intensity A. An intensity- or A-dependent dye normalization method thus may be preferable to global methods. A local A-dependent normalization was performed using the robust scatter plot smoother Lowess from the statistical software package R[7].

$$\log_2(R/G) \rightarrow \log_2(R/G) - c(A) = \log_2\{R/[k(A)G]\} \qquad (7.17)$$

where $c(A)$ is the lowess fit to the M versus A plot. The Lowess (·) function is a scatter plot smoother that was found to perform robust locally linear fits. The Lowess (·) function will not be affected by a small percentage of differentially expressed genes, which will appear as outliers in the M versus A plot. The user-defined parameter f is the fraction of the data used for smoothing at each point; the larger the f value, the smoother is the fit. The M versus A plot amounts to a 45-degree counterclockwise rotation of the log(G), log(R) coordinate system. Within the print-tip group, normalization is simply a (print tip + A)–dependent normalization that is

$$\log_2(R/G) \rightarrow \log_2(R/G) \rightarrow c_i(A) = \log_2[R/k_i(A)G] \qquad (7.18)$$

where $c_i(A)$ is the Lowess fit to the M versus A plot for the ith grid only, $I = 1, 2, \ldots, i$ represents the number of print tips.

Paired-slides normalization applied dye-swamp experiments, two hybridizations for two mRNA samples with dye assignment reversed in the second hybridization. The normalized log ratios for the first slide are denoted by $\log_2(R/G) - c$ and those for the second slide by $\log_2(R'/G') - c'$. Here, c and c' denote the normalization functions for the two slides. These could be obtained by any of the within-slide normalization methods described earlier. If $c \approx c'$,

$$\begin{aligned} &½\{\log_2(R/G) - c - [\log_2(R'/G') - c']\} \\ &\approx ½[\log_2(R/G) + \log_2(G'/R')] \\ &= ½ \log_2(RG'/GR') = ½(M - M') \end{aligned} \quad (7.19)$$

The relative expression levels for the two slides may be combined for the two slides without explicit normalization by a procedure referred to as *self-normalization.* The validity of the assumption can be checked using housekeeping genes or genomic DNA. Given that the dye assignments are reversed in the two experiments, one expects that the normalized log ratios on the two slides are of equal magnitude and opposite sign, that is,

$$\log_2(R/G) - c \approx \log_2(R'/G') - c' \quad (7.20)$$

Therefore, rearranging the equation and assuming again that $c \approx c'$, the normalization function c can be given by

$$c \approx ½[\log_2(R/G) + \log_2(R'/G')] = ½(M + M') \quad (7.21)$$

In practice, $c = c(A)$ is estimated by the Lowess fit to the plot of $½(M + M') = ½\log_2(RR'/GG')$ versus $½(A + A')$, where this time all the genes are used. Global normalization amounts to a vertical translation in an M versus A plot and does not allow for spatial- or intensity-dependent dye biases.

Summary

Microarray techniques can be used to measure gene expression, understand disease states better, and effect cures by better drug design. Schena used glass substrates with less background fluorescence and developed the enzymatic labeling procedure using fluorescent probes from yeast and plant mRNA. The microarray industry is expected to grow in a similar fashion as the microprocessor industry has grown. A microarray is an ordered array of microscopic elements on a planar substrate that allows the specific binding of genes or gene products. To qualify as a microarray, the analytical device must be ordered, microscopic, planar, and specific. The microarray analysis life cycle consists of five steps: formulation of a biologic question, sample preparation, biochemical reaction, detection, and data analysis and modeling. Ten tips were given to ensure success in microarray analysis. Some of the interesting applications of microarrays are gene expression, drug delivery, genetic screening and diagnostics, gene profiling, understanding mechanism of aging, the study of cancer, etc.

The confocal scanning microscope can be used in microarray detection that uses fluorescence scanning. The sample is excited by laser beam, and fluorescence light is emitted from the probe in the sample and can be detected using the difference in wavelength of 24 nm between excitation and emitted light beams. Epi-illumination is used in the scanning process. The excitation and emitted beams pass through the objective lens to and from the sample but in opposite directions. PMT is used as a detecting element. The instrument performance measures are number of lasers and fluorescence channels, detectivity, sensitivity, crosstalk, resolution, field size, uniformity, image geometry, throughput, and superposition of signal sources. High-quality surfaces are needed for the preparation of microarray samples. An ideal microarray surface has to be dimensional, flat, planar, uniform, inert, efficient, and accessible.

Optimal target concentration occurs at a spacing of 1 DNA target molecule per 20 Å. The probe duplex is approximately 24 Å. Optimal probe concentration is the number of probe molecules per unit volume of sample that provides the strongest signal in a microarray assay. Microarrays of oligonucleotides can be prepared using delivery or synthesis methods. The four steps in the process of oligonucleotide synthesis are deprotection, coupling, capping, and oxidation. The three manufacturing methods used during microarray manufacture are ink-jet printing, mechanical microspotting, and photolithography. Stepwise coupling efficiency can be defined to gauge the quality of microarray synthesis. Linker molecules can be used to increase the efficiency of hybridization and DNA attachment at the surface. The time taken for ink-jet printing when jets or pins are used is compared.

Statistical normalization procedures can be used to remove systematic variation in microarray experiments that affects the measured gene expression levels. Speed developed a normalization procedure using gene expression data from lipid metabolism in mice. He used housekeeping genes that have constant levels of expression across a variety of conditions. Differentially expressed genes were identified by computing t statistics. Global normalization methods, M versus A plot, paired-slide normalization, within-slide normalization, and multiple-slide normalization methods are discussed.

References

[1] M. Schena, *Microrrray Analysis.* New York: Wiley, 2003.
[2] M. Schena, *DNA Microarrays: A Practical Approach.* Oxford, UK: Oxford University Press, 2002.
[3] M. Schena, D. Shalon, R. W. Davis, and P. O. Brown, "Quantitative monitoring of gene expression patterns with complementary DNA microarray," *Science.* 270 (1995), 467–470.

[4] M. Schena, R. A. Heller, T. P. Theriault, K. Konrad, E. Lachenmeier and R. W. Davis, "Microarrays:Biotechnology Discovery Platform for Functional Genomics", *Trends Biotechnol.* 16 (1998), 301.
[5] L. Pauling, H. A. Itano, S. J. Singer, and I. C. Wells "Sickle cell anemia: A molecular disease," *Science.* 110 (1949), 543–548.
[6] K. B. Mullis, "The unusual origin of the polymerase chain reaction," *Sci. Am.* 262 (1990), 56–61.
[7] D. Baltimore, "RNA-dependent DNA polymerase in virions of RNA tumour viruses," *Nature.* 226 (1970), 1209–1211.
[8] M. G. Ormerod, "Flow cytometry: a practical approach". Oxford, UK: IRL Press at Oxford University Press, 1994.
[9] T. P. Speed, *Statistical Analysis of Gene Expression Microarray Data.* Boca Raton, FL: CRC Press, 2002.

Exercises

1.0 Compare the silicon chip with the biochip, including the same three fundamental principles.

2.0 Describe the four basic criteria for microarray.

3.0 Why are there maxima in intensity versus target concentration?

4.0 Show how the DNA gets attached along the entire length of the molecule to an amine glass surface.

5.0 Describe the role of blocking agents.

6.0 Describe with a schematic the four-step process of the oligonucleotide synthesis.

7.0 Discuss the hybridization parameters of probe and target with glass substrate.

8.0 What is the role of the mirror in the confocal scanning arrangement in a microarray scanner?

9.0 Show by schematic a geometric beamsplitter in an epi-illuminated scanner.

10.0 Compare the time taken for jet printing versus pin printing of microarray dots.

11.0 What is the drop size dispensed by a nanocapillary jet?

12.0 Discuss the protocol for SNP array synthesis with a schematic.

13.0 Elaborate the protocol for short tandem repeat array synthesis.

14.0 How is the student *t* test used during normalization?

15.0 What are housekeeping genes?

16.0 Discuss *k* means clustering with an illustration.

17.0 Name two examples of target molecules other than DNA.

18.0 What is the difference between an amine-treated and an aldehyde-treated surface?

19.0 Discuss the first prototype confocal scanning microscope.

20.0 What is a Stokes shift?

21.0 What is the role of a PMT detector?

22.0 What is the role of hydration in microarray sample preparation?

23.0 Who was the first to discover the correlation between disease and gene expression?

24.0 Discuss the binding efficiency during hybridization.

25.0 Why is glass a superior choice for a substrate?

26.0 What is a dichroic lens?

27.0 Discuss the instrument performance measures of a confocal scanning microscope.

28.0 What is the difference between photolithography and ink-jet printing?

29.0 What should be the considerations for printing a nanoarray?

30.0 What is dye bias?

31.0 Discuss T. P. Speed's experiments on the lipid metabolism in mice.

32.0 Discuss the pioneering event in the development of the field of microarray technology and the exploration of the function of transcription factors in the flowering plant *Arabidopsis thaliana.*

33.0 Discuss the study of yeast in Brown's laboratory using microrrrays.

34.0 Discuss the properties of zinc titania glass and the advantages of using it as a substrate.

35.0 Draw a neat schematic of the confocal scanning microscope.

36.0 Distinguish between global normalization and within-slide normalization.

37.0 Discuss the key requirements of the microarray substrate.

38.0 Enumerate the 10 tips for sample preparation.

39.0 Distinguish between in situ synthesis and delivery methods of target preparation.

40.0 What is the optimal probe concentration?

41.0 What are amine- and aldehyde-treated surfaces?

42.0 What is an *M* versus *A* plot?

43.0 What is dendrimer technology?

44.0 Write notes on pharmacogenomics.

45.0 What is the role of the fluorescent probe in sample preparation?

46.0 Discuss the five steps of microarray preparation.

47.0 Name the three methods of manufacture of microarray analysis.

48.0 Why is controlled-pore glass used during the synthesis of phosphoramidite?

49.0 What are optical requirements of excitation and emittance?

50.0 What is the difference between microarray and macroarray?

51.0 The presence of water is required during the hybridization reactions. Evaporation rate also increases with temperature. How can the sample be kept warm and hydrated?

52.0 Given Tuppy's estimate for sequencing proteins, how long will it take to sequence a DNA? What would be the reduction in time if microarray technology were used?

53.0 Why is uniformity among different dots important in microarray analysis?

CHAPTER 8

Electrophoretic Techniques and Finite Speed of Diffusion

Objectives

The objectives of this chapter are to

- Understand the role of electrophoresis in sequence distribution measurement.
- Understand the role of molecular diffusion in electrophoresis.
- Understand the limitations of Fick's laws of diffusion.
- Derive a generalized Fick's law of diffusion.
- Apply a generalized Fick's law of diffusion to standard geometries.
- Apply a generalized Fick's law of diffusion to electrophoretic transport.

8.1 Role of Electrophoresis in the Measurement of Sequence Distribution

As discussed in Chap. 1, the sequence distribution of DNA can be obtained by the method of gel acrylamide electrophoresis. The technique of electrophoresis is not described in detail in the current literature. A similar technique, paper chromatography, is used in the acquisition of the sequence distribution of polypeptides. In both these techniques, diffusion plays an important role. Fick's laws of diffusion have limitations, especially in the time frame of the critical events that take place during measurement of sequence distribution.

The critical events are the migration of molecular fragments over varying distances depending on their molecular sizes. Usually, calibration is used to convert the raw measurements to sequence data. Why are mathematical models not used for interpretation of the electrophoretic pattern or the paper chromatographic pattern? It is being realized increasingly among investigators that at short time scales, Fick's description of transient diffusion is not an adequate representation of all the events. This chapter reviews molecular diffusion principles with particular attention to the limitations of Fick's laws of diffusion. A generalized Fick's law of diffusion is used to account for all the transient time events that occur during a real process. The implications on the electrophoretic techniques and sequence errors and shotgun sequencing cannot be overemphasized.

8.2 Fick's Laws of Molecular Diffusion

Diffusion is the migration of a species from a region of a higher concentration to a region of lower concentration under the driving forces of a concentration gradient in a primary manner. Other forces can cause such movement in a secondary manner, such as a superimposed temperature gradient, as in thermophoresis; a superimposed concentration gradient of a second species, as in diffusophoresis; a superimposed electromotive gradient, as in electrophoresis; an osmotic potential; a steam sweep; a centripetal force; a pressure drop; a surface tension gradient; a surface force; and so on. The term *molecular diffusion* refers to the Brownian motion of molecules from a region of higher concentration to a region of lower concentration. The movement of species from a region of lower concentration to a region of higher concentration would be in violation of the second law of thermodynamics.

The *Clausius inequality* states that heat always will flow from a region of higher temperature to a region of lower temperature. It can never flow from a region of lower temperature to a region of higher temperature in a spontaneous fashion. Not all heat can be converted to work without discarding some heat to the lower temperature region. In an analogous manner, mass cannot diffuse from a region of low concentration or low chemical potential to a region of higher concentration or high chemical potential in a spontaneous manner. The direction of transfer is to equalize the concentration.

By another analogy between heat and mass transfer, the stipulation of the third law of thermodynamics that the lowest attainable temperature anywhere in the universe is 0 K translates into the law that there can exist no negative concentration. The lowest concentration achievable anywhere in the universe is 0 mol/m^3. Diffusion plays a pivotal role in the sequence distribution analysis in genome and proteome projects.

Albert Einstein, one of the best physicists of the twentieth century, observed that a cube of sugar placed in the bottom of a hot cup of tea diffused, and a uniform concentration of sugar throughout the entire cup results at the final state. If a few crystals of potassium permanganate ($KMnO_4$) are placed at the bottom of a tall bottle filled with triple distilled water, the pink color will spread slowly throughout the bottle. At first, the color will be concentrated in the bottom of the bottle. After a day, it will penetrate upward a few centimeters. After several years, the solution will appear homogeneous. The process responsible for movement of the colored material is diffusion. Diffusion is a molecular phenomenon. In gases, diffusion progresses at a rate of about 10 cm/min, in liquids its rate is about 0.05 cm/min, and in solids its rate is about 100 nm/min.

In the middle of the nineteenth century, Fick introduced two differential equations that provide a mathematical framework to describe the otherwise random phenomenon of molecular diffusion. The flow of mass by diffusion across a plane was proportional to the concentration gradient of the diffusant across the plane. The components in a mixture are transported by a driving force during diffusion. The ability of the diffusant to pass through a body depends on the diffusion coefficient D (m^2/s). The solubility of the species in the body is also a salient consideration in determining the permeation rates of the species in the body. Fick stated the first two laws of diffusion in the year 1855. He was the youngest of five children of a civil engineer. He was very much interested in mathematics in his high school and was enamored by the work of Poisson. His brother, a professor of anatomy, persuaded Fick to switch to medicine from mathematics. Ludwig served as Fick's tutor. Fick's thesis was on visual errors caused by astigmatism. He performed outstanding work on mechanics in hydrodynamics and hemorheology and in the visual and thermal functioning of the human body. In his first paper on diffusion, published in 1855, Fick interpreted the experiments of Graham with interesting theories, analogies, and quantitative experiments. He showed that diffusion can be described on the same mathematical basis as Fourier's law of heat conduction and Ohm's law of electricity. Fick's first law of diffusion can be written as [1]

$$J = -AD\frac{\partial C_A}{\partial x} \tag{8.1}$$

where J is defined as the one-dimensional molar flux. The diffusivity is the proportionality constant that depends on the material under consideration, a thin shell of thickness Δx with constant cross-sectional area A across which the diffusion is considered to occur. A mass balance in the incremental volume considered $A\Delta x$ for an incremental time Δt, neglecting any reaction or accumulation of the species, can be written as

$$\text{Mass in} - \text{mass out} \pm \text{mass reacted/generated} = \text{mass accumulated} \tag{8.2}$$

$$\Delta t(J_x - J_{x+\Delta x}) = A\Delta x \Delta C_A \tag{8.3}$$

Dividing Eq. (8.3) throughout by $A\Delta x\Delta t$ and obtaining the limits as Δx and Δt go to zero gives

$$-\frac{\partial J}{\partial x} = A\frac{\partial C_A}{\partial t} \tag{8.4}$$

Combining Eqs. (8.1) and (8.4), the governing equation for the diffusing species when the area across which the diffusion occurs is a constant becomes

$$D\frac{\partial^2 C_A}{\partial x^2} = \frac{\partial C_A}{\partial t} \tag{8.5}$$

Equation (8.5) is Fick's second law of diffusion in one dimension. This is a fundamental equation that described the transient one-dimensional diffusion of the migrating species. When Fick attempted to integrate Eq. (8.5), he was discouraged by the numerical effort needed. He found the second derivative difficult to measure experimentally, and he ran into the effect of experimental errors increase by the second difference. Finally, he demonstrated in a cylindrical cell the steady-state linear concentration gradient of sodium chloride (NaCl). He uses a glass cylinder containing crystalline sodium chloride in the bottom and a large volume of water in the top. By periodically changing the water in the top volume, he was able to establish a steady-state concentration gradient in the cylindrical cell. He confirmed his equation from this steady-state gradient.

The *Skylab* science demonstration was the first in a series of investigations designed by Facimire [2] to study low-gravity diffusive mass transfer. The specific objective of the demonstration was to photographically document the diffusion of tea in water in spacecraft. In preparation for the experiment, *Skylab* pilot Jack Lousma filled a ½-in-diameter, 6-ft-long transparent tube three-quarters full with water. A highly concentrated tea solution then was delivered to the water surface (via a 5-cc syringe) through a synthetic fiber wad. The tube was then capped. The fiber pad was employed to try to bring the tea and water in contact without entrapped air. Three attempts to produce the wad were unsuccessful. During the fourth attempt, an "a good bubble-free interface" was realized. The next day, Lousma reported that no diffusion of the tea in the liquid had occurred. Thus the experiment was initiated again. During this new experimental run, the wad was removed, and the tea was delivered on top of the water. After an air bubble between the tea and water was removed via the syringe, a "smooth, continuous interface" was achieved. The tea was allowed to diffuse over the next 3 days. After the flight, 16-mm photographs of the diffusion were analyzed. In 51.15 hours, the visible diffusion front advanced 1.96 cm. It was noted that the diffusion front become increasingly parabolic during the demonstration. It also was noted that very little diffusion occurred near the container wall. A similar ground-based experiment was performed for comparison with the space investigation. After 45.5 hours, three different zones were visible: (1) a dark area, (2) an area of medium darkness, and (3) a very light area. The medium-colored area had advanced 1.6 cm in 45.5 hours.

8.3 Generalized Fick's Law of Diffusion

Fick's model of molecular diffusion is not universal. It is analogous to Fourier's law of heat conduction, as Fick proposed in his stated laws. There are seven reasons to seek a generalized Fick's law of molecular diffusion:

1. The theory of Onsager and the contradiction of Fick's law of molecular diffusion by the theory of microscopic reversibility [3].
2. Nernst [4] found that heat can have inertia in good thermal conductors at low temperatures, which can lead to oscillatory discharge. This cannot be fully explained by Fourier and, by analogy, Fick.
3. Events at high mass flux rates cannot be described using Fick's parabolic equations.
4. Landau and Lifshitz [5] noted that the speed of heat and, by analogy, the speed of mass cannot be greater than the speed of light. That the speed of a moving object has to be less than the speed of light was examined by Kelly [6] for diffusion.
5. Singularities can be found in the solutions to the Fick parabolic model for industrially important cases:
   a. Blowup of surface flux as time goes to zero during transient molecular diffusion in a semi-infinite medium subject to a constant wall concentration in Cartesian coordinates [7].
   b. Surface flux during transient molecular diffusion within a finite slab of width $2a$ subject to a step change in surface concentration.

c. The concentration term in the constant wall flux problem in cylindrical coordinates in infinite medium is solved for using the Boltzmann transformation [8], leading to a solution in an exponential integral [9–15].

d. In the short time limit, solutions to the parabolic equations by Boltzmann transformation for an infinite sphere blowup can be found.

6. Fick's law was developed from empirical observations at steady state, and when used in transient applications, it is an extrapolation that is not confirmed adequately by systematic experimental study or molecular theories.

7. Overpredictions of the theory to experimental observations were found in gel electrophoresis [16], restriction mapping [17], adsorption [18], nuclear fuel rods [19], drug delivery systems [20], and heat transfer systems [21,22] when the Fick model is used. This indicates that there is another mechanism that has not been accounted for.

Boley [23] found that addition of the second derivative in time of temperature to the governing equation is the only way to remove the singularities in the solution to parabolic heat conduction equations. Thus a generalized Fick's law of mass diffusion can be written as

$$J = -DA\frac{\partial C_A}{\partial x} - \tau_r A\frac{\partial J}{\partial t} \tag{8.6}$$

This is a damped-wave diffusion and relaxation equation or a generalized Fick's law of molecular diffusion. When the relaxation time τ_r is zero, Eq. (8.6) will revert to Fick's model of molecular diffusion. Reference to the use of this equation was found in heat conduction and can be traced back to Maxwell [24], Morse and Feshbach [25], Cattaneo [26], and Vernotte [27], who postulated this equation independently. This equation can be used to account for the finite speed of molecular diffusion and removes the infinite speed implied in Fick's model of molecular diffusion. Experimental evidence in heat conduction has been found, and relaxation times on the order of 20 seconds have been reported by Mitra and colleagues [28]. During drying of solids, the relaxation time can be on the order of few thousand seconds [29]. Tzou [30] found the relaxation times to be on the order of a few nanoseconds in heat conduction in stainless steel. A table of relaxation time values is not available in the literature. More research is needed to tabulate the relaxation-time values for molecular diffusion for different species that migrate.

The diffusion coefficient depends on a number of parameters, including the temperature of the medium of migration [31]. One method of deriving an expression for diffusion coefficient for liquids has been reported, and the derived expression is called the *Stokes-Einstein equation* to calculate diffusion coefficients.

8.3.1 Derivation of a Generalized Fick's Law of Diffusion

The Stokes-Einstein equation can be used to calculate diffusion coefficients in liquids:

$$D = \frac{k_B T}{f} = \frac{k_B T}{6\pi\mu R_0} \tag{8.7}$$

where k_B is the Boltzmann constant, f is the frictional drag coefficient, T is the temperature, μ is the viscosity of the surrounding medium, and R_0 is the radius of the solute that is diffusing. Equation (8.7) can be derived as follows: A rigid solute sphere is assumed for the molecule diffusing in a common solvent. The frictional drag force acting on the molecule opposing its motion is proportional to the velocity of the sphere:

$$\text{Drag force} = f v_1 \tag{8.8}$$

where v_1 is the velocity of the molecule. From Stokes law [32] for a sphere moving in a fluid, $f = 6\pi\mu R_0$. The driving force was taken by Einstein [33] to be the negative of the chemical potential gradient ($-\nabla\mu_A$) defined per molecule:

$$-\nabla\mu_A = (6\pi\mu R_0) v_A \tag{8.9}$$

Equation (8.9) is valid when the molecule reaches a steady-state velocity. This occurs when the net force acting on the molecule is zero. The solution is assumed to be ideal and dilute:

$$\mu_A = \mu_A{}^0 + k_B T \ln(x_A) = \mu_1{}^0 + k_B T \ln C_A - k_B T \ln C_B \tag{8.10}$$

For dilute solutions, the concentration of the second species C_B far exceeds the solute concentration and can be taken as constant. The gradient at constant temperature, then, is

$$\nabla\mu_1 = k_B T \frac{\nabla C_A}{C_A} = -(6\pi\mu R_0) v_A \tag{8.11}$$

$$\frac{-k_B T}{6\pi\mu R_0} \nabla C_A = C_A v_A = J/A \tag{8.12}$$

Comparing Eq. (8.12) with Fick's model of molecular diffusion given in Eq. (8.4), the Stokes-Einstein relationship of Eq. (8.7) results.

Equation (8.4) is valid only at steady state. Often, in transient applications, a sudden step change in concentration, i.e., the driving force, is imposed on the system. The molecule will experience an accelerating regime prior to reaching steady state. During the accelerating regime,

$$-\nabla\mu_A - (6\pi\mu R_0)v_A = m\frac{dv_A}{dt} \tag{8.13}$$

where m is the mass of the molecule. Then

$$mC_A\frac{dv_A}{dt} = -(6\pi\mu R_0)C_A v_A \tag{8.14}$$

or

$$-\frac{k_B TA}{6\pi\mu R_0}\nabla C_A = \frac{m}{6\pi\mu R_0}\frac{\partial J}{\partial t} + J \tag{8.15}$$

Equation (8.15) is a generalized Fick's law of diffusion that accounts for the acceleration regime of the molecule as well as the steady-state regime. An expression for the relaxation time for molecular diffusion falls out of the analysis, that is,

$$\tau_r = \frac{m}{6\pi\mu R_0} = \frac{mD}{k_B T} \tag{8.16}$$

In terms of P_{tot}, the system pressure for ideal gas, the relaxation time can be written as

$$\tau_r = \frac{MD\rho_m}{P}$$

where ρ_m is the molar density of the migrating species. The velocity of mass diffusion is given by

$$v_m = \sqrt{\frac{D}{\tau_r}} = \sqrt{\frac{k_B T}{m}} \tag{8.17}$$

Equation (8.17) can be rewritten in terms of the molar gas constant and molecular weight as

$$v_m = \sqrt{\frac{D}{\tau_r}} = \sqrt{\frac{RT}{M}} \tag{8.18}$$

The kinetic representation of pressure can be written after observing that a molecule moving in a cube of dimensions l with a velocity of v_x undergoes a momentum change of $2mv_x$ on one collision with the wall. The number of collisions on the wall can be estimated by first calculating the time taken by the molecule to move the round trip from the wall after a collision to the opposite wall and back as $2l/v_x$. The number of collisions undergone by a molecule is $v_x/2l$. The rate of transfer of momentum to the surface from the molecular collisions then is mv_x^2/l. The total force exerted by all the molecules colliding can be obtained by summing the contributions from each molecule, and the pressure is obtained by dividing the sum by the area of the wall and is given by [34]

$$P_{tot} = \frac{m}{l^3}\,(v_{x1}^{\;2} + v_{x2}^{\;2} + v_{x3}^{\;2} + \cdots) \tag{8.19}$$

Let N_m be the number of molecules in the system and n the number of molecules per unit volume. Then Eq. (8.19) can be rewritten after multiplying the numerator and denominator by N_m:

$$P_{tot} = mn\langle v_x^2\rangle = \rho\langle v_x^2\rangle = \tfrac{1}{3}\rho\langle v^2\rangle \tag{8.20}$$

Since the molecules treated as particles move in random, there is no preferred direction in the box. Hence $v^2 = v_x^2 + v_y^2 + v_z^2$. The square root of v^2 is called the *root mean squared speed* of the molecule and is a widely accepted average molecular speed. From the ideal gas law, $P_{tot} = \rho RT/M$. Combining this with Eq. (8.18) gives

$$\frac{1}{3\langle v^2\rangle} = \frac{RT}{M} = \frac{A_N k_B T}{M} \tag{8.21}$$

Comparing Eqs. (8.21) and Eq. (8.17), it can be seen the velocity of mass is one-third the root mean square velocity. This could be due to the fact that only one-dimensional diffusion has been considered. When all three dimensions are considered, these two velocities would be identical, although derived from different first principles.

The concentration in Cartesian, cylindrical, and spherical coordinates, taking into account the generalized Fick's law of mass diffusion and relaxation, is given by the following equations:

$$\tau_{mr}\left[\frac{\partial^2 C_A}{\partial t^2} + v_x\frac{\partial^2 C_A}{\partial x\partial t} + v_y\frac{\partial^2 C_A}{\partial y\partial t} + v_z\frac{\partial^2 C_A}{\partial z\partial t}\right] + \frac{\partial C_A}{\partial t} + \frac{\partial C_A}{\partial x}\left[\tau_{mr}\frac{\partial v_x}{\partial t} + v_x\right]$$

$$+\frac{\partial C_A}{\partial y}\left[\tau_{mr}\frac{\partial v_y}{\partial t} + v_y\right] + \frac{\partial C_A}{\partial z}\left[\tau_{mr}\frac{\partial v_z}{\partial t} + v_z\right]$$

$$= D\left[\frac{\partial^2 C_A}{\partial x^2} + \frac{\partial^2 C_A}{\partial y^2} + \frac{\partial^2 C_A}{\partial z^2}\right] + R_A \tag{8.22}$$

$$\tau_{mr}\left[\frac{\partial^2 C_A}{\partial t^2}+v_r\frac{\partial^2 C_A}{\partial r\partial t}+\frac{v_\theta}{r}\frac{\partial^2 C_A}{\partial\theta\partial t}+v_z\frac{\partial^2 C_A}{\partial z\partial t}\right]+\frac{\partial C_A}{\partial r}\left[\tau_{mr}\frac{\partial v_r}{\partial t}+v_r\right]$$

$$+\frac{1}{r}\frac{\partial C_A}{\partial\theta}\left[\tau_{mr}\frac{\partial v_\theta}{\partial t}+v_\theta\right]+\frac{\partial C_A}{\partial z}\left[\tau_{mr}\frac{\partial v_z}{\partial t}+v_z\right]+\frac{\partial C_A}{\partial t}$$

$$=D\left[\frac{1}{r}\frac{\partial}{\partial r}\left(r\frac{\partial C_A}{\partial r}\right)+\frac{1}{r^2}\frac{\partial^2 C_A}{\partial\theta^2}+\frac{\partial^2 C_A}{\partial z^2}\right]+R_A \tag{8.23}$$

$$\tau_{mr}\left[\frac{\partial^2 C_A}{\partial t^2}+v_r\frac{\partial^2 C_A}{\partial r\partial t}+\frac{v_\theta}{r}\frac{\partial^2 C_A}{\partial\theta\partial t}+v_\varphi\frac{1}{r\sin\theta}\frac{\partial^2 C_A}{\partial\varphi\partial t}\right]+\frac{\partial C_A}{\partial r}\left[\tau_{mr}\frac{\partial v_r}{\partial t}+v_r\right]$$

$$+\frac{1}{r}\frac{\partial C_A}{\partial\theta}\left[\tau_{mr}\frac{\partial v_\theta}{\partial t}+v_\theta\right]+\frac{1}{r\sin\theta}\frac{\partial C_A}{\partial\varphi}\left[\tau_{mr}\frac{\partial v_\varphi}{\partial t}+v_\varphi\right]+\frac{\partial C_A}{\partial t}$$

$$=D\left[\frac{1}{r^2}\frac{\partial}{\partial r}\left(r^2\frac{\partial C_A}{\partial r}\right)\right]+\frac{1}{r^2\sin\theta}\frac{\partial}{\partial\theta}\left[\sin\theta\frac{\partial C_A}{\partial\theta}\right]+\frac{\partial}{r^2\sin^2\theta}\frac{\partial^2 C_A}{\partial\varphi^2}+R_A \tag{8.24}$$

8.3.2 Taitel Paradox and Final Time Condition

Previous reports by Taitel [34] and Barletta and Zanchini [36] have raised some concerns about the second law of thermodynamics and the Cattaneo and Vernotte equation. The Cattaneo and Vernotte equation is an analogous equation in heat transfer to the generalized Fick's law of molecular diffusion. Taitel considered heat conduction in an infinitely wide parallel slab with thickness $2L$ such that the thermal conductivity k, the thermal diffusivity α, the specific heat at constant volume, and the thermal relaxation time τ_r of the slab can be considered constant. He notes that at time zero, $\partial T/\partial t = 0$ and uses it as one of the time conditions and $T = T_0$ at time zero as the second time condition. For times greater than zero, the temperature distribution on the two sides of the slab is kept uniform with a value $T_w \neq T_0$. By symmetry, at the center of the slab, $\partial T/\partial x = 0$ is the fourth space condition. A second-order hyperbolic partial differential equation (PDE) can be completely described by two space and two time conditions. On obtaining the transient temperature, Taitel points out that the absolute value of the temperature change $(T - T_0)$ may exceed $|T_w - T_0|$. Barletta and Zanchini develop a solution for the finite slab problem by the method of separation of variables. They show by a plot of $1 - u$ versus X for Vernotte number 1 $(\alpha\tau_r/4L^2)$ and Fourier number 0.7 $(\alpha t/4L^2)$ that $|T - T_0|$ may exceed $|T_w - T_0|$, as pointed out by Taitel. In another plot of $1 - u$ versus X for Vernotte number 1 and Fourier number 0.25, the equilibrium value for the temperature was attained by an oscillatory process. The parabolic conduction predicts a continuous increase in temperature from 0 to 1 at any internal position. The solution obtained by Taitel for the centerline temperature of the finite slab is given below. He considered a constant wall temperature, and the initial time conditions included a $\partial T/\partial t = 0$ term in addition to the initial temperature condition. The exact solution presented by Taitel is as follows:

$$u = \sum_0^\infty b_n \exp\left(\frac{-\tau}{2}\right)\exp\left(\frac{-\tau}{2}\sqrt{1 - \frac{4(2n+1)^2\pi^2\alpha\tau_r}{a^2}}\right)$$
$$+\sum_0^\infty c_n \exp\left(\frac{-\tau}{2}\right)\exp\left(\frac{+\tau}{2}\sqrt{1 - \frac{4(2n+1)^2\pi^2\alpha\tau_r}{a^2}}\right) \tag{8.25}$$

Multiplying both sides of the Eq. (8.25) by $\exp(\tau/2)$ gives

$$u\exp\left(\frac{\tau}{2}\right) = w = \sum_0^\infty \exp\left(\frac{-\tau}{2}\sqrt{1 - \frac{4(2n+1)^2\pi^2\alpha\tau_r}{a^2}}\right)$$
$$+\sum_0^\infty \exp\left(\frac{\tau}{2}\sqrt{1 - \frac{4(2n+1)^2\pi^2\alpha\tau_r}{a^2}}\right) \tag{8.26}$$

At infinite times, the left hand side (LHS) of Eq. (8.26) is 0 times ∞ and is 0. The right hand side (RHS) does not vanish. Thus the expression given by Taitel and later discussed as a temperature overshoot may be a result of the growing exponential term in the preceding expression.

Sharma [7] considered a finite slab of width $2a$ with an initial concentration at C_0. The sides of the slab are maintained at constant concentration of C_{AS}. The governing equation in the dimensionless form is then

$$\frac{\partial u}{\partial \tau} + \frac{\partial^2 u}{\partial \tau^2} = \frac{\partial^2 u}{\partial X^2} \tag{8.27}$$

where $$u = \frac{(C_A - C_{AS})}{(C_{A0} - C_{AS})};\tau = \frac{t}{\tau_{mr}};X = \frac{x}{\sqrt{D\tau_{mr}}}$$

The initial condition is given as

$$t = 0,\ u = 1 \tag{8.28}$$

The boundary conditions in space are given by

$$t > 0,\ X = 0,\ \partial u/\partial X = 0 \tag{8.29}$$

$$t > 0,\ X = \pm X_a,\ u = 0 \tag{8.30}$$

The fourth and final condition in time is

$$t = \infty,\ u = 0 \tag{8.31}$$

The governing equation was obtained by a one-dimensional mass balance (in – out + reaction = accumulation). This is achieved by eliminating J'' between the damped-wave diffusion and relaxation equation and the equation from mass balance ($-\partial J''/\partial x = \partial C/\partial t$). This is achieved by differentiating the constitutive equation with respect to x and the mass balance equation with respect to t and eliminating the second cross-derivative of J'' with respect to x and time. This equation is then nondimensionalized. The solution is obtained by the method of separation of variables. Let

$$u = V(\tau)\phi(X) \tag{8.32}$$

Equation (8.27) becomes

$$\phi''(X)/\phi(X) = [V'(\tau) + V''(\tau)]/V(\tau) = -\lambda_n^2 \tag{8.33}$$

$$\phi(X) = c_1 \sin(\lambda_n X) + c_2 \cos(\lambda_n X) \tag{8.34}$$

From the boundary conditions, at $X = 0$,

$$\partial\phi/\partial X = 0, \qquad \text{so } c_1 = 0 \tag{8.35}$$

$$\phi(X) = c_1 \cos(\lambda_n X) \tag{8.36}$$

$$0 = c_1 \cos(\lambda_n X_a) \tag{8.37}$$

$$(2n - 1)\pi/2 = \lambda_n X_a \tag{8.38}$$

$$\lambda_n = (2n - 1)\pi \sqrt{(\alpha\tau_r)/2a} \qquad n = 1, 2, 3, \ldots \tag{8.39}$$

The time domain solution would be

$$V = \exp\left(\frac{-\tau}{2}\right)\left[c_3 \exp\left(\tau\sqrt{\frac{1}{4} - \lambda_n^2}\right) + c_4 \exp\left(\tau\sqrt{\frac{1}{4} - \lambda_n^2}\right)\right]$$

$$\text{or} \quad V\exp\left(\frac{\tau}{2}\right) = \left[c_3 \exp\left(\tau\sqrt{\frac{1}{4} - \lambda_n^2}\right) + c_4 \exp\left(\tau\sqrt{\frac{1}{4} - \lambda_n^2}\right)\right] \tag{8.40}$$

from the final condition $u = 0$ at infinite time. So $V\phi\exp(\tau/2) = W$, the wave concentration at infinite time. The wave concentration is that portion of the solution that remains after dividing the damping component from either the solution or the governing equation. For any nonzero ϕ, it can be seen that at infinite time the LHS of Eq. (8.40) is a product of zero and infinity and a function of x and is zero. Hence the RHS of Eq. (8.40) is also zero, and hence in Eq. (8.40), c_3 needs to be set to zero. Hence

$$u = \sum_{1}^{\infty} \exp\left(\frac{-\tau}{2}\right)\exp\left[-\tau\sqrt{\frac{1}{4} - \lambda_n^2}\right]\cos(\lambda_n X) \tag{8.41}$$

where λ_n is described by Eq. (8.39). C_n can be shown using the orthogonality property to be $4(-1)^{n+1}/(2n-1)\pi$. It can be seen that Eq. (8.41) is bifurcated. As the value of the thickness of the slab changes, the characteristic nature of the solution changes from monotonic exponential decay to subcritical damped oscillatory. For $a < \pi\sqrt{(D\tau_r)}$, even for $n = 1$, $\lambda_n > ½$. This is when the argument within the square root sign in the exponentiated time domain expression becomes negative, and the result becomes imaginary. Using Demovrie's theorem and taking the real part for small width of the slab,

$$u = \sum_{1}^{\infty} c_n \exp\left(\frac{-\tau}{2}\right)\cos\left(\tau\sqrt{\lambda_n^2 - \frac{1}{4}}\right)\cos(\lambda_n X) \tag{8.42}$$

Equations (8.41) and (8.42) can be seen to be well bounded. Equation (8.42) becomes zero after some time. This would be the time taken to reach steady state. Thus, for $a \geq \pi\sqrt{(D\tau_{mr})}$, the transient concentration is described by Eq. (8.41), where $c_n = 4(-1)^{n+1}/(2n-1)\pi$ and $\lambda_n = (2n-1)\pi\sqrt{(D_{AB}\tau_r)}\,/2a$.

The centerline concentration is shown in Fig. 8.1. Eight terms in the infinite series given in Eq. (8.41) were taken, and the values were calculated on a 1.9-GHz Pentium IV desktop personal computer. The number of terms was decided on the incremental change or improvement obtained by doubling the number of terms. The number of terms was arrived at a 4 percent change in the dimensionless temperature. The subcritical damped oscillations can be seen in the figure. The time taken to steady state can be read from the x intercept. The figure shows a parametric study of the relaxation time. A small slab of thickness of 1 cm and binary diffusivity of 10^{-5} m^2/s is considered. Twelve terms were taken in the infinite series solution, and four different relaxation times were calculated. The accuracy of the data was less than 4 percent. For the case where the relaxation time was small, i.e., when Eq. (8.41) was applicable for the solution,

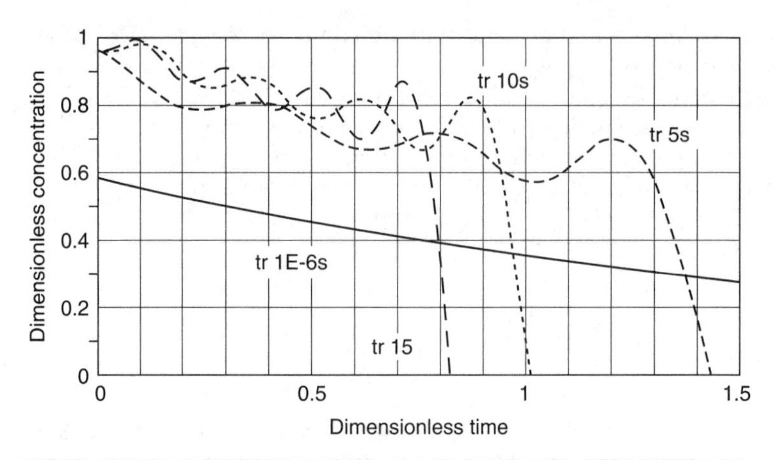

FIGURE 8.1 Dimensionless concentration for a finite slab at different relaxation times.

the centerline concentration decayed monotonically with the x axis as its asymptote. When the relaxation time considered was large in such a fashion that Eq. (8.42) is applicable, the subcritical damped oscillations can be seen. The time taken to steady state can be read from the x intercept in such cases. This happens when

$$\tau_r > \frac{a^2}{\pi^2 D} \tag{8.43}$$

At infinite relaxation time, the governing equation will revert to the wave equation [7], and the D'Alambert solution will result. For a wide range of mass relaxation times, this approach can be seen to be viable.

The Taitel paradox is obviated by examining the final steady-state condition and expressing the state in mathematical terms. The W term, which is the dimensionless concentration on removal of the damping term, needs to go to zero at infinite time. This resulted in a well-bounded solution. Use of the *final* condition may be what is needed for this problem to be used extensively in engineering analysis without being branded as violating the second law of thermodynamics. The conditions that were touted as violations of the second law are not physically realistic. A bifurcated solution results. For small slab width, $a < \pi\sqrt{(D\tau_{mr})}$, the transient concentration is subcritical damped oscillatory.

An exact well-bounded solution that is bifurcated depending on the width of the slab is provided. The transient solution to the damped-wave non-Fick hyperbolic wave propagative and relaxation equation is obtained by the method of separation of variables.

A well-bounded infinite series expression is provided. The temperature overshoot identified by Taitel [35] is obviated by examining the final steady-state condition and expressing the state in mathematical terms. A bifurcated solution results. For small slab width, $a < \pi\sqrt{(D\tau_r)}$, the transient concentration is subcritical damped oscillatory. In both Taitel [35] and Barletta and Zanchini [36], four conditions were used for initial and boundary constraints. The two in the space domain are retained here. The initial concentration at time zero is also retained. However, the slope with the time domain of the concentration at time zero is replaced with the final condition for the time domain, i.e., at steady state, the transient concentration will decay out to a constant value or to zero in the dimensionless form. This consideration is shown to change the nature of the solution considerably to a well-bounded expression that is bifurcated. For small values of the slab, the transient concentration is subcritical damped oscillatory. For other values, the Fourier series representation is augmented by a modification to the exponential time domain portion of the solution. In this section, use of the final condition at steady state as the fourth condition to give a bounded solution in obeyance of the Clausius inequality was achieved.

8.3.3 Relativistic Transformation of Coordinates

The semi-infinite medium is considered to study the spatiotemporal patterns that the solution of the non-Fick damped-wave diffusion and relaxation equation exhibit. This kind of consideration has been used in the study of Fick mass diffusion. The boundary conditions can be different, such as constant wall concentration, constant wall flux (CWF), pulse injection, and convective, impervious, and exponential decay. The similarity or Boltzmann transformation worked out well in the case of parabolic PDE, where an error function solution can be obtained in the transformed variable. The conditions at infinite width and zero time are the same. The conditions at zero distance from the surface and infinite time are also the same.

Baumeister and Hamill [37] solved the hyperbolic heat conduction equation in a semi-infinite medium subjected to a step change in temperature at one of its ends using the method of Laplace transformation. The space-integrated expression for the temperature in the Laplace domain had the inversion readily available within the tables. This expression was differentiated using Leibniz's rule, and the resulting temperature distribution was given for $\tau > X$ as

$$u = \frac{(C_A - C_0)}{(C_{AS} - C_0)} = \exp\left(\frac{-X}{2}\right) + X\int_X^{\tau} \exp\left(\frac{-p}{2}\right)\frac{I_1\sqrt{p^2 - X^2}}{\sqrt{p^2 - X^2}}dp \qquad (8.44)$$

The method of relativistic transformation of coordinates is evaluated to obtain the exact solution for the transient temperature. Consider a semi-infinite slab at initial concentration C_0 imposed by a constant wall concentration C_s for times greater than zero at one of the ends. The transient concentration as a function of time and space in one dimension is obtained. Obtaining the dimensionless variables,

$$u = \frac{(C_A - C_{A0})}{(C_{AS} - C_{A0})}; \tau = \frac{t}{\tau_{mr}}; X = \frac{x}{\sqrt{D\tau_{mr}}}; J^* = \frac{J''}{\sqrt{\frac{D}{\tau_r}}(C_{AS} - C_{A0})} \tag{8.45}$$

The mass balance on a thin spherical shell at x with thickness Δx is written in one dimension as $-\partial J^*/\partial X = \partial u/\partial \tau$. The governing equation can be obtained in terms of the mass flux after eliminating the concentration between the mass balance equation and the non-Fick expression:

$$\frac{\partial J^*}{\partial \tau} + \frac{\partial^2 J^*}{\partial \tau^2} = \frac{\partial^2 J^*}{\partial X^2} \tag{8.46}$$

It can be seen that the governing equation for the dimensionless mass flux is identical in form with that of the dimensionless concentration. The initial condition is

$$\tau = 0, \; J^* = 0 \tag{8.47}$$

The boundary conditions are

$$X = \infty, J^* = 0 \tag{8.48}$$

$$X = 0, C = C_s, u = 1 \tag{8.49}$$

Let us suppose that the solution for J^* is of the form $w \exp(-n\tau)$, for $\tau > 0$, where W is the transient wave flux. Then, when $n = ½$, Eq. (8.46) becomes

$$\frac{\partial^2 w}{\partial \tau^2} - \frac{w}{4} = \frac{\partial^2 w}{\partial x^2} \tag{8.50}$$

Equation (8.50) can also be generated from Eq. (8.46) by multiplying Eq. (8.46) throughout with $\exp(n\tau)$ and realizing that $w = u\exp(n\tau)$ at $n = ½$. The solution to Eq. (8.50) can be obtained by the following relativistic transformation of coordinates for $\tau > X$. Let $\eta = (\tau^2 - X^2)$. Then Eq. (8.50) becomes

$$\frac{\partial^2 w}{\partial \tau^2} = 4\tau^2 \frac{\partial^2 w}{\partial \eta^2} + 2\frac{\partial w}{\partial \eta} \tag{8.51}$$

$$\frac{\partial^2 w}{\partial X^2} = 4X^2 \frac{\partial^2 w}{\partial \eta^2} - 2\frac{\partial w}{\partial \eta} \tag{8.52}$$

Combining Eqs. (8.51) and (8.52) into Eq. (8.50) gives

$$4(\tau^2 - X^2)\frac{\partial^2 w}{\partial \eta^2} + 4\frac{\partial w}{\partial \eta} - \frac{w}{4} = 0 \tag{8.53}$$

$$\eta^2 \frac{\partial^2 w}{\partial \eta^2} + \eta\frac{\partial w}{\partial \eta} - \frac{\eta w}{16} = 0 \tag{8.54}$$

Equation (8.54) can be seen to be a special differential equation in one independent variable. The number of variables in the hyperbolic PDE thus has been reduced from two to one. Comparing Eq. (8.54) with the generalized form of Bessel's equation [35], it can be seen that $a = 1$, $b = 0$, $c = 0$, $s = ½$, and $d = -1/16$. The order of the solution is calculated as 0, and the general solution is given by

$$w = c_1 I_0\left[\frac{\sqrt{\tau^2 - X^2}}{2}\right] + c_2 K_0\left[\frac{\sqrt{\tau^2 - X^2}}{2}\right] \tag{8.55}$$

The wave flux w is finite when $\eta = 0$, and hence it can be seen that c_2 is zero. c_1 can be solved from the boundary condition given in Eq. (8.49). The expression for the dimensionless mass flux for times $\tau > X$ is thus

$$J^* = c_1 \exp\left(\frac{-\tau}{2}\right) I_0\left[\frac{1}{2}\sqrt{\tau^2 - X^2}\right] \tag{8.56}$$

For large times, the modified Bessel's function can be given as an exponential and reciprocal in the square root of time by asymptotic expansion. Consider the surface flux, i.e., when in Eq. (8.56) X is set as zero:

$$J^* = c_1 \exp\left(\frac{-\tau}{2}\right) \frac{\exp\left(\frac{\tau}{2}\right)}{\sqrt{\pi\tau}} = \frac{c_1}{\sqrt{\pi\tau}} \tag{8.57}$$

For times when $\exp(\tau)$ is much greater than the mass flux, it can be seen that the second derivative in time of the dimensionless flux in Eq. (8.46) can be neglected compared with the first derivative. The resulting expression is the familiar expression for surface flux from the Fourier parabolic governing equation for constant wall concentration in a semi-infinite medium and is given by

$$J^* = \frac{1}{\sqrt{\pi\tau}} \tag{8.58}$$

Comparing Eqs. (8.58) and (8.57), it can be seen that c_1 is 1. Thus the dimensionless heat flux is given by

$$J^* = \exp\left(\frac{-\tau}{2}\right) I_0\left(\frac{\sqrt{\tau^2 - X^2}}{2}\right) \tag{8.59}$$

The solution for J^* needs to be converted to the dimensionless concentration u and then the boundary conditions applied. From the mass balance,

$$-\frac{\partial J^*}{\partial X} = \frac{\partial u}{\partial \tau} \tag{8.60}$$

Thus, differentiating Eq. (8.59) with respect to X, substituting in Eq. (8.60), and integrating both sides with respect to τ, for $\tau > X$,

$$u = \int \exp\left(\frac{-\tau}{2}\right)\left[\frac{I_1 \frac{1}{2}\sqrt{\tau^2 - X^2}}{\sqrt{\tau^2 - X^2}}\right] d\tau + c(X) \tag{8.61}$$

It can be left as an indefinite integral, and the integration constant can be expected to be a function of space. The $c(X)$ can be solved for by examining what happens at the wave front. At the wave front, $\eta = 0$, and time elapsed equals the time taken for a mass disturbance to reach the location x given the wave speed $\sqrt{D/\tau_{mr}}$. The governing equations for the dimensionless mass flux and dimensionless concentration are identical in form. At the wave front, Eq. (8.53) reduces to

$$\frac{\partial w}{\partial \eta} = \frac{w}{16} \quad \text{or} \quad w = c' \exp\frac{\eta}{16} = c' \tag{8.62}$$

$$u = c' \exp\frac{-\tau}{2} = c' \exp\frac{-X}{2} \tag{8.63}$$

Thus $c(X) = c' \exp(-X/2)$. Thus

$$u = \int X \exp\frac{-\tau}{2} \frac{I_1 \frac{1}{2}\sqrt{\tau^2 - X^2}}{\sqrt{\tau^2 - X^2}} d\tau + c' \exp\frac{-X}{2} \tag{8.64}$$

From the boundary condition in Eq. (8.49) it can be seen that $c' = 1$. Thus, for $\tau > X$, *Eq. (8.64) gives the exact solution for dimensionless concentration.*

It can be seen that the boundary conditions are satisfied by the Eq. (8.64) and describe the transient concentration as a function of space and time that is governed by the hyperbolic wave diffusion and relaxation equation. The flux expression is given by Eq. (8.59).

It also can be seen that expressions for dimensionless mass flux and dimensionless concentration given by Eqs. (8.59) and (8.64) are valid only in the open interval for $\tau > X$. When $\tau = X$, the wave front condition results, and the dimensionless mass flux and concentration are identical and are

$$J^* = u = \exp\frac{-X}{2} = \exp\left(\frac{-\tau}{2}\right) \tag{8.65}$$

When $X > \tau$, the transformation variable can be redefined as $\eta = X^2 - \tau^2$. Equation (8.50) becomes

$$\eta^2 \frac{\partial^2 w}{\partial \eta^2} + \eta \frac{\partial w}{\partial \eta} + \eta \frac{w}{16} = 0 \tag{8.66}$$

The general solution for this Bessel equation is given by

$$w = c_1 J_0\left[\frac{\sqrt{\eta}}{2}\right] + c_2 Y_0\left[\frac{\sqrt{\eta}}{2}\right] \tag{8.67}$$

The wave temperature W is finite when $\eta = 0$, and hence it can be seen that c_2 is zero. c_1 can be solved from the boundary condition given in Eq. (8.49). The expression in the open interval or the dimensionless heat flux for times τ smaller than X is thus

$$J^* = c_1 \exp\left(\frac{-\tau}{2}\right) J_0\left[\frac{\sqrt{X^2 - \tau^2}}{2}\right] \tag{8.68}$$

On examining the Bessel function in Eq. (8.68), it can be seen that the first zero occurs when the argument becomes 2.4048. Beyond that point, the Bessel function will take on negative values, indicating a reversal of heat flux. There is no good reason for the mass flux to reverse in direction at short times. Hence Eq. (8.68) is valid from the wave front down to where the first zero of the Bessel function occurs. Thus the plane of zero transfer explains the initial condition verification from the solution.

By using the expression at the wave front for the dimensionless mass flux, c_1 can be solved for and is found to be 1. Equation (8.68) also can be obtained directly from Eq. (8.56) by using $I_0(\eta) = J_0(i\eta)$. The expression for temperature in a similar vein for the open interval $X > \tau$ is thus

$$u = \int X \exp\left(\frac{-\tau}{2}\right) \frac{J_1\left[\frac{\sqrt{\tau^2 - X^2}}{2}\right]}{\sqrt{\tau^2 - X^2}} d\tau + \exp\left(\frac{-X}{2}\right) \tag{8.69}$$

Consider a point X_p in the semi-infinite medium. Three regimes can be identified in the mass flux at this point from the surface as a function of time. Series expansion of the modified Bessel composite function of the first kind and zeroth order was accomplished using a Microsoft Excel spreadsheet on a Pentium IV desktop computer. The three regimes and the mass flux at the wave front are summarized as follows:

1. The first regime is a thermal inertia regime when there is no transfer.
2. The second regime is given by Eq. (8.68) for the mass flux and

$$J^* = \exp\left(\frac{-\tau}{2}\right) J_0\left[\frac{\sqrt{X^2 - \tau^2}}{2}\right] \tag{8.70}$$

The first zero of the zeroth-order Bessel function of the first kind occurs at 2.4048. This is when

$$2.4048 = \frac{\sqrt{X^2 - \tau^2}}{2} \quad \text{or} \quad \tau_{\text{lag}} = \sqrt{X^2 - 23.132} \tag{8.71}$$

Thus τ_{lag} is the inertial lag that will ensue before the mass flux is realized at an interior point in the semi-infinite medium at a dimensionless distance X from the surface. As a demonstration, one value of X is used, i.e., 5. Thus, for points closer to the surface, the time lag may be zero. Only for dimensionless distances greater than 4.8096 is the time lag finite. For distances *closer than 4.8096* $\sqrt{\alpha\tau_r}$, the thermal lag experienced *will be zero*. For distances

$$x > 4.8096\sqrt{\alpha\tau_{mr}} \tag{8.72}$$

the time lag experienced is given by Eq. (8.71) and is $\sqrt{X^2 - 4\beta_1^2}$, where β_1 is the first zero of the Bessel function of the first kind and zeroth order and is 2.4048. In a similar fashion, the penetration distance of the disturbance for a considered instant in time beyond which the change in initial temperature is zero can be calculated as

$$X_{pen} = \sqrt{23.132 + \tau_i^2}$$

3. The third regime starts at the wavefront and is described by Eq. (8.59):

$$J^* = \exp\left(\frac{-\tau}{2}\right) I_0\left(\frac{\sqrt{\tau^2 - X^2}}{2}\right) \tag{8.73}$$

4. At the wave front, $J^* = u = \exp(-X/2) = \exp(-\tau/2)$.

The expressions for transient concentration derived above need integration prior to use. More easily usable expressions can be developed by making suitable approximations. Realizing that for PDE, a set of functions instead of constants as in the case of ODE needs to be solved from the boundary conditions, the c in Eq. (8.68) is allowed to vary with time. This results in an expression for transient concentration that is more readily available for direct use of the practitioner. Extensions to three dimensions in space are also straightforward in this method.

In this section, the exact solution for the constant wall concentration problem in semi-infinite medium in one dimension is revisited because of the discussion of the method of Laplace transforms by Baumeister and Hamill. In this section I attempt to derive an expression that does not need further integration. Consider a semi-infinite slab at initial concentration C_0 subjected to a sudden change in concentration at one of the ends to C_s. The mass propagative velocity is $V_m = \sqrt{D_{AB}/\tau_r}$. The initial conditions are

$$t = 0,\ V_x,\ C = C_0 \tag{8.74}$$

$$t > 0,\ x = 0,\ C = C_s \tag{8.75}$$

$$t > 0,\ x = \infty,\ C = C_0 \tag{8.76}$$

Obtaining the dimensionless variables

$$u = \frac{(C - C_0)}{(C_s - C_0)};\ \tau = \frac{t}{\tau_{mr}};\ X = \sqrt{D\tau_{mr}} \tag{8.77}$$

the mass balance on a thin spherical shell at x with thickness Δx is written. The governing equation can be obtained after eliminating J'' between the mass balance equation and the derivative with respect to x of the flux equation and introducing the dimensionless variables.

$$\frac{\partial u}{\partial \tau}+\frac{\partial^2 u}{\partial \tau^2}=\frac{\partial^2 u}{\partial X^2} \tag{8.78}$$

Suppose that $u = \exp(-n\tau)w(X, \tau)$. By choosing $n = ½$, the damping component of the equation is removed. Thus, for $n = ½$, the governing equation becomes

$$\frac{\partial^2 w}{\partial \tau^2}-\frac{w}{4}=\frac{\partial^2 w}{\partial x^2} \tag{8.79}$$

The solution to Eq. (8.79) can be obtained by the following relativistic transformation of coordinates for $\tau > X$. Let $\eta = (\tau^2 - X^2)$. Then Eq. (8.50) becomes

$$\frac{\partial^2 w}{\partial \tau^2}=4\tau^2\frac{\partial^2 w}{\partial \eta^2}+2\frac{\partial w}{\partial \eta} \tag{8.80}$$

$$\frac{\partial^2 w}{\partial X^2}=4X^2\frac{\partial^2 w}{\partial \eta^2}-2\frac{\partial w}{\partial \eta} \tag{8.81}$$

Combining Eqs. (8.80) and (8.81) into Eq. (8.79) gives

$$4(\tau^2-X^2)\frac{\partial^2 w}{\partial \eta^2}+4\frac{\partial w}{\partial \eta}-\frac{-w}{4}=0 \tag{8.82}$$

$$\eta^2\frac{\partial^2 w}{\partial \eta^2}+\eta\frac{\partial w}{\partial \eta}-\frac{\eta w}{16}=0 \tag{8.83}$$

Equation (8.83) can be seen to be a special differential equation in one independent variable. The number of variables in the hyperbolic PDE thus has been reduced from two to one. Comparing Eq. (8.83) with the generalized form of Bessel equation [35], it can be seen that $a = 1$, $b = 0$, $c = 0$, $s = ½$, and $d = -1/16$. The order of the solution is calculated as 0, and the general solution is given by

$$w=c_1 I_0\left[\frac{\sqrt{\tau^2-X^2}}{2}\right]+c_2 K_0\left[\frac{\sqrt{\tau^2-X^2}}{2}\right] \tag{8.84}$$

The wave temperature w is finite when $\eta = 0$, and hence it can be seen that c_2 is zero. c_1 can be solved from the boundary condition given in Eq. (8.75). For $X = 0$, u is 1. Writing the expression for u at $X = 0$,

$$1=c_1\exp\left(\frac{-\tau}{2}\right)I_0\left(\frac{\sqrt{\eta}}{2}\right) \tag{8.85}$$

c_1 can be eliminated by dividing Eq. (8.84) after setting $c_2 = 0$ by Eq. (8.85) to yield in the open interval of $\tau > X$

$$u = \frac{I_0\left[\frac{\sqrt{\tau^2 - X^2}}{2}\right]}{I_0\left[\frac{\tau}{2}\right]} \tag{8.86}$$

In the open interval $X > \tau$,

$$u = \frac{J_0\left[\frac{\sqrt{X^2 - \tau^2}}{2}\right]}{I_0\left[\frac{\tau}{2}\right]} \tag{8.87}$$

It can be inferred that an expression in time is used for c_1. A domain-restricted solution for short and long times may be in order.

8.3.4 Periodic Boundary Condition

Consider a semi-infinite slab at initial concentration C_0 imposed by a periodic concentration at one of the ends by $C_0 + C_1 \cos(wt)$. The transient concentration as a function of time and space in one dimension is obtained. Obtaining the dimensionless variables

$$u = \frac{(C - C_0)}{C_1}; \quad \tau = \frac{t}{\tau_{mr}}; \quad X = \frac{x}{\sqrt{D\tau_{mr}}}; \quad u = (C - C_0)/(C_1);$$

$$\tau = t/\tau_r; \quad X = x/\sqrt{D\tau_r} \tag{8.88}$$

The mass balance on a thin shell at x with thickness Δx is written. The governing equation is obtained after eliminating J between the mass balance equation and the derivative with respect to x of the flux equation and introducing the dimensionless variables. The initial conditions are

$$t = 0, C = C_0, u = 0 \tag{8.89}$$

The boundary conditions are

$$X = \infty, C = C_0, u = 0 \tag{8.90}$$

$$X = 0, C = C_0 + C_1 \cos(\omega t); \quad u = \cos(\omega^* \tau) \tag{8.91}$$

Let us suppose that the solution for u is of the form $f(x) \exp(-i\omega^*\tau)$, for $\tau > 0$, where ω is the frequency of the concentration wave imposed on the surface and the C_1 is the amplitude of the wave. Then

$$(-i\omega^*)f \exp(-i\omega^*\tau) + (i^2\omega^{*2})f \exp(-i\omega\tau) = f'' \exp(-i\omega^*\tau) \tag{8.92}$$

$$i^2 f(\omega^{*2} + i\omega^*) = f'' \qquad f(X) = c\exp(-iX\omega^* \sqrt{\omega^* + i}) \tag{8.93}$$

Then d can be seen to be zero as at $X = \infty$, $u = 0$.

$$u = c\exp(-iX\omega^* \sqrt{\omega^* + i})\exp(-i\omega^* \tau) \tag{8.94}$$

From the boundary condition at $X = 0$,

$$\cos(\omega^*\tau) = \text{real part}[c\exp(-i\omega^*\tau)] \quad \text{or} \quad c = 1 \tag{8.95}$$

$$\begin{aligned} u &= \exp[-X\omega^*(A + iB)\exp(-i\omega^*\tau)] \\ &= \exp(-A\omega^*X)\exp[-i(BX\omega^* + \omega^*\tau)] \end{aligned} \tag{8.96}$$

where

$$A + iB = i\sqrt{\omega^* + i}\,. \tag{8.97}$$

Squaring both sides gives

$$A^2 - B^2 + 2AiB = i^2(\omega^* + i) = -\omega^* - i \tag{8.98}$$

$$A^2 - B^2 = -\omega^* \qquad 2AB = -1 \quad \text{or} \quad B = -½A$$

or

$$A^2 - ¼A^2 = -\omega^* \tag{8.99}$$

$$A^2 = (-\omega^* \pm \sqrt{\omega^{*2} + 1})/2 \qquad B = -½A \tag{8.100}$$

Obtaining the real part

$$u = \exp(-A\omega^*X)\cos[\omega^*(BX + \tau)] \tag{8.101}$$

The time lag in the propagation of the periodic disturbance at the surface is captured by the preceding relation. Thus the boundary conditions can be seen to be satisfied by Eq. (8.101). In a similar vein to the supposition of $f(x)\exp(-i\omega^*\tau)$, the mass flux J'' can be supposed to be of the form $J^* = g(x)\exp(-i\omega^*\tau)$. Thus

$$g = \frac{f'}{(1 - i\omega^*)} \tag{8.102}$$

Combining the f from Eq. (8.93) into Eq. (8.102) gives

$$\begin{aligned} J^* &= -\omega^*(A + iB)\exp[-X\omega^*(A + iB)]\exp(-i\omega^*\tau) \\ &= -\omega^*(A + iB)\exp(-A\omega^*X)\exp[-i(BX\omega^* + \omega^*\tau)] \\ &= -\omega^*(A + iB)\exp(-A\omega^*X)[\cos(BX\omega^* + \omega^*\tau) + i\sin(BX\omega^* + \omega^*\tau)] \end{aligned} \tag{8.103}$$

Obtaining the real part

$$J'' = \sqrt{\frac{D}{\tau_{mr}}}\,\omega^* \exp(-A\omega^* X)\{B \sin[\omega^*(BX + \tau)]$$
$$- A \cos[\omega^*(BX + \tau)]\} \tag{8.104}$$

8.4 Electrophoresis Apparatus

The term *electrophoresis* refers to the movement of a solid particle through a stationary fluid under the influence of an electric field (Fig. 8.2). The constituent that migrates under the field can be large molecules, colloids, fibers, clay particles, and latex spheres. Electrophoresis is often applied to polymeric and biologic samples. It is applied frequently in the analysis of proteins and DNA fragment mixtures. The differences in mobility of different species under an electric field are used to obtain a separation between two or more

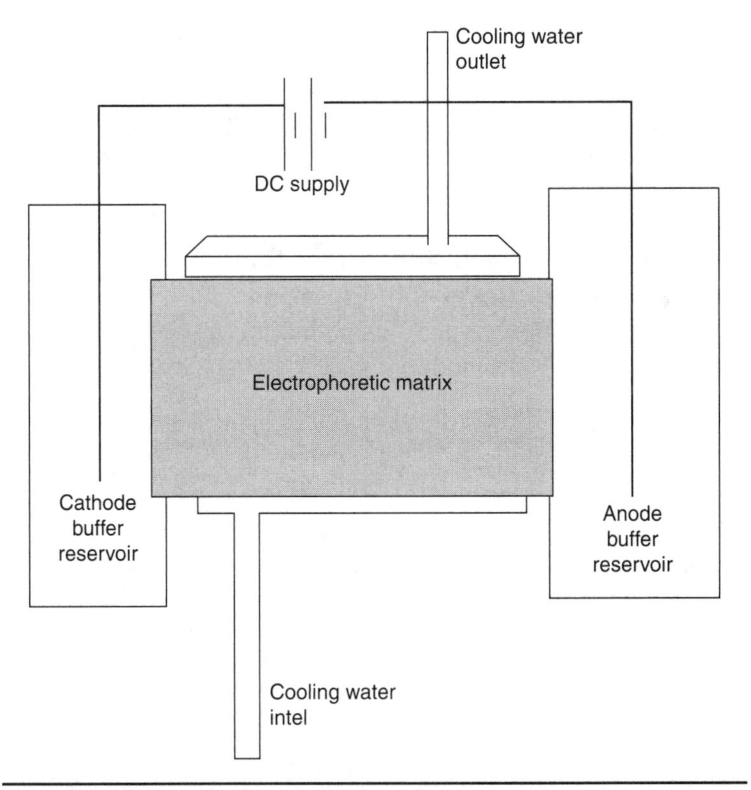

Figure 8.2 Schematic of an electrophoresis apparatus.

species. The advancement of biotechnology was in some measure due to electrophoresis. Variations of this method are used in obtaining the nucleic acid sequences of DNA; isolating active biologic factors associated with diseases such as cystic fibrosis, sickle-cell anemia, myelomas, and leukemia; and establishing immunologic reactions between samples on the basis of individual compounds. The technique is sensitive to small differences in molecular charge and mass. It does not interfere with the species under investigation during the investigation.

The charge separation between the surface of the particle and the fluid surrounding it is tapped into in the electrophoresis technique. The particle is caused to move by the electric force it experiences from the electric field and resulting charge on the particle. The electric field also generates heat.

Different types of gel matrices can be employed. These are agarose, polyacrylamide, paper, capillaries, and flowing buffers. The gel and capillary modes can be used alone or in combination in the different matrices listed to achieve the target objectives for a given application. Over a period of time, a number of different types of electrophoresis methods have been developed. Some of them are (1) disk electrophoresis, (2) zone electrophoresis, (3) native zone electrophoresis, (4) reduced sodium dodecyl sulfate (SDS) electrophoresis, (5) pulsed-field gel electrophoresis, (6) isoelectric focusing, (7) isotachophoresis, (8) agarose electrophoresis, (9) polyacrylamide electrophoresis, (10) paper electrophoresis, (11) capillary electrophoresis, and (12) force-flow electrophoresis.

8.5 Electrophoretic Term, Ballistic Term, and Fick Term in the Governing Equation

The molar flux after taking into account the electric field effects and the finite speed of molecular diffusion effects can be written as

$$-j_A = D\frac{\partial C_A}{\partial z} + \left(\frac{zFm}{RT}\right)C_A + \tau_{mr}\frac{\partial j_A}{\partial t} \tag{8.105}$$

$z_A F$ is the charge per molecule, and F is the Faraday's constant in coulombs per gram. z_A is the valency of the species A, and m is its mass. Lumping the electrophoretic effects as $\varepsilon = zFm/RT$ and combining with the mass balance equation in transient diffusion, the governing equation for concentration of species A under transient conditions can be written as

$$D\frac{\partial^2 C_A}{\partial z^2} + \varepsilon\frac{\partial C_A}{\partial z} = \tau_{mr}\frac{\partial C_A}{\partial t^2} + \frac{\partial C_A}{\partial t} \tag{8.106}$$

Equation (8.106) is a hyperbolic partial differential equation with four terms in the governing equation. Analytical general solutions to the equation are not reported in the literature. There have been some attempts to solve this equation numerically. However, given what is known about the transient nature of the process, the nature and salient characteristics of the solution can be examined. This is done as follows: Equation (8.106) is made dimensionless by the following substitutions:

$$u = \frac{(C_A - C_{AS})}{(C_{A0} - C_{AS})}; \quad \tau = \frac{t}{\tau_{mr}}; \quad X = \frac{z}{\sqrt{D\tau_{mr}}}; \quad Pe_{\text{elec}} = \frac{\varepsilon}{\sqrt{D/\tau_{mr}}} \tag{8.107}$$

Then the governing equation transforms from Eq. (8.106) into

$$\frac{\partial^2 u}{\partial X^2} + Pe_{elec}\frac{\partial u}{\partial X} = \frac{\partial^2 u}{\partial \tau^2} + \frac{\partial u}{\partial \tau} \tag{8.108}$$

The Peclect number (electric) Pe_{elec} is given by the ratio of the electrophoretic velocity and the velocity of mass propagation. Consider a slab with length l maintained at concentration C_{AS} at one end at all times. At length l, the gel is impervious to any further migration, and hence the adiabatic boundary condition of zero flux at the boundary can be assumed. The time and space conditions are then

$$X = 0,\ u = 0 \tag{8.109}$$

$$X = X_l,\ \frac{\partial u}{\partial X} = 0 \tag{8.110}$$

$$\tau = 0,\ u = 1 \tag{8.111}$$

$$\tau = \infty,\ u = 0 \tag{8.112}$$

It is generally known that problems in transient diffusion have an exponential decaying time component to their solution. Further, since the problem is driven by a surface concentration maintained at a higher concentration at $X = 0$, and further, since the end of the apparatus is impervious to diffusion, the species concentration will have a decaying component in space as well. In order to examine the salient characteristics of the solution to Eq. (8.108), let the solution be assumed to take the form

$$u = \exp(-n\tau)\exp(-mX)w \tag{8.113}$$

where w is the wave concentration, which is a function of space and time. It can be shown that at n = ½ and m = ½Pe_{elec}, Eq. (8.108) becomes

$$\frac{\partial^2 w}{\partial X^2} = \frac{\partial^2 w}{\partial \tau^2} + \frac{w}{4}\left[\frac{1}{Pe_{elec}} - 1\right] \tag{8.114}$$

It can be seen from Eq. (8.114) that when the Peclect number (electric) equals 1, Eq. (8.114) reverts to the wave equation. The solution then would be D'Alembert's solution, as discussed in Sharma [35]. When the Peclect (electric) number is equal to 1, the electrophoretic velocity and the velocity of molecular diffusion are equal. Equation (8.114) can be solved by the method of separation of variables. Let

$$w = g(X)V(\tau) \tag{8.115}$$

Substituting Eq. (8.115) into Eq. (8.114) and separating the variables in space and time gives the two differential equations that govern the solution in space and time:

$$\frac{d^2 g}{dX^2} = -\lambda_n^2 \tag{8.116}$$

$$\frac{d^2 V}{d\tau^2} = -\left[\lambda_n^2 + \frac{1}{4}\left(\frac{1}{Pe_{elec}} - 1\right)\right] \tag{8.117}$$

The solution to Eq. (8.116) can be seen to be

$$g = c_1 \sin(\lambda_n X) + c_2 \cos(\lambda_n X) \tag{8.118}$$

From the boundary conditions at $X = 0$, it can be seen that $c_2 = 0$. From the boundary conditions at $X = X_1$, the eigenvalues can be solved for as

$$\lambda_n X_1 = \frac{(2n-1)\pi}{2} \qquad n = 1, 2, 3, \ldots \tag{8.119}$$

It can be seen that the solution to Eq. (8.117) depends on the relaxation time and other parameters of the system. It is a bifurcated solution. For small eigenvalues,

$$V = c_3 \exp\left(-\tau\sqrt{\frac{1}{4} - \frac{1}{4Pe_{elec}} - \lambda_n^2}\right) + c_4 \exp\left(\tau\sqrt{\frac{1}{4} - \frac{1}{4Pe_{elec} - \lambda_n^2}}\right) \tag{8.120}$$

In a similar fashion, as discussed in Sec. 8.2.1, at steady state or infinite time, or from the time condition stated in Eq. (8.112), it can be seen that $w = u\exp(\tau/2)\exp(X/2Pe_{elec})$ will become 0 times infinity and equal to 0. Hence the c_4 in Eq. (8.120) can be set to 0. Thus the general solution to the transient concentration can be written as

$$u = \sum_1^\infty c_n \sin(\lambda_n X)\exp\left[\frac{-X}{2Pe_{elec}}\right]\exp\left[\frac{-\tau}{2}\right]\exp\left(-\tau\sqrt{\frac{1}{4} - \frac{1}{4Pe_{elec}} - \lambda_n^2}\right) \tag{8.121}$$

The c_n can be solved for from the initial condition using the orthogonality property and be shown to be

$$\frac{8l}{(2n-1)^2\pi^2\sqrt{D\tau_{mr}}}$$

Further, it can be seen that for large eigenvalues, the solution for the concentration given in Eq. (8.121) becomes damped oscillatory. Thus, when

$$\lambda_n^2 + \frac{1}{4Pe_{elec}} > \frac{1}{4} \tag{8.122}$$

$$u = \sum_1^\infty c_n \sin(\lambda_n X)\exp\left[\frac{-X}{2Pe_{elec}}\right]\exp\left[\frac{-\tau}{2}\right]\cos\left(\tau\sqrt{\lambda_n^2 + \frac{1}{4Pe_{elec}} - \frac{1}{4}}\right) \tag{8.123}$$

The implications of Eq. (8.122) can be seen that when $n = 1$,

$$\frac{\pi^2 D\tau_{mr}}{l^2} + \frac{\sqrt{\frac{D}{\tau_{mr}}}}{\varepsilon} > 1 \tag{8.124}$$

or

$$l < \sqrt{\frac{\varepsilon\pi^2 D\tau_{mr}}{\left(\varepsilon - \sqrt{D/\tau_{mr}}\right)}}$$

Thus, when the length of the electrophoretic apparatus is less than a critical length, the concentration of species A will undergo subcritical damped oscillations. This would be the case for gel matrices with high relaxation times and low molecular diffusion velocities.

Summary

The sequence distribution of deoxyribonucleic acid (DNA) is obtained by the method of gel acrylamide electrophoresis. The sequence distribution of polypeptide (protein) is obtained by using paper chromatography. In both these techniques molecular diffusion phenomena is a important consideration. At short time scales, such as those associated with critical events during electrophoresis and paper chromatography, Fick's laws of diffusion is not adequate to represent the transient events. There are seven reasons to seek a generalized Fick's law of molecular diffusion. These are the contradiction of Fick's law with the theory of microscopic reversibility of Onsager, observation of Nernst that heat and hence mass possess inertia and in good conductors and hence in good diffusing media lead to oscillatory discharge, Landau and Lifshitz's observation that light possess speediest velocity, high mass rate applications cannot be described by Fick's laws, singularities were found in the description of surface flux in cartesian, cylindrical, of and spherical coordinates using Fick's representation of transient concentration, Fick's laws were developed from empirical observations, overprediction of theory to experiment were found in a number of important industrial applications of Fick's laws.

The generalized Fick's law of diffusion is analogous to the Cattaneo and Vernotte hyperbolic heat conduction equation. It was derived by considering the acceleration of a moving molecule under the Sotkes-Einstein formulation of chemical potential and drag. Expressions for relaxation time of mass was developed in terms of the diffusion coefficient of mass. The Taitel paradox of a temperature overshoot and hence a implied concentration overshoot during damped wave diffusion and relaxation was re-examined. The use of the final condition in time leads to well bounded infinite series solution. No overshoot was found in the solution to damped wave diffusion and relaxation in a finite slab subject to constant wall concentration boundary condition. At large relaxation times, the solution is found to exhibit subcritical, damped oscillations in concentration of migrating species.

The method of relativistic transformation of coordinates was developed to obtain physically realistic solutions to the semi-infinite medium problem subject to constant wall concentration boundary condition. Three different regimes of solution were identified. A inertial regime characterized with zero transfer, a second regime characterized by Bessel composite function of space and time of the zeroth order and first kind, and a third regime of a modified Bessel composite function of space and time of the zeroth order and first kind. Expressions for penetration distance and inertial lag time were developed. The characteristics of the solution to the damped wave diffusion and relaxation subject to a periodic boundary condition were studied using the method of complex temperature.

A schematic of electrophoresis apparatus was provided. Different methods of electrophoresis techniques were discussed. The governing equation for the migrating species subject to finite speed diffusion and electrophoretic force was developed. A dimensionless group, Peclect number (electric) was defined. This was the ratio of the electrophoretic velocity to the finite speed of mass. Well bounded infinite series solution for the migrating species were developed for a finite slab subject to a constant concentration boundary condition. The conditions were the concentration is expected to undergo subcritical, damped oscillations were derived. The mathematical model can be used in place of the calibration method used in electrophoresis methods. This may lead to less errors in the sequence distribution of DNA and protein molecules.

References

[1] A. E. Fick, "Uber Diffusion", *Poggendorff's Annelen der Physik.* 94 (1855), 59.

[2] B. Fascimire, NASATechnical Reports, Skylab Project, NASA Marshall Flight Center, Alabama, 1973.

[3] L. Onsager, "Reciprocal relations in irreversible processes," *Phys. Rev.* 37 (1931), 405–426.

[4] W. Nernst, *Die Theoretischen Grundalgen des n Warmestazes.* Knapp Halle, Frankfurt, DE, 1917.

[5] L. Landau and E. M. Liftshitz, *Fluid Mechanics.* New York: Pergamon Press, 1987.

[6] D. C. Kelly, "Diffusion: A relativistic appraisal," *Am. J. Phys.* 36 (1968), 585–591.

[7] K. R. Sharma, *Damped Wave Transport and Relaxation.* Amsterdam: Elsevier, 2005.

[8] K. R. Sharma, "Damped Wave Conduction and Relaxation in Spherical and Cylindrical Coordinates," *J. Thermophys. Heat Transfer.* 21 (2007), 688–693.

[9] K. R. Sharma, "Manifestation of acceleration during transient heat condution," *J. Thermophys. Heat Transfer.* 20 (2006), 799–808.

[10] K. R. Sharma, "Analytical solution of damped wave conduction and relaxation equation for a finite sphere and cylinder," *J. Thermophys. Heat Transfer* (accepted).

[11] K. R. Sharma, "Temperature solution in semi-infinite medium under CWT for Cattaneo and Vernotte non-Fourier heat conduction," 225th ACS National Meeting, New Orleans, LA, March 23–28, 2003.

[12] K. R. Sharma, "On the use of final condition in time to obtain solution of Cattaneo and Vernotte damped wave transport and relaxation equation," 231st ACS National Meeting, Atlanta, GA, March 26–30, 2006.

[13] K. R. Sharma, "Finite speed diffusion effects in electrophoresis in a finite slab at large electrophoretic velocities," AIChE Spring National Meeting, New Orleans, LA, April 6–10, 2008.

[14] K. R. Sharma, "On the temperature overshoot problem in a cylinder and use of final condition in time to obtain a bounded solution." 235th ACS National Meeting, New Orleans, LA, April 6–10, 2008.

[15] K. R. Sharma, "On the violations of laws of thermodyanics by laws of heat condution," AIChE Spring National Meeting, New Orleans, LA, April 6–10, 2008.

[16] K. R. Sharma, "Confounding effect of charge on gel electrophoresis measurements," 59th Northwest Regional Meeting of the American Chemical Society, NORM/RMRM, Utah State University, UT, June 2004.

[17] K. R. Sharma, "On the use of cutrices in restriction mapping of three enzymes," 231st ACS National Meeting, Atlanta, GA, March 26–30, 2006.

[18] K. R. Sharma, "Removal of arsenic from drinking water by molecular sieve adsorption," 226th ACS National Meeting, New York, September, 2003.
[19] K. R. Sharma, "Critical radii neither greater than the shape limit nor less than cycling limit," AIChE Spring National Meeting, New Orleans, LA, March 30–April 3, 2003.
[20] K. R. Sharma, "Acceleration effects during controlled drug delivery to the brain," 235th ACS National Meeting, New Orleans, LA, April 6–10, 2008.
[21] K. R. Sharma, "Storage coefficient of substrate in a 2 GHz microprocessor," 225th ACS National Meeting, New Orleans, LA, March 23–28, 2003.
[22] K. R. Sharma and R. Turton, "Mesoscopic approach to correlate surface heat transfer coefficients using pressure fluctuations in dense gas-solid fluidized beds," *Powder Technol.* 99 (1998), 109–118.
[23] A. K. Boley, *Heat Transfer Structures and Materials.* New York: Pergamon Press, 1964.
[24] J. C. Maxwell, "On the dynamical theory of gases," *Philos. Trans. R. Soc.* 157 (1867), 49.
[25] P. M. Morse and H. Feshbach, *Methods of Theoretical Physics.* New York: McGraw Hill, 1953.
[26] C. Cattaneo, "A form of heat conduction which eliminates the paradox of instantaneous propagation," *Comp. Rendu.* 247 (1958), 431–433.
[27] P. Vernotte, "Les paradoxes de la theorie continue del'equation de la chaleur," *C. R. Hebd. Seanc. Acad. Sci. Paris.* 246 (1958), 3154–3155.
[28] K. Mitra, S. Kumar, A. Vedavarz, and M. K. Moallemi, "Experimental evidence of hyperbolic heat conduction in processed meat," *J. Heat Transfer.* 117 (1995), 568–573.
[29] E. Mitura, S. Michalowski, and W. Kaminski, "A mathematical model of convection drying in the falling drying rate period," *Drying Technol.* 6 (1988), 113–137.
[30] D. Y. Tzou, *Macro to Microscale Heat Transfer: The Lagging Behavior.* Boca Raton, FL: CRC Press, 1996.
[31] K. R. Sharma, *Principles of Mass Transfer.* New Delhi: Prentice Hall of India, 2007.
[32] R. H. Stokes, "An improved diaphragm-cell for diffusion studies and some test of the methods," *J Am. Chem. Soc.* 72 (1950), 763-767.
[33] A. Einstein, "On the motion of small particles suspended in liquids at rest required by molecular-kinetic theory of heat," *Annalen der Physik.* 7 (1905), 549.
[34] Y. Taitel, "On the parabolic, hyperbolic and discrete formulation of heat conduction equation," *Int. J. Heat Mass Transfer.* 15 (1972), 369–371.
[35] E. Zanchini, "Hyperbolic heat conduction theories and non-decreasing entropy," *Phys. Rev. [B]* 60 (1999), 991–997.
[36] A. Barletta and E. Zanchini, "Thermal-wave heat conduction in a solid cylinder which undergoes a change of boundary temperature," *Heat Mass Transfer [Warema-und Stoffuebertragung].* 32 (2003), 5383–5386.
[37] K. J. Baumeister and T. D. Hamill, "Hyperbolic heat conduction equation: A solution for the semi-infinite body problem," *ASME J. Heat Transfer.* 93 (1971), 126–128.
[38] K. B. McAfee, *Scientific American*, 199, 1, (1958), 52.

Exercises

1.0 What is the difference between binary and ternary diffusion coefficients?

2.0 During Brownian motion, the molecules follow a random zigzag path and sometimes move in the opposite direction compared with the imposed concentration difference driving the diffusion. Is this a violation of the second law of thermodynamics?

3.0 What is self-diffusivity? Is a concentration difference needed for movement of the species that defines the self-diffusivity?

4.0 What are the differences between multicomponent diffusion and binary diffusion?

5.0 What happens to the formula for total flux during equimolar counterdiffusion compared with that for molecular diffusion?

6.0 Correlations for diffusion in gases, liquids, and solids were discussed. What would be appropriate for liquid diffusing in a solid or for gases diffusing in a liquid?

7.0 Explain the effect of temperature on the mass propagation velocity. What happens to the diffusion coefficient and relaxation time at high pressure?

8.0 Why are insects larger in size in the tropics than insects in the arctic region?

9.0 Is the force of gravity taken into account in the derivation of the Stokes-Einstein relationship for diffusivity coefficients?

10.0 Can you expect a plane of zero concentration or null transfer during drug delivery in the tissue region? How so?

11.0 The diffusion coefficient is a proportionality constant in Fick's first law of diffusion independent of concentration. For concentrated solutions, it is said to vary with concentration. How can this be interpreted?

12.0 State the Onsager reciprocal relations. Show that $D_{12} = D_{21}$.

13.0 What was Landau's observation of infinite speed of propagation?

14.0 What is an overshoot?

15.0 What is the drag force experienced by an electron compared with the acceleration term?

16.0 What is penetration length?

17.0 What is inertial lag time?

18.0 What is the first zero of the Bessel function of the first order? How is this used in the derivation of the penetration length and inertial lag time in a three-dimensional medium?

19.0 What is the physical significance of the maxima in Fig. 8.1?

20.0 What is the physical significance of the x intercept in Fig. 8.1? Can an expression for the time taken to steady state be derived from these x-intercept values?

21.0 Examine $I_0(\tau/2)\exp(-\tau/2)$ in terms of extremas and asymptotic limits, and under what conditions can $I_0(\tau/2)$ be reduced to a simpler expression?

22.0 What is the meaning of a negative mass flux? What happens to the ratio of the accumulation and diffusion terms?

23.0 It was shown that for large relaxation times, the transient concentration in a finite slab exhibits subcritical damped oscillations. What is the critical size of the slab below which the oscillations can be seen? What is the value of the diffusion coefficients when the oscillations can be seen?

24.0 Contrast subcritical damped oscillations with critical and underdamped oscillations. What does resonance mean for this problem?

25.0 Scale the governing equation and show that when the temporal derivative of the dimensionless concentration exceeds $\exp(\tau)$, the hyperbolic PDE reduces to the wave equation. Further, when $\exp(\tau)$ is greater than the temporal derivative, the hyperbolic PDE reverts to the parabolic PDE identical to that of Fick's second law of diffusion.

26.0 Why is there a maxima in the dimensionless flux as a function of time?

27.0 *Estimate of the diffusion coefficient of argon in hydrogen.* Calculate the diffusion coefficient of argon in hydrogen at 1 atm and 195°C. Compare this with the experimental values reported in the literature.

28.0 *Parabolic law of oxidation.* During the corrosion of metals, an oxide layer is formed on the metal. Assuming that oxygen diffuses through the oxide layer, show that the thickness of the oxide layer δ can be given by $(C_{\text{bulk}} D_{AB} t/\rho_m)^{1/2}$ using Fick's law of diffusion. A gentle breeze is blowing at a constant velocity U over the corroded layer. Is this going to increase the rate of corrosion owing to the convection contribution?

29.0 *Krogh tissue cylinder.* Capillaries through which blood flows in the human anatomy are surrounded by tissue space. The oxygen and other drugs that are dissolved in the bloodstream need to diffuse through the capillary walls into the tissues. Write the governing equations for the concentration of solute in the capillaries and in the tissue as

$$-V dC/dz = 2/r_c K_0 (C - C_T|_{rc+tm})$$

Considering the effects of diffusion in the x direction only in the tissue and assuming a zeroth order reaction rate,

$$D_{AB} \partial^2 C_T/\partial x^2 = R_0$$

Integrating and substituting for the boundary conditions

$$x = x_c + t_m, \; C_T = C_T|_{xc+tm}$$

$$x = x_T, \; dC_T/dx = 0$$

show that the concentration profile is

$$C_T - C_T|_{xc+tm} = (R_0/2D_{AB})[x^2 - (x_c + t_m)^2] - R_0 x_T/D_{AB}[x - (x_c + t_m)]$$

Show that the variation in concentration as a function of z can be calculated as

$$C = C_0 - R_0 z A_T/VA$$

Combine the two equations and show that

$$C_T - C_0 = R_0 z A_T / VA + K_0 x_c R_0 A_T / 2A + (R_0/2D_{AB})[x^2 - (x_c + t_m)^2]$$
$$- R_0 x_T / D_{AB}[x - (x_c + t_m)]$$

Show that at a critical distance from the capillary wall the concentration in the solute will become zero. This can be solved for from the preceding equations. At and beyond the critical distance,

$$dC_T/dx = 0 = C_T$$

Replacing x_T with x_{critical} gives

$$0 = C_0 + R_0 z A_T / VA + K_0 x_c R_0 A_T / 2A + (R_0/2D_{AB})[x^2 - (x_c + t_m)^2]$$
$$- R_0 x_{\text{critical}} / D_{AB}[x - (x_c + t_m)]$$
$$x_{\text{critical}}^2(-R_0/2D_{AB}) = C_0 + R_0 z A_T / VA + K_0 x_c R_0 A_T / 2A - (R_0/2D_{AB})(x_c + t_m)^2$$
$$- R_0 x_{\text{critical}} / D_{AB}[x - (x_c + t_m)]$$

The quadratic equation in x_{critical} is then

$$A x_{\text{critical}}^2 + B x_{\text{critical}} + C = 0$$

where $A = -(R_0/2D_{AB})$

$B = +(x_c + t_m) R_0 / D_{AB}$

$C = C_0 + R_0 z A_T / VA + K_0 x_c R_0 A_T / 2A - (R_0/2D_{AB})(x_c + t_m)^2 + R_0(x_c + t_m)/D_{AB}$

When the solutions of the quadratic expression for the critical distance in the tissue are real and are found to be less than the thickness of the tissue, then the onset of zero concentration will occur before the periphery of the tissue. This zone can be seen as the anorexic or oxygen-depleted regions in the tissue.

30.0 *Sacred pond.* Evaporation from ponds is retarded by the introduction of lotus leaves in sacred ponds at temples. Assume that in a pond of area 9 × 9 m, 4130 leaves each with a diameter of 3 in were placed. Calculate the reduction in diffusion rate on account of the reduction in area of the path of evaporation.

31.0 *Diffusion coefficient of tobacco mosaic virus.* Estimate the diffusion coefficient of tobacco mosaic virus that is shaped as a cylinder of 0.35 μm length and 9 nm diameter in water at 20°C. Its molecular weight is 42 million and partial specific volume is 0.53 cm^3/g.

32.0 *Diffusion of oxygen through spiracles.* Many insects breathe through spiracles. Spiracles are open tubes that extend into the insect's body. Oxygen diffuses from the surrounding air, and gas exchange takes place through the walls. For every mole of oxygen diffusing in, there is one mole of CO_2 diffusing out. To prevent water losses, the walls of the spiracles are coated with cuticle of 10-μm thickness. The oxygen concentration outside the cuticle is constant and is 5 percent of the equilibrium concentration. What is the local oxygen flux in the spiracle to the tissue? Derive an oxygen concentration profile within the tissue. Is the spiracle an efficient method of respiration?

Spiracle radius: 100 μm

Spiracle length: 9 mm

Oxygen solubility in tissue $C_t = 0.2$ mmol/L

$D_{0,cuticle}$: –3 E-5 cm²/s

$D_{0,air}$: –0.15 cm²/s

33.0 *Scrubbing of SO_2.* During coal combustion, the emission of sulfur dioxide from power plants can be reduced by using CaO scrubbers. In the scrubber,

$$2CaO + 2SO_2 + O_2 \rightarrow 2CaSO_4$$

Consider the diffusion of SO_2 into a spherical particle of CaO. Show that the governing equation can be derived from the shell balance as

$$D_{AB}[1/r^2\partial/\partial r(r^2\partial C_A/\partial r)] = k'''C_A$$

Show that the concentration profile of SO_2 in the spherical lime particle can be written as

$$C_A/C_{As} = X^{-1/2}I_{1/2}[r(k'''/D_{AB})^{1/2}]/I_{1/2}[R(k'''/D_{AB})^{1/2}]$$

The Thiele modulus is $\phi = R(k'''/D_{AB})^{1/2}$.

34.0 *Coextrusion.* In the manufacture of the casings of a solid rocket motor (SRM), the material requirements are bifunctional. They have to have high hoop strength on one side and high ablation resistance on the other. In order to prepare such materials, the technology of coextrusion is used. In a twin-screw extruder, both the materials are coextruded together. During the residence time of the polymers in the extruder, the interdiffusion of either material in the other occurs. Calculate the interlayer thickness as a function of the extruder residence time and diffusivities of the two materials.

35.0 *Diffusion coefficient of milk in the refrigerator.* Estimate the diffusion coefficient of lactic acid in the refrigerator. Compare this with the value at room temperature and that of milk through a plastic container.

36.0 *Wilting of lettuce.* Lettuce leaves in a salad wilt. The process of wilting is accelerated if the lettuce is salted. The water droplets on the surface of the leaves comes from the interior of the lettuce cells. Consequently, the turgor pressure and internal rigidity of the leaves are lowered, and they wilt. The process of water transport out of the cells caused by increase in external salt concentration is an example of osmosis. Dutrochet made systematic observations of osmotic pressure in the 1800s. He observed that small animal bladders filled with a dense solution and completely closed and plunged in water became turgid and swollen excessively. Water flowed into the bladder so as to dilute the solution inside. Van't Hoff noted that the osmotic pressure was proportional to the product of the solute concentration and the absolute temperature with a constant of proportionality that equaled the molar gas constant R. Darcy's law provided the solvent flux as a function of the pressure gradient and the constant of proportionality called *hydraulic permeability*:

$$J_{solv} = -\kappa\partial(p - \pi)/\partial x$$

where J_{solv} is the solvent flux, κ is the hydraulic permeability, and π is the osmotic pressure, which can be written as RTC_{sol}, where C_{sol} is the solute concentration. For the wilting of lettuce, show that the governing equations can be written assuming the salt can permeate through the lettuce and neglecting the hydraulic pressure gradient at steady state as

$$0 = \kappa RT\partial^2 C_{solv}/\partial x^2$$

with the following space conditions:

$$x = \delta,\ C_{solv} = 0$$

$$x = 0,\ C_{solv} = C_{solv,0}$$

Show that if the lettuce is a semipermeable membrane, at steady state the solvent transport can be given by

$$p - p_0 - (\pi - \pi_0) = -J_{solv}/\kappa(x - x_0)$$

where x_0 is the reference location at which the hydraulic and osmotic pressures are known.

37.0 *Restriction mapping.* Endonucleases or restriction enzymes cut the unmethylated DNA at several sites and restrict their activity. About 300 restriction enzymes are known, and they act on 100 distinct restriction sites that are palindromes. Some cuts leave blunt ends, and others leave them sticky. The restriction fragment lengths can be measured by using the technique of gel electrophoresis. The solid matrix is usually agarose or polyacrylamide gel that is permeated with liquid buffer. Since DNA is a negatively charged molecule, when placed in an electric field, the DNA migrates toward the positive pole. DNA migration is a function of its size. Calibration is used to relate the migration distance as a function of size. Migration distance of DNA under a field for a set time is measured. The DNA molecule is made to fluoresce and made visible under ultraviolet light by staining the gel with ethidium bromide. A second method is to tag the DNA with a radioactive label and then to expose the x-ray film to the gel. Show that the migration under gel electrophoresis can be given by

$$J_{frag} = -(z_A u_A F)^*\partial E/\partial x - D_{frag}\partial C_A/\partial x$$

Show that the governing equation can be written in one dimension as

$$0 = D_{frag}\partial^2 C_A/\partial x^2 + -(z_A u_A F)^*\partial^2 E/\partial x^2$$

38.0 *Pheromones and insect control.* During insect control, controlled release of pheromones is used. Pheromones are sex attractants released by insects. When mixed with an insecticide and applied, it annihilates all of one sex of a particular insect pest. The pheromone sublimation rate in the impermeable holder can be given as

$$S_0 = 9\ \text{E -16}\ (1 - 1\text{E6}\ C_1)$$

where C_1 is the concentration in the vapor. The diffusivity through the polymer is 1.2 E-11 cm^2/s. It can be assumed that the pheromone level outside the chamber is 0. If the polymeric diffusion barrier is 600 μm thick and has an area of 1.6 cm^2, what is the concentration of pheromones in the vapor? How fast is the pheromone released by the device?

39.0 *Oxygen transport in the eye.* The cornea is a unique living tissue and is a transparent window through which light enters the eye to be focused on the retina, thus forming the images of our surroundings and enabling sight. When the eye is open, it receives all its oxygen requirements from the surrounding air. Other nutrients are likely delivered via the tear duct fluid that bathes the outer surface of the cornea or the aqueous humor that fills the chamber behind the cornea and in front of the lens. Some oxygen may enter the aqueous humor from the vasculature in the muscle at the periphery of the lens. When the eye is closed, it is cut off from the O_2 source in the air. There is a rich microvascular bed (well perfused with high vascular density on the inner surface of the eyelid) that supplies the cornea with oxygen (and possibly other nutrients). What is the Po_2 at the surface of the cornea when the eye is closed?

Layer	Thickness, μm	Diffusion Coefficient, cm^2/s	Vo_2, mL O_2/mL/s
Epithelium	40	3.8 E-10	2.0 E-4
Stroma	450	3.8 E-10	1.0 E-5
Endothelium	10	3.8 E-4	2.0 E-4

Table of Model Parameters

40.0 *Loss from beverage containers.* Coca-Cola bottles are made out of plastic. The contents diffuse at a slow rate through the walls of the container and out into the air and result in some losses. It has been suggested to coat the inner wall of the container to reduce the losses. With a coating thickness of 25 μm and a diffusion coefficient in the coating of 1 E-9 m^2/s, what would be the benefit to the manufacturer? Assume a thickness of 1.5 mm for the plastic container and a diffusion coefficient of the contents in the plastic container as 1 E-6 m^2/s.

41.0 *Dasangam.* Dasangam is offered to the gods during special pooja. Idealize a dasangam into a cone. Consider the reaction between oxygen and dasangam on ignition to be of first order. As the reaction proceeds, consider an added ash layer through that the oxygen will have to diffuse. Obtain the concentration profile of oxygen in the ash layer. Make suitable assumptions and estimate the time taken for consumption of two dasangams of a cone height of 3 cm and a diameter of 1.5 cm. Consider the diffusion coefficient of oxygen in the ash layer to be 1 E-12 cm^2/s.

42.0 *Reaction and diffusion in a nuclear fuel rod.* In autocatalytic reactions such as during nuclear fission, the neutrons can be studied by a first-order reaction. The mass balance in a long cylindrical rod with first-order autocatalytic reaction can be written at steady state as

$$1/r\partial(rJ_r)/\partial r + k'''C = 0$$

The long cylindrical rod is at zero initial concentration of autocatalytic reactant A. The surface of the rod is maintained at a constant concentration C_s for times greater than zero. The boundary conditions are

$$r = 0, \partial C/\partial r = 0$$

$$r = R, C = C_s$$

Show that the steady-state solution can be obtained as follows after redefining $u^s = C/C_s$:

$$\partial^2 u^s/\partial X^2 + 1/X \partial u^s/\partial X + k^* u^s = 0$$

$$X^2 \partial^2 u^s/\partial X^2 + X \partial u^s/\partial X + X^2 k^* u^s = 0$$

The preceding equation can be recognized as the Bessel equation. The solution is

$$u^s = c_1 J_0(X\sqrt{k^*}) + c_2 Y_0(X\sqrt{k^*})$$

It can be seen that $c_2 = 0$ because the concentration is finite at $X = 0$. The boundary condition for surface concentration is used to obtain c_1. Thus

$$c_1 = 1/J_0(R\sqrt{k^*/D\tau_r})$$

Thus

$$u^s = J_0(X\sqrt{k^*})/J_0(R\sqrt{k^*/D\tau_r})$$

43.0 *Grooming hair with oil.* In order to keep the hair on the human skull from becoming dehydrated, it is oiled or hair cream is applied every day. During the course of the day, estimate the loss of the oil from the human hair by diffusion. Show that there are two contributions: One is from molecular diffusion from the head to the atmosphere in the vertical direction, and the other is by convection from wind blowing in the horizontal direction. Show that the governing equation can be given by

$$\partial^2 u/\partial z^2 = (Ud_{hair}/D)\, \partial u/\partial x$$

Show that the solution for the concentration profile of the oil in the surrounding region of the human skull at steady state can be given by

$$u = 1 - \text{erf}\, Z(Pe_m/4X)^{1/2}$$

Assuming that the diameter of the hair is 2 μm, the velocity of air is 1 m/s, and the diffusivity is 1 E-5 m²/s, estimate the time taken for the layer of cream of 1 μm to be replaced. Make suitable assumptions, such as a cranial area of 2500 cm² and a length of the hair of 5 cm.

44.0 *Dyeing of the wool.* A dye bath at a concentration C_0 and a volume V is used to dye wool that is bathed in it. The dye diffuses into the wool. Measuring the concentration of the dye in the wool as a function of time, can you (a) estimate the diffusion coefficient of the dye? If so, how? And (b) can you estimate the relaxation time?

45.0 *Dopant profile by ion implantation.* Ion implantation is used to introduce dopant atoms into a semiconductor material to alter its electrical conductivity. During ion implantation, a beam of ions containing the dopant is directed at the semiconductor surface. For example, boron atoms are implanted into silicon wafers by Lucent Technologies, Murray Hill, NJ. Assume that the transfer of boron into the silicon surface is on account of both the convection and diffusion contributions at steady state. Show that the governing equation for the transfer of boron at the gas-solid interface is given by

$$-\partial C_A/\partial z = D_{AB}\partial^2 C_A/\partial z^2$$

Given a characteristic length l, show that the equation can be reduced to

$$-Pe_m \partial u/\partial Z = \partial^2 u/\partial Z^2$$

and the solution is

$$u = 1 - J^*_{ss}/Pe_m \exp(-Pe_m Z)$$

46.0 *Soot from a steam engine.* The steam engine that powers the train that takes you from Chennai to New Delhi in 31 hours discharges coal dust at steady rate of 68 kg · mol/h. The train moves at a velocity of 90 km/h. Estimate the thickness of soot that will deposit on a passenger sitting near the window of S6 during the entire journey. S6 is about 200 ft from the engine. Assume that the diffusion coefficient of the soot in air is 1 E-6 m^2/s. Repeat the analysis for a wind speed of 10 km/h. (*Hint:* Bulk concentration of soot in the surrounding air can be calculated by considering a basis of time as that taken for a passenger to move 600 ft to the discharge point in fixed space, and in that time the discharge amount is calculated from the discharge rate and the dispersed region from the penetration length in all three directions.)

47.0 *Steady diffusion in a hollow sphere.* Develop the concentration profile in a hollow sphere when a species is diffusing without any chemical reaction. Consider the concentration of the species to be held constant at the inner and outer surfaces of the cylinder at C_{Ai} and C_{Ao}, respectively. Show that

$$(C_A - C_{Ai})/(C_{Ai} - C_{Ao}) = (1 - R_i/r)/(1 - R_i/R_o)$$

48.0 *Determination of diffusivity.* Unimolar diffusion can be used to estimate the binary diffusivity of a binary gas pair. Consider the evaporation of carbon tetrachloride (CCl_4) into a tube containing oxygen. The distance between the CCl_4 level and the top of the tube is 16.5 cm. The total pressure in the system is 760 mm Hg, and the temperature –5°C. The vapor pressure of CCl_4 at that temperature is 29.5 mm Hg. The area of the diffusion path in the diffusion tube may be taken as 0.80 cm^2. Determine the binary diffusivity of O_2–CCl_4 when in an 11-hour period after steady state, 0.026 cm^3 of CCl_4 has evaporated.

49.0 *Helium separation from natural gas.* McAfee [38] proposed a method to separate helium from natural gas. He noted that Pyrex glass is almost impermeable to all gases but helium. The diffusion coefficient of helium is 25 times the diffusion coefficient of hydrogen. Consider a Pyrex tubing of length L and inner and outer radii R_i and R_o. Show that the rate at which helium will diffuse through the Pyrex can be given by

$$J_{He} = 2\pi L D_{\mathrm{He,pyrex}}(C_{\mathrm{He,1}} - C_{\mathrm{He,2}})/\ln(R_o/R_i)$$

50.0 *Solid dissolution into a falling film.* A liquid is flowing in laminar motion down a vertical wall. The wall consists of a species that is slightly soluble in the liquid. Show that the governing equation for the species diffusing into the liquid from the wall can be written as

$$\partial^2 u/\partial z^2 = (UL/D)\,\partial u/\partial x$$

Show that an error function solution results for this PDE.

51.0 *Carburizing steel.* Low-carbon steel can be hardened to improve wear resistance by carburizing. Steel is carburized by exposing it to a gas, liquid, or solid that provides a high carbon concentration at the surface. Given the percent carbon versus depth graphs for various times at 930°C, how can the diffusion coefficient be estimated from the graphs?

52.0 *Electrophoretic term*

$$-j_A = D\frac{\partial C_A}{\partial z} - \left(\frac{zFm}{RT}\right)C_A + \tau_{mr}\frac{\partial j_A}{\partial t}$$

For some systems, there is a minus sign in the electrophoretic term, as shown in the equation above. What are the implications of the minus sign in this equation? How will this manifest in applications?

APPENDIX A

Internet Hotlinks to Public-Domain Databases

$99 Genetrack DNA Test	www.genetrackus.com
2D-PAGE Databases for Human Proteome	http://biobase.dk/cgi-bin
Affymetrix	www.affymetrix.com
Alignment tools	www.SuccessFactors.com/FreeTrial
Alscript	www.compbio.dundee.ac.uk/Software/Alscript
American Type Culture Collection	www.atcc.org
Applied Biosystems	www.appliedbiosystems.com
Atlas assembler	www.hgsc.bcm.tmc.edu
ArrayExpress	www.ebi.ac.uk/arrayexpress
AAT	http://genome.cs.mtu.edu/aat.html
AMAS	www.compbio.dundee.ac.uk/Software/Amas/
AMPS	www.compbio.dundee.ac.uk/Software/Amps/
AVID	http://bio.math.berkeley.edu
BASE	http://base.thep.lu.se
Bacillus subtilis	http://pbil.univ-lyon1.fr/nrsub

BankIt	www.ncbi.nlm.nih.gov/BankIt
Baylor College of Medicine, RNA Database	http://mbcr.bcm.tmc.edu/smallRNA
BioConductor	www.bioconductor.org
BioMedNet Library	http://biomednet.com
BioPerl	www.bioperl.org
BLAST	www.ncbi.nlm.nih.gov/blast/Blast.cgi
BLAST2	www.Bork.EMBL-Heidelberg.DE/Blast2e
BLASTZ	http://bio.cse.psu.edu
BLAT	http://genome.ucsc.edu/cgi-bin/hgBlat
BLOCKS	http://blocks.fhcrc.org/
BLOSUM substitution matrix	www.ncbi.nlm.nih.gov/Education/BLASTinfo/
Cambridge Structural Database	www.ccdc.cam.ac.uk
Carbohydrate Databases	www.boc.chem.ruu.nl/sugabase
CDD	www.ncbi.nlm.nih.gov/Stucture/cdd.cdd.shtml
Center for Applied Genomics	www.tcag.ca/
Center for Inherited Disease Research	www.cidr.jhmi.edu/markerset.html
CEPH Genotype Database	www.cephb.fr
CHAOS	www.molecularstation.com/bioinformatics/link
Cholinesterase Gene Server	www.ensam.inra.fr
Chromosomes and karyotypes	www.selu.com/bio/cyto/human/
CIBEX	http://cibex.nig.ac.jp
ClustaL	www-igmc.u-strasbg.fr/BioInfo
CLUSTALW	www.ebi.ac.uk
Cooperative Human Linkage Center	http://gai.nci.nih.gov/CHLC
Database of Enzymes and Metabolic Pathways	www.empproject.com
Database of Protein Structure Domains	http://barton.ebi.ac.uk

Database for Rice Transcription Factors	http://drtf.cbi.pku.edu.cn/
Department of Molecular and Cellular Biology	http://golgi.harvard.edu
DAVID	http://david.niaid.nih.gov
DIALIGN	www.gsf.de/biodv/dialign.html
Digital gene expression	www.NanoString.com
DISULFIND	http://disulfind.dsi.unifi.it
DNA Database of Japan	www.ddbj.nig.ac.jp
DNA search	www.genomequest.com
DNA sequence assembly	www.genecodes.com
DNA sequencing software	www.codoncode.com/aligner
Database of Genome Sizes (DOGS)	www.cbs.dtu.dk/databases/DOGS
Dotlet	www.isrec.isb-sib.ch/java/dotlet/Dotlet.html
Dotter	www.cgr.ki.se/cgr/groups/sonhammer/Dotter.html
Dottup	www.emboss.org
DoubleScan	www.sanger.ac.uk/Software/analysis/doublescan
Drosophila melanogaster	http://flybase.bio.indiana.edu
EBI Protein Topology Atlas	www3.ebi.ac.uk/tops
Ebioinformatics	www.ebioinformatics.org
EcoCyc	www.ai.sri.com/ecocyc
EMBO structural databases	http://xray.bmc.uu.se
EMBOSS	http://cbrmain.cbr.nrc.ca
Ensembl	www.ensembl.org
EnteriX	http://bio.cse.psu.edu
Entrez	www.ncbi.nlm.nih.gov/sites/gquery
ENZYME	www.expasy.org/enzyme/

Escherichia coli Database Collection	http://susi.bio.uni-giessen.de
eGenome	http://genome.chop.edu
eShadow	http://eshadow.dcode.org
euGenes	http://iubio.bio.indiana.edu:8089
European Bioinformatics Institute	www.ebi.ac.uk
European Molecular Biology Institute	www.embl-heidelberg.de
ExoFish	www.genoscope.cns.fr/proxy/cgi-bin/exofish.cgi
Exon Annotation Database	http://hollywood.mit.edu
Exon-Intron Database	www.meduohio.edu/bioinfo/eid/
ExPASy	www.expasy.org
FASTA—EBI	www.ebi.ac.uk/fasta33/
FASTA—Virginia	http://fasta.bioch.virginia.edu
fGENEH	www.bioscience.org/urllists/genefind.htm
Finishing standards	www.genome.wustl.edu/Overview/g16stand.php
Flicker 2D gel analysis software	www-lecb.ncifcrf.gov/flicker
FootPrinter	http://bio.cs.washington.edu/software.html
Free bioinformatics tools	www.clcbio.com
GALA	http://bio.cse.psu.edu
GenBank	www.ncbi.nlm.nih.gov/Web/Genbank
Gene expression for mouse brain	www.brain-map.org/welcome.do
Gene Expression Omnibus	www.ncbi.nlm.nih.gov/geo
Gene Ontology Consortium	http://genome-www.stanford.edu
Gene Ontology Project	http://geneontology.org
GeneBuilder	http://125.itba.mi.cnr.it/~webgene/genebuilder.html

GeneCards	http://bioinformatics.weizmann.ac.il/cards
Genedoc	www.psc.edu/biomed/genedoc
GeneExpress	www.mgs.bionet.nsc.ru/mgs/systems/geneexpress/
GeneLoc	http://genecards.weizmann.ac.il/cards
GeneMark.hmm	http://genemark.biology.gatech.edu/GeneMark/hum.cgi
GeneID	www1.imim.es/geneid.html
GeneView	http://125.itba.mi.cnr.it/~webgene/wwwgene.html
GeneWise	www.sanger.ac.uk/Software/Wise2
GENEMARK	http://exon.gatech.edu/GeneMark
GenePaint	www.genepaint.org/
Genetic analysis software	http://linkage.rockefeller.edu/soft
Genie	www.fruitfly.org/seq_tools/genie.html
GenomeScan	http://genes.mit.edu/genomescan
GenomeVista	http://pipeline.lbl.gov
Genome programs of the DOE	http://genomics.energy.gov/
Genome Sequence Database	www.ncgr.org
Genomics Institute of Novartis Research Foundation	http://symatlas.gnf.org/SymAtlas/
GENSCAN	http://genes.mit.edu/GENSCAN.html
GLASS	http://groups.csail.mit.edu/cb/glass/cgi-bin/glass.cgi
GLIMMER	www.cbcb.umd.edu/software/glimmer
GPCRD	www.gpcr.org/

GRAIL	http://compbio.ornl.gov/grailexp
GRAILEXP	http://grail.lsd.ornl.gov/grailexp/
Hemophilia A Mutation Database	http://europium.csc.mrc.ac.uk
Haemophilus influenzae Database	http://susi.bio.uni-giessen.de
HIV Immunology Database—Harvard University	http://hiv-web.lanl.gov/immuno/index.html
HMMER	http://bioweb.pasteur.fr/seqanal/motif/hmmer-uk.html
HMMgene	www.cbs.dtu.dk/services/HMMgene
HMMPRO	www.bio.net/bionet/mm/bio-soft/1999-January
HMMSTR	www.bioinfo.rpi.edu/~bystrc/hmmstr/about.html
Human Genome Project Information	www.ornl.gov/sci/techresources/Human_Genome/
HUPO Proteomics Standards	http://psidev.sourceforge.net
IBM Bioinformatics and Pattern Discovery Group	http://cbcsrv.watson.ibm.com/Tspd.html
Immunogenetics Database	www.ebi.ac.uk
Institute for Genomic Biology	www.igb.uiuc.edu/
Institute of Genomic Research	www.tigr.org/
InterPro	www.ebi.ac.uk
J. Craig Venter Institute, Rockville, MD	http://rsng.nhlbi.nih.gov
JalView	www.jalview.org
Johns Hopkins University OWL Web Server	www.bis.med.jhmi.edu/
JPRED	www.compbio.dundee.ac.uk/~www-jpred/
Kyoto Encyclopedia of Genes and Genomes	www.genome.ad.jp/kegg/

LAGAN	http://lagan.stanford.edu/lagan_web/index.html
Listing of molecular biology databases	gopher://gopher.nih.gov/11/molbio/other
MAFTT	www.biophys.kyoto-u.ac.jp/~katoh/prgrams
Mammalian Genome Size Database	www.unipv.it
MAP-O-MAT	http://compgen.rutgers.edu/mapp,at
MASCOT	www.matrixscience.com
Mauve	http://gel.ahabs.wisc.edu/mauve/documentation.php
MegaBLAST	www.ncbi.nlm.nih.gov/BLAST
Meta-Meme	http://metameme.sdsc.edu
MGC	http://mgc.nci.nih.gov
Microarray Gene Expression Data Society	www.mged.org
Molecular diagnostics	www.G2Reports.com
Molecular Informatics Resource for Analysis	www.ifti.org/Mirage.mirage.html
Molecular modeling servers and databases	www.rsc.org/lap/rsccom/dab
Molecular Probe Database	www.biotech.ist.unige.it
Molecules R Us	http://cmm.info.nih.gov/modeling/net_services.html
Molscript	www.chemie.fu-berlin.de/chemnet/use/suppl/molscript
Mouse Genome Database	http://BioMedNet.com/cgi-bin/mko
Mulan	http://mulan.dcode.ord
MUMer	http://mummer.sourceforge.net/
MUSCLE	www.drive5.com/muscle
MyHits	http://myhits.isb-sib.ch

MZEF	http://argon.cshl.org/genefinder/human.htm
National Institute of Genetics	www.nig.ac.jp/index-e.html
NCBI	www.ncbi.nlm.nih.gov
NCBI BLAST Accelerator	www.BlastStation.com
Nobel Museum	http://nobel.se
NRL_3D, Sequence-Structure Database	www.gdb.org
Of Gene Expression, MIRAGE	www.ifti.org/
O-GLYCBASE	www.cbs.dtu.dk/OGLYCBASE
OMIM	www.ncbi/nlm.nih.gov/omim
OPAL	http://opal.cs.arizona.edu
p53 mutations in human tumors and cell lines	ftp://ftp.ebi.ac.uk/pub/databases/p53
PAH mutation analysis	www.mcgill.ca
Pairwise sequence alignment	http://searchlauncher.bcm.tmc.edu
PEPTIDESEARCH	www.narrador.embl-heidelberg.de
PFAM	www.sanger.ar.uk/Software/Pfam/
PHDSec	www.predictprotein.org/doc/methodsPP.html
PHYLIP	http://evolution.genetics.washington.edu/
PROCLAME	http://prclame.unc.edu
PhyloBLAST	www.pathogenomics.bc.ca/phyloBLAST
Phylogeny programs	http://evolution.genetics.washington.edu/phylip
PipMaker	http://bio.cse.psu.edu/pipmaker
PIR	http://pir.georgetown.edu
PROCRUSTES	www-hto.usc.edu/software/procrustes/qpn.html

Profile HMMs	http://helix.nih.gov/docs/gcg/hmmanalysis.html
PROSITE	http://expasy.hcuge.ch/sprot/prosite.html
Protein kinase resource	www.sdsc.edu/projects/kinases
Protein microarrays	www.proteinbiotechnologies.com
Protein Mutant Database	http://pmd.ddbj.nig.ac.jp/
ProteinProspector	http://prospector.ucsf.edu
Proteins in gene regulation	www.access.digex.net
PROWL	http://mcphar04.med.nyu.edu
PSI-BLAST	www.ncbi.nlm.nih.gov/BLAST
PubMed	www.ncbi.nlm.nih.gov/Genbank/index.html
PubMed Central	www.pubmedcentral.nih.gov
PUZZLEBOOT	www.tree-puzzle.de
QSAR	www.cris.com/~Hyposoft
RCSB PDB	www.rcsb.org/pdb/home/home.do
Rat Genome Database	http://rgd.mcw.edu
ReadSeq	http://dot.imgen.bcm.tmc.edu
REBASE—Restriction Enzymes	www.neb.com/rebase
RepeatMasker	http://repeatmsaker.genome.washington.edu
REPuter	www,genomes.de
Rfam	www.sanger.ac.uk/Software/Rfam
Ribosomal Database Project	http://rdpwww.life.uiuc.edu
Ribosomal Database Project II	www.cme.msu.edu/RDP/html

RNA Modification Database	http://medlib.med.utah.edu
ROSETTA	www.rosettabio.com
Rutgers University Linkage Physical Maps	http://compgen.rutgers.edu/maps
Saccharomyces cerevisiae	www.proteome.com
Sakura	http://sakura.ddbj.nig.ac.jp
SAM	www.cse.ucsc.edu/research/compbio/sam.html
SAM-T99	www.soe.ucsc.edu/compbio/HMM-apps/
Sanger Institute	www.sanger.ac.uk
SCOP	http://scop.berkeley.edu
SEG	ftp://ncbi.nlm.nih.gov/pub/seg
Sequence alignment tool	www.Geneious.com
Sequest	http://fields.scripps.edu/sequest
Sequin	www.ncbi.nlm.nih.gov
Sequences of tRNA	www.uni-bayreuth.de
SLAM	http://bio.math.berkeley.edu/slam
SGP-I	www.1.imim.es/datasets/humanmouse
SNP Consortium GL Maps	http://snp.cshl.org/linkage.maps
Source Database	http://source.stanford.edu
STAMP	www.compbio.dundee.ac.uk/Software/Stamp/
Stanford Microarray Database	http://genome-www5.stanford.edu/
STRAP	www.charite.de/bioinf/strap
Structural classification of proteins	http://scop/mrc-lmb.cam.ac.uk
SWISS-2DPAGE	www.expasy.org/ch2d/
SWISS-MODEL.	http://swissmodel.expasy.org/repository/
Swiss-Prot (EBI)	www.ebi.ac.uk/swissprot

Swiss-Prot (Ex-PASy)	www.expasy.org/sprot
T-Coffee	www.ch.embnet.org/software/TCoffee.html
TM4 Software	www.tigr.org/software/tm4
The Better Bradford Assay	www.piercenet.com
TPF assembly tool	www.ncbi.nlm.nih.gov/projects/zoo_seq
Tree of Life	http://phylogeny.arizona.edu
TreeView	http://taxonomy.zoology.gla.ac.uk/rod/treeview.html
TWINSCAN	http://genes.cs.wustl.edu
UCSC	http://genome.ucsc.edu
UIUC Metabolomics Center	www.biotech.uiuc.edu/centers/MetabolomicsCenter/
UK Human Genome Mapping Project	http://hgmp.mrc.ac.uk
Unimode	www.unimod.org
Uniprot	www.uniprot.org
UP Patent Citation Database	http://cos.gdb.org/repos/pat/
UW-Madison Server for Virology	www.bocklabs.wisc.edu
VBASE	www.mrc-cpe.cam.ac.uk
VISTA	www-gsd.lbl.gov/vista
VISTA Browser	http://pipeline.lbl.gov
WEBPHYLIP	http://sdmc.krdl.org.sg:8080
Washington University Genome Sequencing Center	http://genome.wustl.edu
WHS	www.cladistics.org/education.html
Worldwide Protein Databank	www.wwpdb.org/
Wormbase	www.wormbase.org/
Yeast Homology Databases	www.ch.embnet.org
zPicture and multi-zPicture	http://zpicture.dcode.org

APPENDIX B

PERL for Bioinformaticists

Practical Extraction and Report Language (PERL) was invented by Larry Wall in 1986. It is an excellent pattern-matching scripting language. It is a programming language with good string processing capabilities and can be used for doing such things as sequence analysis, database management, etc. It has few data types. The source code for PERL is free. PERL is distributed under the General Public License (GPU). It is user-friendly for biologists. Small programs can be downloaded easily from the Comprehensive PERL Archive Network (CPAN), as well as from BIOPERL. It is a glue language. It is superb at common gateway interface (CGI) front. The source code of programs can contain few lines. The Internet access is at www.perl.org and www.bioperl.org. PERL is supported on UNIX, MS DOS, VMS, OS/2, Mac, Windows, and LINUX operating systems. PERL is made of sed, awk, UNIX shell, and C. It does not need compiling processes like JAVA. The code can be written in Notepad, available with Windows. The generated file can be saved with a .pl extension. Some of the special features of PERL are

- Hashes (or associated arrays). % is used before hashes.

  % translation = (aug => 'ALA'; caa => 'CYT'; ctt => 'GLU';). One line of PERL as a hash is used to convert the three-letter amino acid codons (AUG, GAA, CAT, etc.) to the amino acids (ALA, GLU, THY, SER, etc.).

  $translation (aug), where PERL will interpret aug as ALA. There are commands for repeating a sequence and for pattern finding.

```
$pattern = "aaaa";
$sequence – agttcgaaaaccggt;
@result = split/$pattern/$sequence;
print @result;
```

The program splits the sentence and finds the pattern, and a gap is inserted to denote its presence.

The function list in PERL is as follows:

• Array	Chomp, join, keys, map, pop, push, reverse, shift, sort, splice, split, unshift, values
• Database	dbmclose, dbmopen
• Directory	chdir, closedir, mkdir, opendir, readdir, rewindir, rmdir, seekdir, telldir
• File	binmode, chdir, chmod, chown, chr close, eof, fnctl, filenxflock, getc, glob, loctl, link, lstat, open, print, printf, read, readdir, readlink, rename, rmdir, seek, select, start, symlink, sysopen, sysread, syswrite, tell, truncate, umask, unlink, write
• Group	endgrent, getgrent, getgrid, getgrname, getpgrp, setgrent, setpgrp
• Hash	delete, each, exists, keys, values
• Host	endhostent, gethostbyaddr, gethostby-name, gethostent
• Input	getc, read, sysread
• Interprocess communication	msgctl, msggct, msgrcv, msgsnd, pipe, semctl, semgct, semop, shmctl, shmget, shmread, shmwrite
• Math	abs, atan2, cos, exp, hex, int, log, oct, rand, sin, sqrt, srand
• Message queues	msgctl, msgget, msgrcv, msgsnd
• Time	gmtime, localtime, time
• Unix	chnwd, chown, chroot, dump, endgrent, endhostent, endnetent, endprotent, endp-went, endservent, fnetl, fork, getgrgrid, getgrname, gethostent, getlogin, getnetent, getpgrp
• Miscellaneous	bless, defined, do, eval, formline, import, ref, scalar, syscall, tie, tred, undf, untie, wantarray
• Network	endnetent, getnetbyaddr, getnetbyname, getnetent, setnetent
• Output	die, print, printf, syswrite, write, wavn

• Password	endpwent, getpvent, getpwname, getpwaid, setpwent
• Process	alarm, die, dump, exec, exit, fork, getlogin, getpgrp, getppid, getpriority, kill, setpriority, sleep, system, times, unmask, wait, waitpid
• Protocol	endprotent, getprotobyname, getprotobynumber, getprotent, getservname, getservbyport, getservent, getprotoent
• Regular expression	grep, pos, quotemeta, rest, split, study
• Scope	local, my, culler
• Service	endservant, getservbyname, getservbyport, getservent, setservent
• Socket	accept, bind, connect, gethostbyaddr, gethostbyname, gethostent, getpeername, getservbyport, getservent, getsocketname, get sockoporet, listen, reev, select, send, setsockoport, shutdown, socket, socketpair
• String	chop, chr, crypt, hex, index, join, le, lcfirst, length, oct, pack, q, qq, quotemetr, qw, reverse, rindex, split, spintf, susbtr, uc, ucfirst, unpack, vec

Index

A

C

G

O

P

S